AF804093

Rethinking Cognitive Enhancement

Rethinking Cognitive Enhancement

Edited by

Ruud ter Meulen

Ahmed Dahir Mohamed

Wayne Hall

OXFORD
UNIVERSITY PRESS

OXFORD
UNIVERSITY PRESS

Great Clarendon Street, Oxford, OX2 6DP,
United Kingdom

Oxford University Press is a department of the University of Oxford. It furthers the University's objective of excellence in research, scholarship, and education by publishing worldwide. Oxford is a registered trade mark of Oxford University Press in the UK and in certain other countries

© Oxford University Press 2017

The moral rights of the authors have been asserted

Chapters 1–12, 14–16, 18–22, and 19 © Oxford University Press, 2017

Chapter 13 © 2008, BMJ Publishing Group Ltd

Chapter 17 © The Hastings Center, 2012 and Taylor and Francis Group, LLC, 2011

Chapter 18 © Taylor and Francis Group, LLC, 2012

Excerpts from "Lessons for Enhancement From the History of Cocaine and Amphetamine Use" from *AJOB Neuroscience*, 3 (2), pp. 24-9 by Stephanie K. Bell, Jayne C. Lucke, and Wayne D. Hall. Copyright 2012 by Taylor & Francis Ltd. Reprinted by permission of the publisher (Taylor & Francis Ltd, http://www.tandfonline.com).

Impression: 1

All rights reserved. No part of this publication may be reproduced, stored in a retrieval system, or transmitted, in any form or by any means, without the prior permission in writing of Oxford University Press, or as expressly permitted by law, by licence or under terms agreed with the appropriate reprographics rights organization. Enquiries concerning reproduction outside the scope of the above should be sent to the Rights Department, Oxford University Press, at the address above

You must not circulate this work in any other form and you must impose this same condition on any acquirer

Published in the United States of America by Oxford University Press
198 Madison Avenue, New York, NY 10016, United States of America

British Library Cataloguing in Publication Data

Data available

Library of Congress Control Number: 2016960549

ISBN 978–0–19–872739–2

Printed and bound by
CPI Group (UK) Ltd, Croydon, CR0 4YY

Oxford University Press makes no representation, express or implied, that the drug dosages in this book are correct. Readers must therefore always check the product information and clinical procedures with the most up-to-date published product information and data sheets provided by the manufacturers and the most recent codes of conduct and safety regulations. The authors and the publishers do not accept responsibility or legal liability for any errors in the text or for the misuse or misapplication of material in this work. Except where otherwise stated, drug dosages and recommendations are for the non-pregnant adult who is not breast-feeding

Links to third party websites are provided by Oxford in good faith and for information only. Oxford disclaims any responsibility for the materials contained in any third party website referenced in this work.

Preface

This multidisciplinary volume, including scientific, ethical, and legal perspectives, offers a much needed reality check to the debate on cognitive enhancement. While there may be potential benefits to human enhancement, and to cognitive enhancement in particular, there is a danger of slipping into a way of speaking about it that plays down potential downsides. It may be implied or even asserted that enhancement is inevitable or that it is by definition an improvement. Even if it *is* inevitable, however, that a person or persons will seek to find means of cognitive enhancement, and even succeed in doing so, that by no means settles the ethical, legal, and policy issues. Also, we should beware of making enhancement an improvement by definition. Whether any particular enhancement intervention is an improvement from an ethical point of view is an open question. For whom will it be an improvement and in what way? We have to have regard to the *respects in which* something is enhanced. As papers in this volume make clear, enhancement in one characteristic may result in worse performance in some other characteristic or characteristics: the associated risks need proper assessment and consideration of the extent to which the trade-off is worthwhile.

While these remarks are true of enhancement in general, where cognitive enhancement is concerned, there are specific issues to consider. On the plus side it might be thought that cognitive capacities are good for whatever life plan an individual might want to follow. Surely, it might be argued, greater cognitive capacities are associated with increased probability of a successful career and all the benefits that flow from that, although it is true that they do not guarantee happiness or well-being. Again, some people may argue that we cannot, and do not think we should, avoid affecting our cognitive capacities by education and other means such as meditation. Contemporary interest in practices such as mindfulness and their effects on neuroplasticity is increasing. On the other hand, there has been concern among paediatricians that some activities which people currently do without a thought may affect our brains in a deleterious way, such as too much passive consumption of screen time, especially in very young children. Given these facts, then why should there be anything wrong with choosing to enhance our cognitive capacities deliberately and in a targeted way, using pharmaceuticals and/or other technologies?

First, however, we need to have regard to what exactly is envisaged in cognitive enhancement, as this volume makes clear: how are we to understand the concept of cognitive enhancement, precisely which characteristics of the brain are to be enhanced and by what methods? Different technologies, including pharmaceutical products, brain stimulation, and genetic technologies, are all candidates, and each of these has associated safety, ethical, and regulatory issues. The fact that manipulation of the brain is involved gives rise to multiple concerns discussed in this collection, including scientific evidence about the significance, in terms of effectiveness, of the baseline cognitive potential in an individual prior to an intervention; safety concerns about possible side effects such as addiction, especially but not exclusively in relation to drugs; philosophical questions about identity and disability; and ethical worries about the space for authentic autonomous choice. The inclusion of the legal dimension in this volume is particularly refreshing, as the potential implications of cognitive enhancement for professional responsibility and the laws of tort, and for other areas of law such as product liability, could be considerable.

So the issues about when, how, and where cognitive enhancement should be introduced require considered thought, and input from different disciplines, in order to address the pertinent questions. In particular what needs to be asked includes the following: What are the purposes of any given enhancement intervention? Is there a moral difference between introducing a cognitive change as a remedial measure as opposed to enhancing someone who already has high cognitive capacity, perhaps even beyond the current limits of human cognition? Among other dimensions to this issue which have been given perhaps less attention elsewhere, the question of potential biological constraints, relating to the way in which the human mind has evolved, is considered in this volume.

Given the importance of the human brain to human identity, the current stage of research into its complexity, and the uncertainty about consequences of some of the enhancement interventions envisaged, it is very timely to have a note of caution injected into the debate, in order to facilitate the introduction of any potential future programmes of intervention in accordance with scientific, ethical, and policy considerations which are in turn informed by rigorous academic debate. This is to be welcomed in this collection.

Ruth Chadwick
University of Manchester

Editorial

This book came about because of unease of the editors, and of many of their colleagues, with the current debate on the possibilities of human enhancement by the use of pharmacological drugs or other technologies impacting on the brain. They felt that the optimistic view of human cognitive enhancement as presented in the bioethics and transhumanist publications was not matched by evidence in the neurosciences about what these drugs could accomplish, and did not consider their harmful effects, including addiction and dangerous overconfidence. Moreover, they were not happy about the way critical views from the field of ethics, law, public health, and social science are ignored or pictured as conservative and Luddite responses that stand in the way of scientific and societal progress.

When the three editors met in Bristol a few years ago at a presentation by Wayne Hall on deep brain stimulation and addiction, they agreed to work together on this edited volume. The ground work was already prepared by Ahmed Dahir Mohamed, who had drafted an initial proposal to Oxford University Press. The editors are grateful to Oxford University Press for their willingness to publish the book and for their support during the process of collecting and editing the chapters. They feel particularly indebted to Martin Baum for believing in this project and for Charlotte Green for her invaluable help during the editing of the book.

The book *Rethinking Cognitive Enhancement* tries to present a critical reflection on the possible benefits and harms of the efforts to enhance the cognitive functioning of human individuals by the use of psychopharmacological drugs. This reflection is led by evidence from neurological and neuropsychological research, philosophical and ethical analysis, legal approaches, and perspectives from public health and drug policy. We hope that this multidisciplinary approach will help to "debunk" the high expectations of these drugs in academic circles but also the hype in the popular press about what these drugs could bring to people. What are needed are not exaggerated fantasies, but plain evidence and critical debate as the basis for sensible policy-making regarding the use of so-called cognitive-enhancing drugs.

The editors would like to acknowledge some people who have been important in the editing of this book. Ahmed Dahir Mohamed is indebted to Anthony Holland and Simon Baron Cohen, who respectively supervised and advised his doctorate in psychology at the University of Cambridge, Marilyn Williams his mentor and undergraduate supervisor, his friend Jenny Lewis, and finally Evianne Van Gijn and Anthony Edward Phillips, who were working alongside him when he was finishing his doctorate at Clare Hall, Cambridge. They have followed his progress with enthusiasm and curiosity and have been supportive all along the way.

Wayne Hall would like to thank his colleagues at the Centre for Youth Substance Abuse Research at the University of Queensland, particularly Stephanie Bell, Jayne Lucke, and Brad Partridge, for helping to refine his thinking about the topic of cognitive enhancement. He also thanks Sarah Yeates for her invaluable assistance in conducting literature reviews and preparing manuscripts for publication over many years.

Ruud ter Meulen would like to thank his colleagues and post-graduate students at the Centre for Ethics in Medicine for their discussions about human enhancement, particularly Sylvie Allouche, Alex McKeown, and Heather Bradshaw. He also thanks the many colleagues he worked with in European projects on human enhancement, like the ENHANCE project and the EPOCH project. Their contributions from the field of ethics, law, social science, and public policy-making have strengthened his belief in a multidisciplinary approach to the ethical issues of human enhancement as the basis for policy-making. Finally, he wants to thank Ruth Chadwick for writing the preface to this volume.

Bristol, Otago, Brisbane
January 2016

Contents

Part III **Ethical, philosophical, legal, and policy issues of cognitive enhancement**

Contributors

Mark Attiah, MD, MS, MBE, neurosurgery resident at Department of Neurosurgery, University of California Los Angeles, USA

Stephanie Bell, PhD, Principal Research Fellow, University of Queensland, Australia

Brendon P. Boot, BA, MBBS, FRACP, Assistant Professor of Neurology, Harvard School of Medicine, Boston MA; Consultant Neurologist, Brigham and Women's Hospital, Boston, MA; Medical Director, Voyager Therapeutics, Cambridge, MA

Heather Bradshaw-Martin, PhD, Research Associate in Bioethics, University of Bristol, UK

Reinoud de Jongh, PhD, lecturer, Department of Medical Psychology, Erasmus Medical Center, Rotterdam, the Netherlands

Imogen Goold, Associate Professor, Faculty of Law, University of Oxford, UK

Wayne Hall, PhD, Professor and Director of the Centre for Youth Substance Abuse Research at the University of Queensland; and Professor of Addiction Policy at the National Addiction Centre, Kings College London, UK

Roy H. Hamilton, MD, MS, Assistant Professor, Laboratory for Cognition and Neural Stimulation, Department of Neurology, University of Pennsylvania

Andreas Heinz, MD, PhD, Professor of Psychiatry and the Head of the Department of Psychiatry and Psychotherapy, Charité Campus Mitte, Charité—Universitätsmedizin Berlin, Germany

Ralph Hertwig, PhD, Director of the Center for Adaptive Rationality and Professor of Psychology, Max Planck Institute for Human Development, Berlin, Germany

Thomas Hills, Professor of Psychology, Department of Psychology, University of Warwick, UK

Jayne Lucke, Professor, College of Science, Health and Engineering School of Psychology and Public Health Australian Research Centre in Sex, Health and Society, La Trobe University, Australia

Alex McKeown, PhD, Teaching Associate in Biomedical Ethics and Law, and Research Associate in Biomedical Ethics, School of Social and Community Medicine, University of Bristol, UK

Charles F. Massie, MDCM, medical resident, Internal Medicine Program, Hôpital du Sacré-Coeur de Montréal, Montreal QC, Canada

Ruud ter Meulen, PhD, Professor of Ethics in Medicine (emeritus) and Psychologist Centre for Ethics in Medicine, School of Social and Community Medicine, University of Bristol, UK

Ahmed Dahir Mohamed, PhD, a registered psychologist in New Zealand and chatered psychologist and associate fellow of the British Psychological Society in the United Kingdom; currently a visiting fellow in Neurscience and Ethics at the Bioethics Centre, University of Otago, New Zealand

Sabine Müller, PhD, MSc, Assistant Professor and leader of the workgroup Neurophilosophy, Neuroethics and Medical Ethics, Department of Psychiatry and Psychotherapy of the Charité, Charité Campus Mitte, Charité—Universitätsmedizin Berlin, Germany

Brad Partridge, PhD, Senior Research Fellow, Centre for Youth Substance Abuse, Faculty of Health and Behavioural Sciences, University of Queensland, Australia

Boris B. Quednow, PhD, Professor of Experimental and Clinical Pharmacopsychology, Department for Psychiatry, Psychotherapy and Psychosomatics at the Psychiatric Hospital of the University of Zurich, Switzerland

Maartje Schermer, Professor of Philosophy, Medicine and Human Enhancement, Medical Ethics and Philosophy of Medicine, Erasmus Medical Center, Rotterdam, The Netherlands

Stephan Schleim, associate professor, Theory and History of Psychology, Faculty of Behavioral and Social Sciences, University of Groningen, The Netherlands; and Research Center for Neurophilosophy and Ethics of Neurosciences, Munich Center for Neurosciences, Ludwig-Maximilians-University, Munich, Germany

Priyanka P. Shah-Basak, PhD, formerly post-doctoral researcher, Laboratory for Cognition and Neural Stimulation Department on Neurology, University of Pennsylvania; currently Research Fellow, Hospital for Sick Children, Toronto, Canada

Dan J. Stein, FRCPC, PhD, DPhil, Professor and Chair, Department of Psychiatry, University of Cape Town, South Africa

John Strang, MD, Professor of Addictions; and Director of the National Addiction Centre, Institute of Psychiatry, Psychology and Neuroscience, Kings College London, UK

Eric M. Yamga, MD, medical resident, Internal Medicine Program, Hôpital du Sacré-Coeur de Montréal, Montreal QC, Canada

Part I

Introduction

Chapter 1

Introduction

Ruud ter Meulen, Ahmed Dahir Mohamed, and Wayne Hall

There has been a recent excitement among some neuroscientists and bioethicists about the possibility of using drugs and other technologies to enhance cognition in healthy individuals (Buchanan 2011; Harris 2007; Naam 2005; Sandberg 2011; Schaeffer et al. 2014). This excitement arises from recent advances in neuroscientific technologies such as drugs that increase alertness and wakefulness in healthy individuals or technologies that can stimulate activity in different parts of the brain—either via the scalp or via electrodes in the brain, all of which raise the possibility of producing cognitive and affective improvements in otherwise healthy individuals. This development has been described using the term "cognitive enhancement," meaning an improvement of the cognitive and intellectual capacities of the brain. It is part of a wider drive to enhance human capacities by way of biotechnologies, including physical enhancement, mood enhancement, and extension of the life span (Savulescu et al. 2011).

This possibility raises important questions: What is meant by "improvement" or, more specifically, "improvement of the brain"? Does it mean merely improvements that result in better college grades or better work performance, or does it mean improvements that result in more well-being and happiness in individuals' personal lives? How can taking a drug improve these functions especially in healthy individuals free from clinical disorders?

While there is an increased interest in cognitive enhancement, and a strong ethical debate on the merits of cognitive enhancement (Bell et al. 2013; Bostrom and Roache 2009; Carter and Hall 2011; Mohamed 2014), there has been limited critical appraisal of (i) what we mean by cognitive enhancement and (ii) whether we can or should aim to achieve this in healthy individuals.[1]

On the basis of evidence from the neurosciences, the book aims to highlight the possibility that humans may face evolutionary, psychological, and social limitations in increasing their cognition function. For example, the idea that healthy individuals are cognitively enhanced in linear fashion has been challenged by evidence that appeals to the inverted U-shaped function relating arousal and performance. Arguments based on the evolutionary limitations of human cognitive function reinforce the implausibility of pharmaceuticals producing a linear increase in cognitive function in healthy individuals. There may also be significant psychological trade-offs in increasing attention in healthy individuals that include impairments in creativity, flexibility of thought, and global thinking. Furthermore, it is not clear whether drugs that are claimed to enhance cognition in some healthy individuals have the capacity to meaningfully enhance cognition in the normal population. For example, there is as yet no evidence that these "cognitive-enhancing" drugs improve well-being, happiness, and real life achievements in healthy neurotypical individuals. There is in fact evidence that they do just the opposite where they induce depression in healthy individuals who take them (Teter et al. 2010).

The book has several objectives. Firstly, it reviews evidence of the neurosciences to critically evaluate and appraise the concept of cognitive enhancement. It challenges the assumption that healthy individuals will be unequivocally enhanced by the use of pharmacological drugs or other neuroscientific technologies. The key value of this book to the readers is that they will learn about the achievements and shortcomings of neuroscientific research on cognitive enhancement and appreciate that cognitive enhancement, as described by some researchers and indeed by the media, is an ambiguous concept. Do these drugs indeed improve wakefulness or memory in a meaningful way? What are the side effects of using these drugs? What is the impact on other cognitive functions when enhancing one particular function? Several contributions to this volume address these questions by reviewing evidence from laboratory studies and other empirical information.

Secondly, the book considers to what extent the ethics of cognitive enhancement might need to be reframed. Given that it is questionable that drugs like modafinil and methylphenidate meaningfully improve general cognitive function in healthy individuals, this work will consider whether some ethical questions, like cheating in the classroom, are even relevant. Instead of an uncritical praise of cognitive enhancement, the book examines possible trade-offs that may arise from the potential risks of the healthy using these drugs. For instance, if a drug is addictive but improves memory, should we allow its use among healthy individuals, particularly young ones whose brain are still in development? Since most healthy individuals do not always have access to balanced and unbiased evidence of the effects of these drugs on the brain and on the body, it might make their decision to take these drugs less informed and, in consequence, they might not be exercising their autonomy authentically. In the field of neuroethics these ethical issues are thus far widely ignored.

Thirdly, the book aims to contribute to discussions about cognitive enhancement and public health. For example, what are the risks posed by enhancement practices in relation to public health, particularly in respect of addiction? To answer such a question, the book includes a chapter by Heinz and Müller analyzing the risks of using stimulant drugs like modafinil and methylphenidate for enhancement purposes (see Chapter 5). Another chapter by Hall and Strang analyzes the question of whether cognitive drugs should be made widely available or whether policies should be restrictive toward their availability in view of the risks to public health (see Chapter 19).

The volume has two main sections: the first section reviews the (experimental and other empirical) evidence regarding the possible improvements of human cognition by the use of neuropharmacological drugs as well as the limitations and possible side effects from using these drugs for enhancement purposes. The second section includes chapters about a range of ethical, philosophical, legal, and policy issues of the use of neuropharmaceutical drugs for cognitive enhancement.

The two sections are preceded by an overview of the debate on the ethical issues of human enhancement in general and cognitive enhancement in particular. This debate is characterized by strong oppositional views about the benefits and risks of the use of cognitive enhancement as well as other ethical perspectives like the respect for individual autonomy and the role of social justice. The chapter by Ruud ter Meulen (Chapter 2) distinguishes between a favorable view and a cautious position that supports more restrictive policies toward human enhancement. He highlights that on the favorable view, there are liberal and utilitarian authors who see nothing wrong in human enhancement which they argue has always been part of human history. An example is John Harris who argues that enhancement is not only an ethical pursuit but one that we have a moral obligation to pursue (Harris 2009). Cautious authors argue that the use of medical technologies for human enhancement will undermine important human values like dignity and solidarity with weaker groups in our society (President's Council on Bioethics 2003). Some critical authors are very skeptical about the possibility of cognitive enhancement and argue that the debate about the ethics of cognitive enhancement is a "phantom debate" (Quednow 2010). Ter Meulen's chapter deals with the concept and moral value of human enhancement as opposed to therapy; enhancement in relation to the goals of medicine; the benefits and risks of (cognitive) enhancement technologies; enhancement from the perspective of justice and access to enhancement technologies; and the relation of enhancement to fundamental values, like human nature, human dignity, human virtues, and authenticity. While many authors in the field of bioethics might be familiar with this debate, this overview might help readers outside the field of bioethics to better appreciate the various positions and empirical claims as well as the various ethical and legal questions that are raised in this volume. Moreover, the volume will help bioethicists to better appreciate the evidence regarding the use of cognitive-enhancing drugs and to balance the claims of their putative positive effects against the limitation and risks of using them.

1.1 Risk and benefits of the use of neuropharmacological drugs for cognitive enhancement

The first main section in this book reviews the evidence on the potential, the limitations, and the possible risks of cognition-enhancing drugs. This section starts with a chapter by de Jongh reviewing the experimental evidence about the possible enhancement of cognition, including functions such as memory, attention, language, perception, and executive functioning (Chapter 3). De Jongh limits his discussion to the study of cognition-enhancing drugs and uses the term "smart drugs." De Jongh argues that the use of drugs to enhance cognition is far from new. The stimulant caffeine, for example, has been used for this purpose, among other motivations, for at least a thousand years. De Jongh evaluates the effects of these drugs on cognition and shows that although there may be some benefits in healthy people, the size of these benefits is small.

The chapter by Massie, Yamga, and Boot calls for better evidence on safety and efficacy on the neuroenhancement use of pharmaceutical drugs (Chapter 4). They define neuroenhancement as the use of medications by healthy people in order to boost cognitive and affective functions. They argue that the lack of evidence on safety and efficacy that motivates the proscription of neuroenhancement for children also applies to adults because, they assert, prescribing drugs for neuroenhancement requires that we re-evaluate the medication risk–benefit calculus. Massie, Yamga, and Boot reason that, in the case of neuroenhancement, because there is no disease to treat, and hence no disease-related harm to weigh against the risk of treatment, we should only accept the use of drugs for enhancement whose risks are well characterized as minor. They call for caution and for better evidence before physicians prescribe drugs to healthy individuals for the purpose of neuroenhancement.

Heinz and Müller argue that the debate about the ethics of cognitive enhancement has exaggerated the benefits and downplayed the risks (Chapter 5). Proponents of cognitive neuroenhancement usually assume either that stimulant drugs are effective neuroenhancers that can be used without serious risks and side effects or that such drugs will be discovered in the near future. Heinz and Müller argue that these assumptions underestimate the risk of addiction to cognitive enhancers, underestimate the medical risks of using cognitive enhancers, and finally overestimate the benefits of putative cognitive enhancers. They make the point that the neuronal mechanisms of learning and memory are fundamentally related to those underlying the development and maintenance of addictive behavior. Given this, it can be anticipated that drugs which modify the mechanisms of learning and memory will increase the risk of becoming addicted to these drugs. In addition to addiction, the authors note, there are significant psychiatric, cardiovascular, and other medical risks of using drugs like modafinil and methylphenidate for cognitive enhancement.

The chapter by Ahmed Dahir Mohamed reports the results from a randomized controlled trial on the effects of modafinil, a drug licenced for narcolepsy, on creativity (Chapter 6). These results are highly relevant to the debate about human enhancement because modafinil is reportedly one of the most popular pharmacological cognitive enhancers used by healthy individuals with no psychiatric disorders. The drug did not improve creative thinking in healthy individuals overall but its effects were (inversely) dependent on the individuals' level of creativity. Modafinil reduced the ability to creatively problem solve, as measured by the Remote Association Test, in participants who were highly creative but increased performance in participants who were low in creativity. Mohamed highlights the impact of modafinil on divergent thinking tasks (i.e., thinking outside of the box which is often seem as a hallmark of creativity). The experimental results by Mohamed's chapter indicates that modafinil reduces convergent thinking in healthy individuals who are highly creative and uniformly reduces divergent thinking in most healthy individuals.

Mohamed's second chapter reviews neuropsychopharmacological evidence on the effects of modafinil on cognition in humans (Chapter 7). He concludes that, similar to his experimental findings, modafinil improves cognition in healthy individuals low in cognitive function, but it impairs cognition in healthy individuals who are high in cognitive function. His chapter shows that the cognitive and attention-enhancing effects of modafinil are mediated by effects on other forms of cognition such as motivational reinforcement and salience of pleasure. Modafinil is beneficial for narcolepsy but, as with amphetamines and psychostimulants, there is emerging evidence that modafinil has a potential for abuse. Because of these mixed effects, modafinil might have both positive and negative impacts on healthy individuals and on society. The cognitive-enhancing effects of modafinil are, however, small or at best moderate and there is a lack of ecological validity on its cognitive-enhancing effects in the real world. The chapter also summarizes evidence concerning modafinil's adverse effects and presents its safety information. Finally, the effects of modafinil on social cognition and ethical and moral reasoning are currently unknown and merit further rigorous research.

Shah-Basak and Hamilton provide an analysis of the opportunity, feasibility, and risks of pursuing cognitive enhancement using noninvasive brain stimulation (Chapter 8). They discuss two emerging questions. Firstly, they ask whether noninvasive brain stimulation can reliably enhance cognition in healthy individuals. Secondly, they explore the possible risks in using noninvasive brain stimulation. In addressing the first question, they review experimental data from cognitive neuroscience supporting the notion that noninvasive brain stimulation, and specifically transcranial direct current stimulation, can transiently enhance some aspects of cognition. In addressing the possible risks of the enhancement use of noninvasive brain stimulation, they consider the social environments in which the demand for optimal performance may prompt healthy individuals to use noninvasive brain stimulation. In regard to the future of noninvasive brain stimulation they argue that experts in neuroscience, public health, and public policy have an obligation to find an appropriate balance between ensuring public safety and respecting the autonomy of individuals who wish to use noninvasive brain stimulation.

In his chapter Attiah addresses the use of brain stimulation technology for cognitive enhancement and the potential for addiction (Chapter 9). Brain stimulation technologies are currently used for several therapeutic purposes, but they also have the potential for enhancing those without an illness. The phrase "brain stimulation" conjures a vast range of emotions from different segments of society, with fear or apprehension being a common and understandable reaction. The brain reigns as the control center for breathing, eating, and moving, to relating, feeling, and understanding. Changing these functions with electricity or magnetism can fundamentally change how we interact with our environment and one another. Even if this change is beneficial, there can still be a cause for concern. Enjoying the advantages that enhancement might bring could be intoxicating, as can be the case with having great wealth, prestige, beauty, or athletic ability. This chapter explores the implications of such possible enhancement uses, as well as the notion that it could create a dependence on the stimulation akin to an addiction.

The first section ends with a critical review by Schleim and Quednow on the benefits and risks of cognitive-enhancing drugs (Chapter 10). Like other contributors in this volume, they point out that we are still uncertain about the safety of the long-term use of stimulant drugs by healthy individuals of the most commonly discussed cognitive enhancers such as methylphenidate, modafinil, and amphetamines. Schleim and Quednow argue that the vigilance-enhancing effects of these drugs are strongly baseline dependent; that is, they ameliorate impaired cognitive and affective functioning in people with low baseline levels of functioning while impairing cognitive and affective functioning in people with high baseline levels of functioning. They also point out that the use of these drugs entails trade-offs in which improvements in one cognitive domain often comes at a cost of impairments in other cognitive domains. It is possible that the same trade-offs occur in the enhancement of affective functions. Schleim and Quednow argue that the use of stimulant drugs as performance enhancers is neither new nor more common than decades ago. Their analysis of scientific sources in the 1960s–1980s shows that stimulant consumption for enhancement purposes was present and investigated before the "new" neuroenhancement debate. They conclude that the ethical significance of neuroenhancement has been exaggerated and that a more cautious stance would be more appropriate.

1.2 Ethical, philosophical, legal, and policy issues of cognitive enhancement

The second main section of this volume contains a number of chapters on conceptual and other theoretical issues. This section starts with a chapter by Hertwig and Hills on the evolutionary limits of neuroenhancement (Chapter 11). They argue that there are evolutionary limits to how much we can neuroenhance. Their chapter, which focuses on the pharmacological enhancement of cognitive traits, asks why traits such as focus and memory have not already been enhanced through evolution. They argue that the current understanding of human cognitive evolution is at odds with the assumption that more-is-better which underlies the claim that we need to use drugs to improve cognitive functioning in normal persons. Using examples from memory and attention, they demonstrate that evolution of the mind has produced a delicate balance between too much and too little of the cognitive trait in question (e.g., attention, memory, alertness). They highlight the evidence of trade-offs exemplified in the inverted U-shaped performance curves commonly associated with pharmacological interventions. This phenomenon, known as the Yerkes–Dodson law, describes an empirical relationship between arousal and performance in which performance increases with physiological or mental arousal up to a point at which further arousal produces a decline in performance. Enhancements—even routinely used ones, such as coffee—have side effects on other traits and their effects seem to follow the same inverted U-shaped curves. Hertwig and Hills present several examples of this trade-off in the context of memory and reasoning, highlighting the point that trying to enhance healthy individuals with psychostimulants may achieve the opposite effects, that is, may impair rather than improve performance.

McKeown asks whether is it possible to draw a line between enhancement and therapy in using putative pharmaceutical methods for cognitive enhancement such as modafinil, Adderall, and Ritalin (Chapter 12). He also considers this question in relation to the potential for genetic cognitive enhancements, should they become available in future. McKeown makes three interconnected arguments. Firstly, he argues that the distinction between therapy and enhancement is ambiguous and logically unstable. Secondly, he asserts that despite this instability there is a relatively simple theoretical solution. This solution could if implemented, he asserts, negotiate the difficulties raised concerning the distinction between the two concepts and protect the just allocation of scarce medical resources according to need. Thirdly, he argues that contemporary medicine in the developed countries such as the UK is not institutionally ready to implement his proposed solution because of its use of "normality" to define the boundary of appropriate medical practice. McKeown concludes that we should limit our expectations about what can practically be achieved via the widespread use of cognitive enhancement drugs until the institutional assumptions of health care, and the training of medical professionals whose practice is informed by them, have undergone substantial reorientation.

In case enhancements are proven effective in improving cognitive functioning, a frequently used argument is that such a use is a form of cheating, particularly in the context of education and exams. The chapter by Schermer analyzes the enhancement-is-cheating argument by comparing sports and education, and by evaluating how the argument can be interpreted in both contexts (Chapter 13). If cheating is understood as breaking the rules in order to gain an unfair advantage over others, it can be argued that some enhancements are a form of cheating. A further analysis of the intuitions behind the enhancement-is-cheating argument, however, shows that if sports and education are understood as "practices," with their own internal goods and standards of excellence, some potential problems of enhancement can be articulated. These concern the internal goods and standards of excellence that are characteristic of specific practices (i.e., working hard, being honest, studying for the exams, and competing fairly). Seen from this perspective, the important question is how enhancement technologies might be embedded in specific practices—or how they might corrode them.

Although some drugs may improve a patient's functioning, including their cognition, it is unclear whether these drugs enhance cognition or ameliorate a debilitating clinical disorder. There are certain cases where therapy is required, as in the case of patients who are suffering from Parkinson's disease who require treatments rather than enhancement per se. Hence, for many clinicians psychiatric classification is the key in deciding whether to use cognitive enhancers. This is an issue that Stein addresses in his chapter on the relevance of psychiatric nosology to cognitive enhancement (Chapter 14). Stein considers to what extent psychiatric diagnostic manuals can assist clinicians when to decide about treatment in case of patients who are just under the threshold to be diagnosed with an illness and who might in fact ask for drugs to enhance their cognition or psychological well-being. He argues that there has been renewed interest in psychiatric classification, with the recent development of the DSM-5, the ICD-11, and the RDoC framework. He argues that from a DSM-5 perspective, the clinical significance criterion delineates normality from disorder. This suggests that clinical judgment may be the key in making decisions about the diagnosis of mild symptoms. From an ICD-11 and global mental health perspective, Stein argues that the clinical significance criterion may be pseudo-precise. He suggests that instead the focus of clinical attention should be on evidence-based treatments for serious mental disorders. Finally, Stein thinks that the RDoC framework has emphasized that behaviors lie on dimensions, and that psychiatrists and physicians need to better account for the physiological mechanisms that underpin these dimensions of behavior. For the foreseeable future, an integrative approach to the assessment and treatment of patients with subthreshold symptoms will need to incorporate DSM, ICD, RDoC, and other constructs, and weigh up a broad range of relevant facts and values in deciding whether to use cognitive enhancers.

Bradshaw asks the questions of whether cognitive enhancement produces more well-being in the case of people with disabilities (Chapter 15). She argues that cognitive enhancements need to be assessed on a case-by-case basis using the morphological identity framework. After defining the terms relevant to morphological identity, she suggests that cognitive enhancement is one example of the wider class of morphological changes humans can undergo. As such, frameworks for assessing the impact on well-being of other morphological changes may also be relevant for cognitive enhancements. One such framework, she argues, arises from work with people with disabilities who have experienced multiple morphological changes. According to Bradshaw, the concept of morphological identity helps us better understand the moral value of cognitive enhancement technologies because it allows us to relate their use to the effects on the well-being of individual people, and to the operation of societies more widely, as we might do for other life choices. Bradshaw's chapter highlights the importance of paying attention to how cognitive enhancement technologies may affect identity in people with physical disabilities and mental disorders. She shows that using technologies to overcome disabilities can have a major impact on how people with disabilities relate to themselves, to others, and to the world.

The focus of the literature on enhancement has mainly been on ethical issues; there has been little discussion of the legal issues. The chapter by Goold examines a range of legal issues that may be raised by putative cognitive enhancement technologies (Chapter 16). She focuses on pharmaceuticals such as modafinil, which some studies have suggested can reduce the impacts of tiredness and fatigue, and improve attention and focus in those who are well rested. Similar issues may arise with new technologies that claim to improve other cognitive abilities, such as transcranial direct current stimulation devices, which are marketed as a means of improving a person's capacity to concentrate for long periods of time and to improve memory, learning, and facial recognition. Goold's chapter presents some of the essential legal principles that may be relevant to these putative cognitive enhancement technologies. Her chapter, for example, examines how product liability rules might apply to the sale and supply of enhancement products, particularly the growing nonmedical use of devices. Related to this are questions about sale of goods and fitness for purpose, where drugs or devices are marketed with claims about what they may do for users. According to Goold, cognitive enhancement technologies also pose legal challenges for tort law, most particularly in negligence, where the availability of cognitive enhancement may affect the standard of care that is expected of persons in some professions. Goold argues that if enhancement enables us to improve our capacities, and if that enhancement becomes normalized or widespread (whether it is safe or not), this might influence how we define "reasonable care" and a "reasonable person." Goold asks the question: If drugs can improve our reaction times or our capacity to maintain our attention, should the law apply different standards to the enhanced and the unenhanced? She also suggests that questions may arise as to whether we should oblige some professionals to enhance themselves. This raises issues in the areas of negligence and employment law. Goold's chapter also touches briefly on the criminal law implications of enhancement. She focuses on the mental element of crimes and issues of consent and explores the relevance of enhancement to what it may mean to form an intention. Finally, Goold's chapter pays attention to the implication of enhancement for human rights law in a discussion that touches on privacy, particularly the emerging idea of mental or psychological privacy, and how we should protect it.

The chapter by Partridge critically analyzes the enthusiasm for cognitive enhancement shown by some bioethicists (Chapter 17). It challenges the evidence for claims commonly made in the bioethics literature on the cognitive enhancement use of stimulants, namely that it is common and increasing among college students, and that these drugs do in fact enhance cognitive function in normal persons. Partridge argues that the prevalence of enhancement use of stimulant drug use is much lower than some bioethicists claim and much of it is nonmedical use rather than use for cognitive enhancement; that controlled studies find it difficult to find evidence of the putatively cognitive-enhancing effects of stimulant drugs; and that bioethicists have underestimated the challenges in assessing the safety and efficacy of putatively cognitive-enhancing drugs. We should be aware of the potential risks to health from the nonmedical use of prescription drugs: uncritical appraisals about the prevalence and risk–benefit profile of cognitive-enhancing drugs could give rise to unwarranted policy decisions about the practice. For example, facilitating the practice by removing laws that prohibit the use of stimulants without a prescription assumes that cognitive enhancement is likely to be beneficial to the user and society. But this would appear to ignore the public health imperative that underpins regulation of these drugs in the first place. Conversely, calls for tighter regulations can also be unwarranted if not grounded in evidence. One speculated measure is for universities to drug test students prior to examinations, just as professional athletes are dope tested. However, such a policy would assume that currently available "cognitive enhancers" do in fact improve exam performance, giving users an unfair advantage over nonusers, and that there are a large number of users that can be "caught" this way. And yet there is little evidence that any of these things are true. In fact, in some situations, such a policy might only increase the prevalence of cognitive enhancement by giving nonusers the impression that stimulants do improve performance and that many of their colleagues are using them.

In their chapter Bell, Lucke, and Hall argue that the creation of the term "cognitive enhancement" has obscured historical experiences with two medicinal drugs for which similar enhancement claims were made, namely, cocaine in the late nineteenth and early twentieth centuries, and amphetamines in the mid-twentieth century (Chapter 18). These drugs were initially introduced as medicinal agents in Europe and North America before becoming more widely used for a variety of nonmedical purposes, including what would nowadays be called cognitive enhancement. Their trajectory of use conformed to the typical use cycle of psychotropic drugs: an initial steep rise in prescribing for medical use, followed by nonmedical use fueled by enthusiastic descriptions of the drug's effects; increased societal concern as the number of regular users increased and problems related to use (such as addiction) became apparent; and the passage of laws banning nonmedical and, eventually, medical use. This historical experience shows that the adverse effects of enhancement use of pharmaceuticals only becomes apparent with regular, wide-scale use of a drug. Bell et al. highlight the similarities between the historical enthusiasms for cocaine and amphetamines and the contemporary enthusiasm for using prescription stimulants for cognitive enhancement. The authors argue bioethicists should not encourage the cognitive enhancement use of drugs such as methylphenidate in the absence of evidence on the efficacy and safety of their use for cognitive enhancement purposes.

Hall and Strang outline some of the challenges in regulating the enhancement use of stimulant drugs by normal individuals (Chapter 19). They focus on approaches to regulating the use of putatively cognitive-enhancing drugs such as dexamphetamine, methylphenidate, and modafinil because these are the drugs most often discussed in bioethics debates about the cognitive enhancement use of pharmaceuticals. Their key observation is that much of the discussion of possible regulatory regimes for cognitive enhancers in the neuroethics literature ignores a critical fact, namely, that the nonmedical use of stimulant drugs is prohibited in most developed countries under provisions of the 1971 international drug control treaty. This means that the most probable regulatory response to any new neuropharmaceuticals that are (supposed to be) cognitively enhancing will be much the same, especially if their use proves popular among young adults. In the absence of good evidence about their safety and efficacy when used for cognitive enhancement prohibition is the common precautionary response that is justified by the argument that it will minimize the risk of serious adverse health outcomes that may occur if these drugs are used recreationally. These international treaties are currently under challenge in the USA where four states have legalized the recreational use of cannabis, a drug whose nonmedical use is also prohibited under the same treaties. It remains to be seen whether any reconsideration of the way that the treaties regulate cannabis use in the USA will eventually be extended to the use of stimulant drugs for cognitive enhancement.

In view of the high expectations of cognitive enhancement and concerns about the potential risks of using cognitive technologies, this book critically engages with the scientific and ethical issues in cognitive enhancement. The book aims to inform critical readers and the public of the risks as well as the promises of cognitive enhancement. It examines the assumptions made about cognitive enhancement in healthy individuals in recent ethical discussions. A distinguishing feature about this book is that, for the first time, neuroscientists, neuropsychopharmacologists, ethicists, philosophers, public health professionals, and policy researchers work together to offer a multidisciplinary, critical consideration of the neuroethics of the use of psychopharmacological drugs for cognitive enhancement.

We hope that this book makes a valuable and positive contribution to the field of neuroethics and that, as the title suggests, by providing a critical analysis of the neuroscience, as well as the ethical, legal, and social aspects of the use of smart drugs, it provides the reader a chance to rethink about the relevant issues in cognitive enhancement.

Note

1. When we use the term "healthy individuals," we mean neurotypical people who have no psychiatric problems or clinical issues, i.e., neurotypically healthy individuals.

References

Bell, S., B. Partridge, J. Lucke, et al. 2013. Australian university students' attitudes towards the acceptability and regulation of pharmaceuticals to improve academic performance. *Neuroethics* 6(1):197–205.

Bostrom, N., and R. Roache. 2009. Smart policy: Cognitive enhancement and the public interest. In *Enhancing human capacities*, ed. J. Savulescu, R. ter Meulen, and G. Kahane, 138–49. Oxford: Wiley Blackwell.

Buchanan, A. 2011. *Better than human: The promise and perils of enhancing ourselves*. Oxford: Oxford University Press.

Carter, A., and W. Hall. 2011. *Addiction neuroethics: The promises and perils of neuroscience research on addiction*. Cambridge, UK: Cambridge University Press.

Harris, J. 2007. *Enhancing evolution*. Princeton: Princeton University Press.

Harris, J. 2009. Enhancements are a moral obligation. In *Human enhancement*, ed. J. Savulescu and N. Bostrom, 131–54. New York: Oxford University Press.

Mohamed, A. D. 2014. Neuroethical issues in pharmacological cognitive enhancement. *Wiley Interdisciplinary Reviews: Cognitive Science* 5(5):533–49.

Naam, R. 2005. *More than human—embracing the promise of biological enhancement*. New York: Random House-Broadway Books.

President's Council on Bioethics. 2003. *Beyond therapy—Biotechnology and the pursuit of happiness, a report of the President's Council on Bioethics*. Washington, DC: US Government Office. Available at http://www.bioethics.gov.

Quednow, B. 2010. Ethics of neuroenhancement: A phantom debate. *BioSocieties* 5:153–6.

Sandberg, A. 2011. Cognition enhancement: Uploading the brain. In *Enhancing human capacities*, ed. J. Savulescu, R. ter Meulen, and G. Kahane, 71–91.Oxford: Wiley Blackwell.

Savulescu, J., R. ter Meulen, and G. Kahane, eds. 2011. *Enhancing human capacities*. Oxford: Wiley Blackwell.

Schaeffer, G. Owen, G. Kahane, and J. Savulescu. 2014. Autonomy and enhancement. *Neuroethics* 7:123–36.

Teter, C. J., A. E. Falone, J. A. Cranford, C. J. Boyd, and S. E. McCabe. 2010. Nonmedical use of prescription stimulants and depressed mood among college students: Frequency and routes of administration. *Journal of Substance Abuse Treatment* **38**:292–8.

Chapter 2

The ethical debate on human enhancement and cognitive enhancement by way of biotechnologies

Ruud ter Meulen

2.1 Introduction

In the past two decades there has been a lively ethical debate about the use of medical and biomedical technologies beyond traditional medical goals. While many of these technologies are developed to heal and restore health, they can also be used to improve or *enhance* human capacities beyond what is considered normal levels of human functioning (Savulescu et al. 2011). The ethical debate focuses on the question whether the use of medical technology for such enhancement can be justified from a moral point of view and whether doctors or other health care professionals should contribute to such a practice. In general one can distinguish two main positions in this debate (Schermer 2012). On one side, there are liberal and (partly) utilitarian authors who cannot see anything wrong in such efforts and even argue for a moral duty of individuals to enhance themselves (Harris 2007). On the other side, there are authors who take a more cautious and conservative position, arguing that the use of medical technologies for human enhancement may lead to a decline of important human values like dignity (Kass 2002) and solidarity with weaker groups in our society (Fukuyama 2003).

The purpose of this chapter is to provide the reader with a better understanding of the ethical implications of the empirical findings reported in this volume. At the same time we will wade into the ethical debate on cognitive enhancement by referring in various sections to the outcomes of empirical studies about the risks and benefits of cognitive drugs (and other neurotechnologies).

The chapter will start with a discussion about how to understand human enhancement, including its moral value. The chapter will continue with a discussion of enhancement in relation with the goals of medicine. This section will be followed by a discussion of the benefits and risks and, after that, a discussion of justice and access to enhancement technologies. We will then move to a discussion of more fundamental topics in relation to human enhancement, like human nature and human dignity. We will finish with a discussion about enhancement and the importance of authenticity.

2.2 The moral value of human enhancement

Before discussing the moral value of human enhancement, including cognitive enhancement, it will be important to understand what human enhancement actually means. According to Buchanan (2011) an enhancement is an "intervention … that improves some capacity (or characteristic) that normal beings ordinarily have or, more radically, that produces a new one" (Buchanan 2011, 5). Cognitive enhancement is the improvement of cognitive capacities, including various kinds of memory, attention, reasoning, and executive function (the ability to monitor direct and coordinate various mental operations (ibid.).

According to Harris (2007), enhancement has been part of human history. From "our first beginnings" (Harris 2007, 16) there has been a continuous effort to improve our functioning by education, health care, housing, language, cultivation, cooking, farming, etc. These are all ways to improve human life and can all be considered human enhancements. Harris mentions this as an argument in favor of the moral value of human enhancement by technological means:

> If the goal of enhanced intelligence, increased powers and capacities, and better health is something that we might strive to produce through education, including of course the more general health education of the community, why should we not produce these goals, if we can do so safely through enhancement technologies or procedures? (Harris 2007, 2)

As mankind has always tried to improve its capacities, what is wrong when we do this by technological means? Actually, improving by way of technology is morally superior as it is more efficient and leads to quicker results than waiting for evolutionary or cultural processes to reach a better level of functioning.

However, one can argue that there is a difference between enhancement by way of cultural and evolutionary processes and enhancement by the use of technology (Schermer 2012, 8). The direct changes in bodily and psychological functioning by means of technological interventions are different, because this process represents a new and different methodology (ibid.). Perhaps helpful is the definition by the Science and Technology Office of the European Parliament (STOA): "an enhancement is a modification aimed at improving individual human performance and brought about by science-based or technology-based interventions in the human body" (STOA 2009, 13). We shall use this definition of human enhancement and use the term *biomedical* enhancement (Buchanan 2011; Schermer 2012) to distinguish these interventions from the cultural, social, and evolutionary processes mentioned by Harris in the previous paragraph.

As mentioned in the introduction, there are different views on the moral value of enhancement. Harris, for example, argues that enhancements are unequivocally good; otherwise, we would not call them enhancements:

> Enhancements of course are good if and only of those things we call enhancements do good, and make us better, not perhaps by curing or ameliorating our ills, but because they make us better people. (Harris 2007, 2)

Harris continues that, apart from the added value of better memories, better experiences, and better processing and assimilating our experiences, we will be less slave to illness, pain, disability, and premature death. We will have less pain and we will be less dependent on doctors and medical science.

This positive view on human enhancement technologies reflects the ideals of the Enlightenment and its utopian perspective of improving the world and ameliorating human suffering through the use of science and technology (Jotterand 2010a). The utopian and rationalist perspective of the Enlightenment has been criticized by the critical theory of the Frankfurt School (Jay 1973), which has emphasized that instrumental reason has resulted in domination of our lives by technology and in an impoverishment of human relationships.

As opposed to the optimistic view of Harris, conservative authors argue that the use of biotechnologies should be limited to the goals of therapy. The use of biotechnologies for human enhancement will have fundamental consequences for human nature and will limit human freedom. Fukuyama, for example, argues that biotechnology "and a greater scientific understanding of the human brain" will have significant political ramifications. The knowledge of pathways in the brain and the workings of certain psychopharmacological drugs will open possibilities for social engineering and control: "Neuropharmacology has already produced not just Prozac for depression but Ritalin to control the unruly behaviour of young children" (Fukuyama 2003, 16). As we discover the molecular pathways between genes and traits like intelligence, aggression, and alcoholism, "it will inevitably occur to people that they can make use of this knowledge for particular social ends" (ibid.).

The US President's Council on Bioethics was particularly concerned about this possible development, which it saw as more morally problematic than medical therapy. While therapy was considered natural in trying to assist the natural healing process, enhancement was regarded by the council as adding something—possibly detrimental—to the human being that was considered unnatural:

> When a physician intervenes therapeutically to correct some deficiency or deviation from a patient's natural wholeness, he acts a servant to the goal of health and as an assistant to the nature's won powers of self-healing, themselves wondrous products of evolutionary selection. But when a bioengineer intervenes for non-therapeutic ends, he stands not as nature's servant but as her aspiring master, guided by nothing but his own will and serving ends of his own devising. (President's Council 2003, 285–6)

Critics of human enhancement technologies point out the possible eugenic tendencies that may be reinforced by these technologies (Sandel 2007). They fear that people with physical and learning disabilities might be subjected to (further) discrimination by the application of enhancement technologies, much like they were during the Nazi regime and, before that, in sterilization programs in Europe and the United States. However, the proponents of human enhancement technologies argue that there is a big difference from the "old" eugenics because the "new" eugenics emphasizes free choice and autonomy (Agar 2004). Nonetheless, according to the critics, the basic idea is the same, namely, the weeding out of undesirable physical and psychological traits.

A possible way to understand the moral significance of human enhancement is to compare its goals to those of "normal" medical therapy. Yet it is not easy to establish robust distinctions between therapy and enhancement. According to Juengst, "enhancement interventions are any interventions designed to produce improvements in human form or function that do not respond to legitimate medical needs" (Juengst 1998, 31). However, Juengst argues that the distinction between therapy and enhancement is rather complex. There is no consensus about what should be seen as legitimate medical needs or a normal application of medical technology. Normal means a treatment that falls within the goals of medicine, like the treatment of disease and the alleviation of suffering (see section 2.3). Therapy and enhancement are to a certain extent overlapping; all successful therapies are a kind of enhancement of impaired function, even if not all enhancements can be called therapeutic. For example, the improving and regenerating of organs and tissues in the elderly may be seen as enhancement, but it can be considered to be a therapy as well.

According to Harris, the distinction between therapy and enhancement "cannot be coherently or consistently maintained" (Harris 2007, 57). Therapy and enhancement are not mutually exclusive categories: they have the moral imperative both to prevent harm and to confer benefit. In that moral light, Harris argues, it is unimportant whether the protection (for example, by vaccination) or benefit conferred is classified enhancement or improvement, protection or therapy. As there are no differences between the two, it makes no sense to argue for permissibility or impermissibility of one or the other based on such a distinction (ibid., 58). Instead, Harris argues, we should look at the benefits or harms that both therapy and enhancement confer and look at the right balance between these when making decisions about their permissibility.

However, though a certain therapy can be an enhancement, not every enhancement is a kind of therapy. When modafinil is used for the treatment of narcolepsy, it is clearly used for a therapeutic purpose. However, when pilots or surgeons take modafinil to stay awake for a longer period (to cope with a long flight or when performing a long surgical procedure) we cannot speak of a therapy but of enhancement. The benefits and harms have a different and noncomparable meaning in both cases: when using modafinil to treat narcolepsy we will accept a different balance of benefits and harms than when it is used as an enhancement. The benefits and harms in the case of therapy are entirely defined by the interests of the affected individual. In the case of enhancement we look not just at the benefits and harms of the pilot or surgeon, but also at the risks of the patient when undergoing surgery or the risks of the passengers on the long haul flight. For example, there is some evidence that modafinil could increase overconfidence which can put the patient at risk (Baranski and Pigeau 1997; Baranski et al. 2004; Repantis et al. 2010).

2.3 Enhancement and the goals of medicine

An important ethical question is whether doctors and health care professionals should be involved in human enhancement. In other words, does enhancement belong to what is called the *goals of medicine*, meaning which activities should be part of medical practice and what are the professional duties of doctors in relation with these activities? In a special supplement to the Hastings Center Report on the goals of medicine, a group of experts from fourteen countries identified four core values which would help maintain medicine "to maintain its integrity in the face of political or social pressures to serve anachronistic or alien purposes" (Hastings Center Report 1996):

1) The prevention of disease and injury and the promotion and maintenance of health;
2) The relief of pain and suffering caused by maladies;
3) The care and cure of those with a malady and the care of those who cannot be cured; and
4) The avoidance of premature death and the pursuit of a peaceful death.

The discussion about what constitutes the goals of medicine has for an important part to do with the boundaries of medical practice and of the health care system: when we have a shared understanding of what doctors (and other health care professionals) should consider as their core values, we have a better idea of which activities should be reckoned to be part of our health care system and, in accordance with this, which activities should be funded by the public health care system.

A definition of the goals of medicine might help to limit the tendency of medicine to stretch the limits of the medical domain and engage in activities that cannot be considered appropriate in relation to its core values and goals. The discussion about the goals of medicine is particularly meant to put a halt to the *medicalization* of society, a concept used to describe the increasing influence of medicine in various areas of our society. The concept of medicalization was launched in the early 1970s by Zola to describe the process in which modern medicine has become one of the most important mechanisms of social control, taking the place of religion and law (Zola 1972). By defining certain phenomena like alcoholism, aggression, ageing, or reproduction from a medical perspective, society is becoming more able to control such phenomena. Illich further developed the thesis of Zola in his book *Medical Nemesis* (1975) in which he argued that medicine and health care do not improve health but in many cases are responsible for a worsening of people's health, a process which he called *iatrogenesis*. According to Illich, medicalization means increased power of doctors and an increasing dependency of individuals on the medical system.

An example, which might be relevant for this volume, is the rapidly increasing use of Ritalin to treat attentional problems in young people who are diagnosed as having attention deficit hyperactivity disorder (ADHD). In a growing number of cases, the diagnosis of ADHD and the use of Ritalin to remedy this disorder can be regarded as medicalization of children's behavior in order to better control it. It includes a definition of what is called appropriate behavior and also an explanation of this behavior as caused by biological processes in the children's brain. Critics of the increased use of Ritalin refer to the social and psychological origins of so-called problematic behavior of young people, which in their view are snowed under the biomedical explanations. Moreover, according to the critics, the indications for ADHD are not clear and there is a tendency to apply the diagnosis to an increasingly wider range of behaviors (Coppock 2002).

Does enhancement fall under the goals of medicine, or is it another manifestation of medicalization, meaning the use of medical technology in areas like education which would normally depend on social and individual factors? It is difficult to answer this question: on the one hand, enhancement cannot be seen as healing a disease, but on the other hand it may promote health and help to avoid premature death. An important problem is that there is no consensus on how to define health and disease. An example is the question whether we should treat children with low height with growth hormone. In some cases there is no biological cause of the low height. According to Daniels (1992, 2000) one can ask whether one can make a clear-cut distinction between treatment of a child with growth hormone deficiency that is the result of a brain tumor and enhancement of one's child's height because of a normal hereditary shortness (Daniels 1992, 2000). Both children have the same height, but it is a height that is seen as inferior according to our cultural standards. However, in the first case, the problem has a medical cause, namely the tumor. In the second case, the child's (future) shortness is part of a normal variation in human height. The big question is, do the two cases fall within the area of medical necessity, and if yes, do they qualify for medical treatment? Or are we on the path of unacceptable medicalization when we think that both cases qualify for medical treatment and reimbursement?

According to Kass, medicine should limit itself to restoring normal functioning and should only treat physical and biological needs (Kass 1985). This biological view on health (and restoring health) can also be found in the works of Christopher Boorse (1977). According to Boorse (1977), health can be described in terms of functions which are typical for a certain species to which an organism belongs. A function is something that contributes to the goals of a certain organism including its natural design. Such a design is different for every species. However, we can determine empirically what this design is and which goals an organism is striving for. To do so, we should research a large number of exemplars of a species to find its "species design," differentiated to age and gender. The result of this biological-statistical analysis is that biological functions are in the end focused on survival and reproduction of the species. On the basis of this analysis, Boorse argues that health equals "normal functional ability." To assess the health of an individual, a doctor should focus whether the body of that individual still works according to the species design.

Boorse's biological-statistical concept of health has been much criticized as being reductionist and failing to account for conditions that, according to his theory, can be considered healthy, but which are treated by the health care system (like caries, prostate enlargement, and arterial problems; Ten Have et al. 2013). Moreover, many mental conditions and physical handicaps are excluded from medical treatment as they have nothing to do with normal species functioning. An important problem with Boorse's (and Kass') approach is that it does not take into account the social and cultural context of specific conditions. A condition like color blindness, for example, is in itself a neutral condition. To call it a disease is dependent on the context in which people with this condition live (ibid.). According to some authors, the distinction between disease or disability and health or normality is, in fact, culturally defined. Engelhardt, for example, argues that concepts of health and disease are guided by value judgments and prejudices that may change over time (Engelhardt 1996). Examples are masturbation, which was for a long time considered a disease (Engelhardt 1974), and homosexuality, which was removed from the *Manual of Psychiatric Disorders* (DSM) in 1974.

One can argue that the distinction between normal and enhanced is basically normative rather than objective or universally valid, as Boorse seems to assume. One can also argue that the distinction between treatment and enhancement is based on a cultural definition of what counts as a disease and what does not. If one follows that argument, the distinction between therapy and enhancement will become highly questionable and nothing will stand in the way of the application of biotechnologies outside the medical domain.

However, it seems that doctors do not feel much interested in going beyond the traditional goals of their clinical practice. This could be concluded from McKeown's interview study reported in his chapter in this volume among renal doctors regarding the use of erythropoietin for physical enhancement (see Chapter 12). The doctors involved did not object to enhancement as such, but thought that the provision of enhancement drugs was far away from their day-to-day clinical work, which was primarily helping sick patients and caring for their immediate needs. They did not regard enhancement as a goal of medicine, but would reconsider their position if enhancement would become a more legitimate goal of medical practice. Nonetheless, in daily clinical practice there are signs that doctors do go beyond the goals of medicine, for example by supplying Propranolol, a beta-blocker used for the treatment of high blood pressure, to professional musicians in order to reduce performance anxiety. Similarly, doctors prescribe psychopharmacological drugs like Prozac for people who feel depressed but do not meet the criteria for clinical depression. With reference to the title of a book by Carl Elliott (2003), one can say, that these people want to feel "better than well."

Though it is difficult to draw the boundaries between normal functioning and what goes beyond it, it is important to continue the discussion on what counts as legitimate goal of medicine and whether so-called enhancement technologies fall within this scope. As Juengst argues, the distinction about enhancement, as distinguished from therapy, is helpful in defining the boundaries of medical practice. Concepts of health and disease, sociological perspectives on medical practice, or a theory about what is the human norm can be considered tools to help define those boundaries. This is important because, as Juengst argues, the line that the treatment versus enhancement distinction draws is the boundary of medical obligation, not the boundary of medical tolerance (Juengst 1998, 44). Doctors should be able to deny prescribing enhancements if they go against their professional judgment, but such enhancements should still be permissible. This distinction is relevant for policy-making regarding access to care and public funding of what from a medical point of view is necessary and what should be left for private arrangements and funding in respect with self-improvement. This does not mean that in private arrangements for self-improvement doctors do not have professional obligations: as, for example, in cosmetic surgery, doctors still need to take care of the well-being of their patients and protect them against the risks of medical procedures or drugs.

2.4 Risks and benefits

An important aspect of the ethical assessment of human enhancement technologies is weighing their possible benefits against their risks and burdens. Such an outcome-oriented assessment is typical for consequentialist approaches in ethics of which utilitarianism is the most well-known and dominant type. Utilitarian authors are usually rather positive in their assessment of enhancement technologies which they regard as important improvements of individual human functioning and the functioning of society. Enhancement technologies are considered to fulfill what utilitarians see as the primary moral goal of human action, that of producing the "greatest good for the greatest possible number". As Harris (2007) argues, if we want to make the world a better place, we should not only change the world, but change human nature as well, even if the risk will be that our descendants will not be regarded as human any more. Instead of being dependent on the natural lottery, we should intervene in it and take control of human evolution beyond the point where we will have changed and have developed "perhaps into a new and certainly into a better species" (ibid., 5). Human enhancement, according to Harris, can be defended and justified because it will make people "longer-lived, stronger, happier, smarter, fairer" (ibid., 5). As these are important individual and social goals, he argues that we have duty to enhance ourselves.

Harris argues that enhancements should be promoted if they bring good and if they are safe. However, there is little scientific evidence that this is the case. For example, as many chapters in this volume demonstrate, the benefits of using cognitive drugs like Ritalin, Aderall, and modafinil are hardly proven, while the risks of using these drugs are quite substantial. First of all, it should be made clear that these drugs do not improve "intelligence." Amphetamine, methylphenidate, and modafinil improve primarily motivation and vigilance. There is some evidence that metylphenidate (Ritalin) may improve some aspects of memory (Repantis et al. 2010), executive functions, and working memory (see Chapter 3 by de Jongh in this volume). Smith and Farah (2011) found amphetamine improved long-term memory (particularly factual knowledge). Adderall may improve function in subjects with low IQ, but impairs functioning in subjects with high baseline performance, which is called the "baseline dependent effect" (Mehta et al. 2000; Quednow 2012). In view of these "baseline" or "ceiling effects," the alleged use of cognitive drugs among academic staff and students (Boseley 2014) seems to be pointless, as cognitively high-performing individuals do not benefit from using these drugs.

In general, laboratory studies on the effects of cognitive drugs give mixed results. This may be due to individual (genetic) differences and the baseline dependent effect (see Chapter 3 in this volume). However, there are some other effects which may lead to a less favorable view of these drugs than they currently receive in the media and the general public (Boseley 2014). An important issue is, for example, the overall impact of these drugs on the brain and on cognitive functioning. Research has found out that improvements in one cognitive domain are associated with impairment in other domains (Quednow 2012; Sandberg 2011). In a recent paper Ahmed Dahir Mohamed points at that drugs which promote attention and increased focus may decrease creativity (Mohamed 2014). Modafinil increases wakefulness of sleep-deprived individuals but does not improve their attention and executive control (see also Mohamed and Lewis 2014). This raises the issue of the risk of overconfidence (Baranski and Pigeau 1997; Repantis et al. 2010; Chapter 3 in this volume), which makes it questionable whether, for example, surgeons or airline pilots should take this drug.

There also concerns about the addictive potential of some of the cognitive drugs (modafinil) as they raise the dopamine levels in the brain. The dopamine release is much higher with these drugs (175–1000%) than by food, human communication, and sex (50–100%) (Heinz et al. 2012). The evidence on the potential risks of addictive behavior induced by the use of drugs like modafinil raises questions whether such drugs are really improving individuals' autonomy as some authors argue (Juth 2011; Schaeffer et al. 2014). Besides, these drugs (particularly the drugs containing amphetamine) have serious side effects, like heart and blood pressure problems, heart rate problems, headache, nervousness, insomnia, and mental health problems (Quednow 2012). When these drugs are used outside a medical context the safety standards should be high, certainly if there is no control or monitoring by doctors.

One of the problems for utilitarianism in general is that it is not clear what counts as "benefits" and what counts as "harms." Though this is already difficult in the case of therapy, it is even more problematic in the case of enhancement. This concept is strongly linked to individual views on personal identity and to personal preferences for specific lifestyles. The use of cognitive drugs may seem attractive for those who are engaged in the academic "rat race" or other social arenas, but may seem totally irrelevant for people who are in the creative world or in some caring professions.

2.5 **The perspective of justice**

One of the issues in the ethical debate on human enhancement technologies is to which extent societies should support individuals, particularly from low-income groups, to have access to these drugs: if they are effective, should we as a society make funds available to create a level playing field by helping disadvantaged groups? Or should we allow further inequality in access to important societal functions as some groups may be more able to afford to use cognitive enhancers? This issue relates to the fundamental ethical issue of access to drugs or the *just distribution* of benefits and burdens of access to cognitive drugs.

Justice has to do with our deep understanding that as a society we have an obligation to help those persons who are in need of medical treatment but who do not have the means get these treatments. Though many would agree with this idea, we find it difficult to accept that societal resources are spent on technologies that are only fulfilling individual preferences, for example, to lead a happier life or to enhance physical performance in sport, or to be smarter in the class room and at exams. Certainly when resources are scarce, we tend to limit the access to medical services. In these circumstances we do not think medical assistance should be offered that is not meant to fulfill a need for medical treatment because of illness or disease, but that is meant to fulfill a preference for individuals to perform better than they usually do, or to feel "better than well" (Elliott 2003). Access to care, at least publicly funded care, is narrowly linked with the perspective of the goals of medicine as discussed in section 2.3. For treatments to be funded, most health care systems in the world have as a condition that they should fall within the medical domain. This is a necessary, but not a sufficient condition: there are medical treatments, like advanced cancer drugs or transplantations, which can be considered medical treatments, but which are so expensive that society cannot afford to pay for these treatments for all those who are in need of them.

However, the idea that we should only publicly fund services that treat a disease and that we should not fund enhancements because they are just preferences is not so easy to implement in public policy. As discussed earlier, the distinction between needs and wants is not clear, particularly because there is no consensus on how to define health and what constitutes disease. In section 2.3 we discussed the biological-statistical concept of health as advocated by Boorse. This concept has been widely criticized because it rules out many conditions and is not able to account for cultural and historical variations (Engelhardt 1996). Nonetheless, Boorse's concept of normal functioning plays an important role in the approach of Daniels to distributive justice in health care based on Rawls' theory of justice (Daniels 1985, 2008; Rawls 1972). According to Daniels, normal functioning health should be defined in terms of the biological functioning that is typical for individuals of a certain species. Typical functions are those that contribute to the goals of the organism, and these goals can be inferred from the "natural design" of that organism. Health means "normal functional ability," i.e., able to function according to the natural design of the species. Disease is then any state that interferes with "normal functioning." In case of disease, normal functioning is blocked or diminished, and herewith the access of individuals to the range of opportunities that are open to them according to their abilities. According to Daniels, there is a moral obligation on society to restore normal functioning because it helps individuals to have access to their normal opportunity range. By restoring normal functioning, health care preserves people's ability to participate in political, social, and economic life. Buchanan et al. (2001) argue that there is no moral obligation on society to fund enhancement technologies as they are not focused on the restoration of normal functioning. In their view, the distinction between enhancement and therapy is relevant as the basis for decision-making about the kind of medical and biotechnical services that should be supplied by society (Buchanan et al. 2001).

This view is challenged by some authors who argue that in Daniels' view there will still exist differences among individuals in their possession of capabilities in society, which hinders some of them from reaching good positions in society. For example, intelligence and good memory might be essential for success in the competition for social positions. The application of enhancement technologies might result in a more even distribution of these capabilities among individuals, which would be more just from an egalitarian point of view. For example, Sen (1990) argues from the perspective of "resource egalitarianism" that, if resources ought to be distributed equally and natural resources are resources, then we ought to intervene in the natural lottery, because whenever doing so would be the best way of equalizing resources (Sen 1990). Scanlon, who advocates a "brute luck theory," argues that persons should not have fewer opportunities as a result of factors that are beyond their control. These circumstances do not only include the effects of the social lottery, like poverty or ethnic discrimination, but also the consequences of the natural lottery like genetic differences (Scanlon 1989).

Both resource egalitarianism and brute luck theory are committed to the thesis that justice may require interventions to counteract natural inequalities by means of biotechnological interventions, regardless of whether they constitute diseases. The findings reported earlier that some cognitive drugs are helping individuals who are at a lower level of cognitive functioning (while individuals on a higher base level do not respond favorably to these drugs) could be considered as an argument in favor of an egalitarian approach to access to cognitive drugs for enhancement purposes (see Chapter 3 in this volume).

Such proposals are contested by liberal authors: they admit that in a market-oriented society, we can expect that genetic technologies and other biotechnologies will be used by the rich and powerful beyond the medical domain in order to enhance their capacities and opportunities in social life. From the viewpoint of distributive justice, it can be argued that society should prevent a widening gap between the better-off and the worse-off in respect to social opportunities. However, that is no reason to equalize the differences in capacities by funding access to enhancement technologies for all members of our society. Such a policy would seriously affect the possibility of competition in our society and the principle of liberty which is a major principle in Rawls' theory of justice:

> Just as we do not owe it to our friends or others in general to contribute our resources to making them happy when they are unhappy because they have developed extravagant tastes, we do not owe it to others to improve any and every capability that they judge to be disadvantageous to them, given their plans of life. (Buchanan et al. 2001, 136)

Egalitarianism and radical egalitarians do not see a difference between enhancement and therapy, as they both affect the chances of individuals in society. Contrary to this position, liberal or liberal-egalitarian authors like Daniels, Buchanan, and others want to stick to this distinction because failure to do so would seriously affect liberty in our society. However, they do advocate access to biotechnologies where they restore normal functioning. Access to technologies for the purpose of enhancement of capabilities above normal functioning fall outside the domain of health care and should be regulated within the broader social context.

2.6 Enhancement and human dignity

A fundamental value that is sometimes referred to in the context of cognitive enhancement, and human enhancement in general, is human dignity. Though the concept is said to lack a clear definition and is dismissed by some as a meaningless perspective in bioethics (Macklin 2003), it is seen by others as an important concept in exploring the moral significance of human enhancement technologies (Ashcroft 2005; Beyleveldt and Brownsword 2001; President's Council 2003). Human dignity is a key concept in the influential, but often contested, report *Beyond Therapy* (2003) by the US President's Council on Bioethics on human enhancement technologies. The report argued that biotechnical interventions may help improve excellence, for example, by improved attention and memory in the area of human cognition. Drugs that decrease drowsiness, increase alertness, sharpen memory, or reduce distraction may help people in the actual pursuit of learning or performing (President's Council 2003, 291–2). However, there is a risk that drugs or biotechnologies take over the activities or performance which may then lose by consequence their "naturalness." According to the report, in such cases there is a "danger of violating or deforming the nature of human agency and the dignity of the naturally human way of activity" (ibid., 292). By using cognitive enhancement technologies the human subject becomes separated from the human significance of the activities so altered: "He can at best *feel* their effects *without understanding their meaning in human terms*" (ibid., italics report). Instead of risking "self-alienation" the Council argues for the recognition and acceptance of limits to our abilities.

Similarly, Fukuyama (2003) argued that the reduction of human qualities for utilitarian purposes will affect our dignity, which he defined as a range of human qualities and abilities that connect us to other human beings. The utilitarian goal of minimizing suffering leads to a destruction "of the way we react to, confront, overcome, and frequently succumb to pain, suffering and death" (Fukuyama 2003). In the absence of these human evils, we lose qualities like sympathy, compassion, courage, heroism, solidarity, or strength of character which are essential for human dignity.

However, some *transhumanists* see no reason why human enhancement technologies will affect human dignity in a negative sense. Transhumanism is an international movement to transform humanity by the development and use of enhancement technologies (Bostrom 2005a). Proponents of this idea claim that enhancement of existing human capacities will make us overcome our current vulnerability, and will result in a post-human state in which we are no longer dependent on our natural surroundings (ibid.). The possibility, for example, of uploading our brains like computers will eventually make us more able to deal with human mortality. According to Bostrom, human enhancement technologies, instead of losing or diminishing human dignity, may in fact help to increase our dignity since they will empower post-human beings to control the risks and uncertainties of their existence. This process will certainly not lead to the disappearance of agency and individuality as the council argues: instead enhancements may help individuals to develop their creativity and originality and thus to further develop their dignity (Bostrom 2005b, 2008).

Bostrom's views were criticized by Jotterand (2010b), who argued that the idea of dignity as a quality would result in the view that some people may have more dignity than others. This idea goes against the original idea of dignity, which is an expression of the idea that all human beings have the same intrinsic value. This idea has religious roots in Christianity, but received a secular foundation in the philosophy of Immanuel Kant. According to Kant, persons have intrinsic worth or dignity because of their rational and moral capacities (Kant 1788, 2004). This idea was the background of Kant's categorical imperative which said that one should always act so as to treat humanity, whether in your own person or in others, always as an end, and never merely as a means.

Respect for human dignity has become a key issue in various documents and international declarations on bioethics, like the Universal Declaration of Bioethics and Human Rights (Jotterand 2010b). According to Jotterand, the respect for human dignity and the fundamental equality of individuals as an intrinsic value is at odds with the idea that one can augment dignity or increase it by enhancement technologies. Jotterand argues that Bostrom's views may lead to a dehumanizing of society as it goes against the idea of equality and uniqueness of every individual. Dignity as quality means that dignity becomes a contingent value which some people may have and others not.

Fukuyama and other authors are afraid that the use of human enhancement technologies, particularly when use for genetic enhancement, may create new divisions in our society. Fukuyama (2003) is worried about the emergence of new genetic classes, which could increase inequality in our society. Fukuyama thinks that genetic enhancement, if and when it comes possible, may lead to a more egalitarian society, particularly when access to these technologies is funded for everybody, but there is also a serious danger that it will divide our society into different genetic classes. According to Fukuyama, genetic enhancement will

> not be threats to the dignity of normal adult human beings but rather to those who possess something less than the full complement of capabilities that we have defined as characterising human specificity. The largest group of beings in this category are the unborn, but it could also include infants, the terminally sick, elderly people with debilitating diseases, and the disabled. (ibid., 174)

Genetic enhancements will put increasing emphasis on intelligence, cognitive capacities, and sensitive emotions as defining characteristics for dignity and humanity. Humans that do not have these (enhanced) capacities will be seen as inferior and as possessing fewer human rights.

Fukuyama's analysis is shared by Sandel (2007), who raises serious concerns about the quality of our future relationships in the "posthuman society." If genetic enhancement and other kinds of enhancement are not integrated in a broader moral worldview on human personhood and dignity, Sandel believes that they might indeed increase individualization in our society, deny the dignity of nonenhanced groups, and decrease our solidarity with weaker and more vulnerable groups.

The discussion above illustrates that the concept of dignity can be understood in different ways and can lead to different views when used in the ethical evaluation of enhancement technologies. Some authors reject dignity as a useful concept in bioethics, including the ethical analysis of human enhancement technologies. Others think that enhancement may increase dignity (Bostrom), while others argue that enhancement will threaten human dignity (Fukuyama). Despite these different opinions, the ethical debate should include the concern about the impact of a nonrestricted or nonregulated use of enhancement technologies on the vulnerable groups in our society. A defense of the principle of dignity and the unique value of each human individual may help to prevent use of enhancement technologies producing a less favorable view of nonenhanced individuals and possibly a division in society between those who are enhanced, genetically or by other technological and cognitive interventions, and those who are not.

2.7 **Enhancement and human nature**

The idea of respecting human dignity is to some extent connected with concerns about human nature which might be changed by the application of enhancement technologies. For example, if we implant devices in human brains or replace limbs which add physical strength or mobility, are we not changing the nature of humanity? Can we still speak of human beings or are they transformed to post-human entities?

The President's Council argued that underlying enhancement is a "hyper-agency," a "Promethean desire to remake nature, including human nature, to serve our purposes" (President's Council 2003, 286). The Council makes a plea to appreciate and respect the "giftedness" of the world. In his book *The Case against Perfection* (2007) Sandel, who was a member of the President's Council when they produced the report on human enhancement, argues that we have become *too* responsible for our own fate. "Parents become responsible for choosing, or failing to choose, the right traits for their children. Athletes become responsible for acquiring, or failing to acquire, the talents that will help their team win" (Sandel 2007, 87). This "explosion of responsibility" has come at the expense of an acceptance of our limitations and the notion of "giftedness." Sandel argues that there is a connection between the notions of "giftedness" and "solidarity": as soon as we are aware of the contingency of our gifts, we will develop our capacity of seeing ourselves as sharing a common fate.

In opposition to this cautious position, proponents of enhancement claim that the desire and pursue of enhancement using new technologies is part of human nature and must be considered morally good. Because of his nature, man is enabled and even morally obliged to enhance his capacities. We may even be considered as having a moral obligation to do so. They argue that, as opposed to other living beings, only human beings possess the potential to master their physical and cognitive capacities to an extent that goes beyond the element of natural chance:

> far from being unnatural, the drive to alter and improve on ourselves is a fundamental part of who we human are. As a species we've always looked for ways to be faster, stronger, and smarter and to live longer. Many past enhancements that we now take for granted—from blood transfusions to vaccinations to birth control—were called unnatural or immoral when they were first introduced. Yet over time we've become accustomed to these new levels of control over our minds and bodies, and have used them for the betterment of ourselves, our families, and our world. (Naam 2005, 9)

A radical elaboration of this approach is represented in the idea of *transhumanism* or *post-humanity* (see section 2.6). Proponents of these ideologies claim that enhancement of existing human capacities will overcome our current vulnerability, and result in a post-human state in which we are no longer dependent on our natural surroundings (Bostrom 2005a). The possibility, for example, of uploading our brains like computers and growing new organs using human stem cells will eventually make us able to deal with human vulnerability to disease and damage. Such radical alterations will change the core nature of Homo sapiens and turn us into post-humans. As mentioned earlier, this argument is criticized by Fukuyama (2003) and others, as this development would lead to divisions in our society between enhanced or post-human beings, on the one hand, and non-enhanced human individuals, on the other hand.

Enhancement can be interpreted both as part of human nature and as a potential violation of human nature. Some would claim that nature and natural process are fragile, and that any interference in them must be approached with caution. According to the more cautious position, the unknown long-term effects of the new enhancement technologies make our own desire for enhancement of human capacities a potential danger for present and future generations. Optimistic authors claim that the desire to enhance ourselves is a natural part of being human. "To turn our backs on this power would be to turn our back on our true nature" (Naam 2005, 10).

According to Edgar, post-humanists and enhancement supporters refuse to see the (existing) human body as part of our fate (or what is "given" to us) but as "brute matter" which we can manipulate and change without any impact on our self-understanding (Edgar 2009). Authors who are critical of human enhancement argue that intervening in the body with the purpose of enhancement is not a morally neutral procedure: It is an attempt to control the contingency of our nature with important consequences for self-understanding of the individual and interpersonal relations in our society.

Human nature is a complex concept and can lead to rather different interpretations and moral evaluations of human enhancement technologies. However, that is not a reason to dismiss it as an invalid concept in the ethical debate. The fact that it is often referred to in the debate may be considered as a sign that the concept expresses an important moral value that we should take into consideration when evaluating the impact of the various enhancement technologies, including cognitive enhancement.

2.8 Easy solutions and virtues

Human enhancements are sometimes regarded as easy solutions or "quick fixes" for educational or learning problems. This can displace efforts to resolve these problems by natural, social, or psychological means. Sandberg, for example, refers to the use of Ritalin and other ADHD medications which are used not only as cognitive enhancers by healthy individuals but also are widely used to treat so-called difficult behavior among young people to make them calmer and more cooperative. This development can be seen as an example of the growing power of medicine (see section 2.3) but also as a quick technological solution to an educational or cognitive problem as well as social problems. An important moral question then is whether we are losing the capacity to develop ourselves and our personalities by taking resource to "easy solutions" like cognitive drugs and whether, for example, giving drugs to subdue children with "difficult" behaviors in the classroom is a substitute to good parenting and early emotional and cognitive enrichment.

The moral issue is whether we should value struggle and effort or whether we should not feel bad about taking the easy route to our goals and aims. Shickle (2000) argues that genetic enhancement technologies "bypass" the struggle involved in many activities. Following the ideas of Joseph Amato about the value of suffering, Shickle argues that struggle is involved in many activities and that human activities are valued in relation to the amount of suffering that they involve:

> Struggle is a measure of human activity, and even pleasurable activities require efforts to that result in pain .... The more that we have to struggle to attain a thing, the stronger the claim to possession. It is the process of striving that makes things worthwhile. (ibid., 349).

By eliminating struggle in our existence, human enhancement technologies decrease our ability to develop a sense of value and of meaningful relationships with others: "individuals using enhancements with a view to gaining a positional advantage are attempting to circumvent (struggle). The individual and a society made up of such individuals may be poorer as a consequence" (ibid., 349–50).

Other ethicists have no problem at all with the use of enhancement technologies as an easy route towards our goals. Caplan, for example, asks why we should feel bad about taking a pharmacological intervention to improve our attention, while we do not so when taking a coffee every morning. Are the cognitive drugs just "cheap thrills that you don't really deserve? ... Should you run around the block a few times instead of taking these shortcuts?" (Caplan 2006, 37).

> Some people do think that the only way to get to the top of the mountain is to hike up there. I don't have a problem with that, that's fine. Me, I like a helicopter. View's the same. I don't care. I get to the top. I get to see it. It's faster. I leave. (ibid.)

Against such an outcome-oriented perspective, critics argue that the use of enhancement technologies, like taking a cognitive drug to pass an exam, makes people not only dependent on these drugs, but makes their accomplishments less authentic. It is the drug that does the work and not the individual. This argument also underlies the concerns of the President's Council about the loss of dignity and the increase of non-naturalness of human efforts when using enhancement technologies. According to the council the use of cognitive drugs like Ritalin for tests lessens our regard for the achievements of the doer. At the same time, we can question to which extent the achievements express the individuality of the doer (President's Council 2003, 294).

According to Juengst, the question is indeed whether the student earned the grade, that is whether the grade is serving its usual function of signaling the disciplined study and active learning of the student. That is what being a student requires. If not, the grade is a "hollow accomplishment" (Juengst 1998, 39). The use of cognitive drugs like Ritalin in exams does not only raise questions about the intrinsic nature and value of the accomplishments of students taking these drugs, but also about the fairness of the outcomes of the tests (ibid.). Did the individuals using the enhancing short-cuts not have an undeserved advantage over the nonenhanced pupils? In other words, have they not been cheating? In a paper by Schermer included in this volume, she argues that strictly speaking taking cognitive drugs is not cheating because there are no rules regarding the use of these drugs and there are by consequence no rules broken. However, she points at the idea that taking these drugs may break internal values and practices that are inherently part of the academic achievement process such as working hard, honestly studying for the exams, and fair play. With reference to the defense of virtue ethics of MacIntyre (1984), Schermer argues that we should better look at the internal values of education as a practice and of the virtues and promotion of virtues to sustain such a practice. It is not clear whether cognitive-enhancing drugs are contributing to virtues and personal development or whether they just help pupils to compete in the academic and societal rat race.

According to Elliott, some individuals use these technologies to make their existence more authentic (Elliott 2003). With reference to Charles Taylor (1991), Elliott argues that authenticity has become a powerful ideal in our modern western culture. Authenticity means nowadays that we have to get in contact with our own inner nature, our inner voice, particularly when it threatens being lost because of external pressure to conform. It is strongly linked to the idea of originality; each voice has something of its own to say. According to Elliott, this ideal of authenticity is driving the language of individuals using biotechnologies and other technologies to enhance their functioning. From Prozac to face-lifts, using these technologies or drugs make people feel more authentic, more "being themselves" (Elliott 2003).

Some communitarian philosophers argue that such a search for the self has resulted in narcissism and estrangement from others, a process reinforced by the use of enhancement technologies. In its discussion of the role of Prozac and other antidepressant drugs in mood elevation, the President's Council stated that individuals become so preoccupied with their state of mind "that they remove themselves increasingly from active participation in civic life, discarding those attachments without which they cannot achieve the happiness they seek and without which the community cannot survive and flourish" (President's Council 2003, 264). The danger of a widespread use of mood-enhancing drugs, according to the council, involves the "solipsistic self worried only about the state of his feelings, who uses psychopharmacology to ensure a flat and shallow self-regarding psychic pleasure" (ibid., 264).

However, according to other authors the President's Council on Bioethics may be too pessimistic about the way enhancement supposedly reduces social and civic life. It is debatable whether the use of antidepressants leads to "affective blunting" (Furlan et al. 2004). Other authors argue that the use of these drugs can help people to relate better with their spiritual and social world. In his book *Listening to Prozac*, Peter Kramer gives examples of how patients by using Prozac felt more able to cope with ordinary life, overcoming shyness and social inhibition (Kramer 1994). It is not clear whether mood-enhancing drugs and cognitive drugs indeed affect the identities of people, their authenticity, and their relations with others. Nonetheless, it will be important to look at identity and authenticity as important issues in the context of the use of cognitive drugs.

2.9 **Conclusion**

In this chapter we have presented an overview of the main issues in the ethical debate on the use of biotechnologies to enhance human functioning. We have distinguished two main positions in this debate: on the one hand, there are favorable or even enthusiastic views, particularly among utilitarian authors, about the possibilities of human enhancement in relation to cognitive, educational, and physical capacities. On the other hand, there are more cautious and conservative authors who are concerned about the impact of enhancement technologies on the personal development of individuals and on fundamental values like human dignity, solidarity, and human nature. It will be difficult to find a consensus on the potential contribution of enhancing technologies due to the very different philosophical and partly ideological positions that are guiding the debate.

Interestingly, both the optimistic and the cautious positions hardly refer to empirical evidence about the impact of enhancement technologies: The optimistic view is not very interested in the question whether enhancement technologies really work or at the possible harmful side effects and risks of the use of these technologies. Similarly the gloomy scenarios pictured by cautious authors are not based on empirical data about the impact of human enhancement technologies on our societies. This is particularly true of the discussion of cognitive enhancement technologies, like psychopharmacological drugs and other neurological technologies to improve cognitive functions and abilities like alertness, memory, and attention. Do these drugs indeed improve wakefulness or memory? What are the risks of using these drugs? What is the impact on other cognitive functions when enhancing one particular function? Various contributions to this volume are trying to answer these questions by presenting evidence about laboratory studies and other empirical information. Although empirical data cannot in the end decide the outcome of the moral debate, this debate will have little value if it is not informed by empirical evidence about what such technologies and drugs accomplish: Do they work and what are the consequences of using them on physical and psychological health and on the well-being of individuals and the functioning of society as a whole?

References

Agar, N. 2004. *Liberal eugenics: In defence of human enhancement.* Oxford: Blackwell.

Ashcroft, R. 2005. Making sense of dignity. *Journal of Medical Ethics* **31**:679–82.

Baranski, J., and R. Pigeau. 1997. Self-monitoring cognitive performance during sleep deprivation: effects of modafinil, d-amphetamine and placebo. *Journal of Sleep Research* **6**:84–91.

Baranski, J., R. Pigeau, P. Dinich, and I. Jacobs. 2004. Effects of modafinil on cognitive and metacognitive performance. *Human Psychopharmacology: Clinical and Experimental* **19**(5):323–32.

Beyleveld, D., and R. Brownsword. 2001. *Human dignity in bioethics and biolaw.* Oxford: Oxford University Press.

Boorse, C. 1977. Health as a theoretical concept. *Philosophy of Science* **44**:542–73.

Boseley, S. 2014. £200,000 "smart" drugs seizure prompts alarm over rising UK sales. *The Guardian* 24 October 2014.

Bostrom, N. (2005a). A history of transhumanist thought. *Journal of Evolution and Technology* **14**(1):1–27.

Bostrom, N. (2005b). In defense of posthuman dignity. *Bioethics* **19**:202–14.

Bostrom, N. (2008). Dignity and enhancement. In *Human dignity and bioethics: Essays commissioned by the President's Council on Bioethics*, 173–206. Washington, DC: US Government Office. Available at http://www.bioethics.gov.

Buchanan, A., D. Brock, N. Daniels, and D. Wikler. 2001. *From chance to choice. genetics and justice.* Cambridge: Cambridge University Press.

Buchanan, A. 2011. *Better than human: The promise and perils of enhancing ourselves.* Oxford: Oxford University Press.

Caplan, A. 2006. Is it wrong to try to improve human nature. In *Better humans? The politics of human enhancement and life extension*, edited by P. Miller, and J. Wilsdon, 31–9. London: Demos.

Coppock, V. 2002. Medicalising children's behaviour. In *The new handbook of children's rights*, edited by B. Franklin, 139–54. London and New York: Routledge.

Daniels, N. 1992. Growth hormone therapy for short stature. *Growth: Genetics and Hormones* **8**(supplement):46–8.

Daniels, N. 1985. *Just Health Care*. Cambridge: Cambridge University Press.

Daniels, N. 2000. Normal functioning and the treatment-enhancement distinction. *Cambridge Health Ethics Quarterly 2000* (9): 309–322.

Daniels, N. 2008. *Just health: Meeting health needs fairly.* Cambridge: Cambridge University Press.

Edgar, A. 2009. The hermeneutic challenge of genetic engineering. *Medicine, Health Care and Philosophy* **12**:157–67.

Elliott, C. 2003. *Better than well: American medicine meets the American dream.* New York and London: Norton & co.

Engelhardt, H. T. 1974. The disease of masturbation: values and the concept of disease. *Bulletin of the History of Medicine* **48**:234–48.

Engelhardt, H. T. 1996. *The foundations of bioethics*, 2nd ed. Oxford: Oxford University Press.

Fukuyama, F. 2003. *Our posthuman future: Consequences of the biotechnology revolution.* London: Profile Books.

Furlan, P. M., M. J. Kallan, and T. T. Have. 2004. SSRIs do not cause affective blunting in healthy elderly volunteers. *American Journal of Geriatric Psychiatry* **12**(3):323–30.

Harris, J. 2007. *Enhancing evolution.* Princeton: Princeton University Press.

Hastings Center Report. 1996. *The goals of medicine: Setting new priorities.* Hastings Center Report Special Supplement **26**(6):S1–27.

Have, H. ten, R. ter Meulen, and E. van Leeuwen. (2013). *Medische Ethiek* (*Medical ethics*), 4th ed. Houten: Bohn Staff and Van Lochum.

Heinz, A., R. Kipke, S. Mueller, and U. Wiesing. 2012. Cognitive enhancement: false assumptions in the ethical debate. *Journal of Medical Ethics* **38**(6):372–5.

Illich, I. 1975. *Medical nemesis: the expropriation of health.* London: Marion Boyars.

Jay, M. 1973. *The dialectical imagination.* Boston: Little, Brown.

Jotterand, F. 2010a. At the roots of transhumanism: from the Enlightenment to a post-human future. *Journal of Medicine and Philosophy* **35**:617–21.

Jotterand, F. 2010b. Human dignity and transhumanism: Do anthro-technological devices have moral status? *American Journal of Bioethics* **10**(7):45–52.

Juengst, E. (1998) What does enhancement mean? In *Enhancing human traits: Ethical and social implications*, edited by E. Parens. Washington, DC: Georgetown University Press.

Juth, N. 2011. Enhancement, autonomy and authenticity. In: *Enhancing human capacities*, edited by J. Savulescu, R. ter Meulen, and G. Kahane, 34–48. Oxford: Wiley Blackwell.

Kant, I. 1788; 2004. *Kritik der praktischen Verninft* (*Critique of practical reason*). In: *Immanuel Kant, De drie kritieken*. Nijmegen: SUN.

Kass, L. 2002. *Life, liberty and the defense of dignity*. San Francisco, CA: Encounter Books.

Kass, L. 1985. *Toward a More Natural Science*. New York: The Free Press.

Kramer, P. 1994. *Listening to Prozac*. London: Fourth Estate.

MacIntyre, A. 1984. *After virtue: A study in moral theory.* Notre Dame, IN: Notre Dame University Press.

Macklin, R. 2003. Dignity is a useless concept. *British Medical Journal* **327**:1419–20.

Mehta, M., A. Owen, B. Sahakian, N. Mavaddat, J. Pickard, and T. Robbins. 2000. Methylphenidate enhances working memory by modulating discrete frontal and parietal lobe regions in the human brain. *Journal of Neuroscience* **20** (RC65):1–6.

Mohamed, A. D. 2014. The effects of modafinil on convergent and divergent thinking of creativity: a randomized controlled trial. *Journal of Creative Behavior* 15 September 2014:1–21.

Mohamed, A. D., and C. R. Lewis. 2014. Modafinil increases the latency of response in the Hayling sentence completion test in healthy volunteers: a randomised controlled trial. *PLoS ONE,* **9**:e110639.

Naam, R. 2005. *More than human—embracing the promise of biological enhancement.* New York: Random House—Broadway Books.

President's Council on Bioethics (2003). *Beyond therapy—biotechnology and the pursuit of happiness, a report of the President's Council on Bioethics.* Washington, DC: US Government Office. http://biotech.law.lsu.edu/research/pbc/reports/beyondtherapy/beyond_therapy_final_report_pcbe.pdf

Quednow, B. 2012. *Efficacy of pharmacological neuroenhancement. Facts and fallacies.* Presentation EPOCH Workshop, Karlsruhe 29–30 March 2012.

Rawls, J. 1972. *A theory of justice.* Oxford: Oxford University Press.

Repantis, D., P. Schlattman, O. Laisney, and I. Heuser. 2010. Modafinil and methylphenidate for neuroenhancement in healthy individuals: a systematic review. *Pharmacological Research* **62**(3):187–206.

Sandberg, A. 2011. Cognition enhancement: uploading the brain. In: *Enhancing human capacities,* edited by J. Savulescu, R. ter Meulen, and G. Kahane, 71–91. Oxford: Wiley Blackwell.

Sandel, M. 2007. *The case against perfection. Ethics in the age of genetic engineering* Cambridge, MA: Belknap Press of Harvard University Press.

Savulescu, J., R. ter Meulen, and G. Kahane, eds. 2011. *Enhancing human capacities.* Oxford: Wiley Blackwell.

Scanlon, T. 1989. A good start: reply to Roemer. *Boston Review* **20**(2):8–9.

Schaeffer, G. Owen, G. Kahane, and J. Savulescu. 2014. Autonomy and enhancement. :123–36.

Schermer, M. 2012. *Van genezen naar verbeteren.* Inaugural lecture. Rotterdam: Erasmus Medical Centre.

Science and Technology Options Assessment European Parliament (STOA). 2009. *Human Enhancement.* Brussels: European Parliament.

Shickle, D. 2000. Are "genetic enhancements" really enhancements? *Cambridge Health Ethics Quarterly* 9:342–52.

Sen, A. 1990. Justice: means versus freedoms. *Philosophy and Public Affairs* **19**:111–21.

Smith, M. E., and M. J. Farah 2011. Are prescription stimulants "smart pills"? The epidemiology and cognitive neuroscience of prescription stimulant use by normal healthy individuals. *Psychological Bulletin* **137**(5):717–41.

Taylor, C. 1991. *The Ethics of Authenticity.* Cambridge, MA: Harvard University Press.

Zola, I. 1972. Medicine as an institution of social control. *Sociological Review* **20**(4):487–504.

Part II

Risk and benefits of the use of neuropharmacological drugs for cognitive enhancement

Chapter 3

Overclocking the brain? The potential and limitations of cognition-enhancing drugs

Reinoud de Jongh

3.1 Introduction

Enhancement can be defined as "interventions designed to improve human form or functioning beyond what is necessary to sustain or restore good health" (E. T. Juengst; in Parens 1998, 29). In this chapter, we will look at the enhancement of cognition, which includes functions such as memory, attention, language, perception, and executive functioning. While there are numerous approaches to cognitive enhancement, for example exercise, nutrition, brain training, brain stimulation, and mnemonic strategies (see Dresler et al. 2013), we will limit ourselves in this chapter to the study of cognition-enhancing drugs (CEs), or "smart drugs" as they are more popularly known. The use of drugs to enhance cognition is far from new. The stimulant caffeine, for example, has been used for this purpose, amongst other motivations, for at least a thousand years (Mehlman, 2004).

In *Why I want to be a posthuman when I grow up* (2008), philosopher and transhumanist Nick Bostrom paints us a picture of what our future cognitively enhanced lives might look like:

> You also discover a greater clarity of mind. You can concentrate on difficult material more easily and it begins making sense to you. You start seeing connections that eluded you before. You are astounded to realize how many beliefs you had been holding without ever really thinking about them or considering whether the evidence supports them. You can follow lines of thinking and intricate argumentation farther without losing your foothold. Your mind is able to recall facts, names, and concepts just when you need them (p. 111).

For Bostrom, then, there is no doubt that people feel the need to enhance themselves:

> People also seem to be keen on improving cognition. Who wouldn't want to remember names and faces better, to be able more quickly to grasp difficult abstract ideas, and to be able to "see connections" better? Who would seriously object to being able to appreciate music at a deeper level? The value of optimal cognitive functioning is so obvious that to elaborate the point may be unnecessary (p. 116).

Are people indeed keen on improving cognition? Riis et al. (2008) studied the consumer demands for CEs in undergrads. Participants read the following instruction:

> We would like your opinions about some of the enhancements that are expected to be possible. In each case, assume that a new kind of prescription pharmaceutical has been developed that can

> safely alter specific parts of the brain. Think of these as a sort of magic pill. They only have to be taken once and the effect is permanent. Imagine that the pills are completely safe with no long-term or short-term side effects. They only alter the specific part of the brain that they target (p. 498).

Then they were asked if they would take a pill to enhance themselves on each of 19 different traits. Some of those traits were rated to be more fundamental to the self, such as empathy, kindness, and self-confidence, while others were rated as less fundamental to self-idenity, such as memory, wakefulness, concentration, and foreign language ability. Results showed that more than half of the students were willing to take a safe pill to improve memory, wakefulness, and foreign language ability, but that they were much more reluctant to enhance traits considered fundamental to self-identity, such as empathy and kindness.

Are people also prepared to take the CEs that are currently available to reach their goal, drugs that are far from completely safe and have clear side effects? Among college students, Teter et al. (2006) found a lifetime prevalence rate of 8.3% for the illicit use of prescription stimulants, with a past-year prevalence rate of 5.9%. Although there are few empirical studies on the use of CEs outside of the USA, the prevalence in European countries appears to be lower (Ragan et al. 2013). Thus, while some people are prepared to take current CEs to enhance their cognition, it is far from common practice, even in a population of students presumably well motivated to do so.

In *Aphorismen zur Lebensweisheit* (1851), Arthur Schopenhauer already warned us that, while free of suffering, it is our desire and craving for a better life that will inevitably lead to suffering. One could argue, for example, that our memory is good enough as it is. In *Moonwalking with Einstein*, science journalist and former US memory champion Joshua Foer notes: "For all of our griping over the everyday failings of our memories—the misplaced keys, the forgotten name, the factoid stuck on the tip of the tongue—their biggest failure may be that we forget how rarely we forget" (p. 27). Even so, he still wonders: "What would it mean to have all that otherwise-lost knowledge at my fingertips? I couldn't help but think that it would make me more persuasive, more confident, and, in some fundamental sense, smarter" (p. 7). He asks himself: "But how many worthwhile ideas have gone unthought and connections unmade because of my memory's shortcomings?" (p. 7).

In the case of CEs, the suffering that Schopenhauer warned us about might arrive in the form of addiction and side effects. As Ragan et al. (2013) write: "the history of the development of medicines tells us that there is no such thing as a safe drug, only a drug whose benefits outweigh its drawbacks" (p. 592). In this chapter, we will consider those drawbacks, as well as focus on "suffering" in the form of trade-offs, where pharmacological enhancement of one task is associated with impairment in another area.

While the striving to enhance cognition, to liberate ourselves from our biological constraints, might be far from new, the possibilities for (pharmacological) enhancement of mental functioning, however, are said to have increased as a result of recent developments in the fields of neuroscience and psychopharmacology. Let us therefore start with the question: What are, at present, the possibilities for pharmacological enhancement of cognition?

3.2 The effectiveness of current putatively cognitive-enhancing drugs

In 2008, we reviewed and critically examined the available evidence for the possibility of pharmacological enhancement of cognition (de Jongh et al. 2008). We concluded that several CEs were indeed capable of improving episodic and working memory, sustained attention, and executive functioning on laboratory tasks, although a number of factors limit the practical usefulness of these drugs. Since then, a lot of studies have added to the research literature on CEs, some taking new directions, such as studying the effects of CEs on creativity (Farah et al. 2009; Mohamed 2014; Müller et al. 2013). A number of excellent and thorough reviews have also been published (Repantis et al. 2010a,b; Smith and Farah 2011) which have further tempered the enthusiasm for the possibility of CE. Let us look again at the most promising CEs—donepezil, methylphenidate, amphetamine, and modafinil—in light of these new findings.

3.2.1 Donepezil

Donepezil (Aricept) is an acetylcholinesterase inhibitor indicated for mild to moderate Alzheimer's disease. Acetylcholinesterase inhibitors exert their effects by inhibiting the breakdown of acetylcholine. This increases the amount of acetylcholine in the synaptic cleft that can bind to muscarinic and nicotinic acetylcholine receptors (Mumenthaler et al. 2003). Based on the studies available in 2008, we concluded that donepezil might enhance the ability to retain a practiced skill (although see section 3.4 for a discussion of this finding) and enhances episodic memory. Negative effects on cognition had also been reported, which might be explained by suboptimal changes in the cholinergic system as a result of the fact that the drug was administered for 14 days, instead of 21 days in the studies that found an enhancing effect (de Jongh et al. 2008).

In their review of acetylcholinesterase inhibitors, Repantis et al. (2010b) found three new studies on the effects of repeated administration of donepezil. In elderly particpants, donepezil administered for 6 weeks improved immediate and delayed word recall (FitzGerald et al. 2008). In young participants, donepezil administered for 17 days had no effect when subjects were well rested, but reduced the decline in memory and attention that resulted from 24 hours of sleep deprivation (Chuah and Chee 2008; Chuah et al. 2009).

Repantis et al. (2010b) also found five studies on the acute effects of a single dose of donepezil. In these studies, acute administration of donepezil improved information processing (Hutchison et al. 2001; Zaninotto et al. 2009), verbal episodic memory, spatial memory, working memory (Zaninotto et al. 2009), visuospatial working memory (Snyder et al. 2005), and procedural memory (Hornung et al. 2007). However, null results were also reported on attention, working memory (Nathan et al. 2001), and declarative memory (Hornung et al. 2007).

3.2.2 Methylphenidate and amphetamine

Methylphenidate and amphetamine are stimulants prescribed to treat attention deficit hyperactivity disorder (ADHD) that increase the synaptic concentrations of dopamine and norepinephrine by blocking their reuptake. Amphetamine also increases catecholamine levels by increasing neurotransmitter release (Smith and Farah 2011). When surveying the literature in 2008, we found that methylphenidate enhanced spatial working memory (Mehta et al. 2000), and in another study, enhanced spatial working memory and planning on novel tasks, but impaired previously established performance (Elliott et al. 1997). In healthy elderly volunteers, methylphenidate was without effect on working memory, response inhibition, and sustained attention (Turner et al. 2003b). For amphetamine, we found evidence for an improvement of working memory and executive function (Barch and Carter 2005; Mattay et al. 2000, 2003).

In 2010, Repantis et al. (2010a) reviewed 19 randomized controlled trials on the effects of methylphenidate. They only found effects of a single dose of methylphenidate on one outcome, memory, on which methylphenidate had a large positive effect, most prominently on spatial working memory. No significant effects were found on attention, executive functioning, and mood.

Smith and Farah (2011) reviewed 28 studies on the effects of psychostimulants, methylphenidate, and amphetamine on learning, working memory, cognitive control, and other executive functions. They found enhanced recall and recognition of verbal material, but only when subjects were tested after longer delays, from 1 hour up to 1 week. In general, no benefits were seen immediately following learning, which is when experimenters generally test memory for learned material (Smith and Farah 2011). The finding that stimulants help with the consolidation of declarative memory, especially when longer periods intervene between learning and testing, is especially interesting for discussions on the ethical concern of fairness. Greely et al. (2008) suggest that cognitive enhancement would be unfair if it just temporarily boosts exam performance, as it would prevent a valid measure of the competency of the examinee. If, however, it enhances long-term learning, as seems to be the case, we might be more willing to accept it.

For working memory, the results were mixed, with some studies reporting enhancement and others null results. Some studies found greater improvement for less able subjects. Smith and Farah (2011) concluded that "stimulants probably do enhance working memory, at least for some individuals in some task contexts, although the effects are not so large or reliable as to be observable in all or even most working memory studies" (p. 15). For cognitive control, a process which is needed for the inhibition of inappropriate actions or the shift from one kind of action or mental set to another, results were also mixed, but now with slightly more null results than positive findings and evidence for impairment in one task. Again, effects were found to be greatest for the least able participants (Smith and Farah 2011).

In summary, Smith and Farah (2011) conclude that the evidence is consistent with enhanced consolidation of long-term declarative memory, but effects on executive functioning (working memory and cognitive control) are less reliable, although they do appear on some tasks and for some individuals.

More recently, Ilieva et al. (2013) took baseline cognitive performance and catechol-O-methyl transferase (COMT) genotype into account when studying the effects of mixed amphetamine salts (MAS; brand name Adderall) on 13 measures of cognitive ability. They did not find enhancement of cognition for participants in general, but below-median baseline performers did improve on word recall and embedded figures, and there was a trend toward improved performance for these less able participants on the Raven's Progressive Matrices, a test of nonverbal intelligence. We will look more closely at these kinds of baseline dependent effects in section 3.5.2.

3.2.3 **Modafinil**

Modafinil is a wake-promoting agent that is FDA approved for the treatment of excessive daytime sleepiness associated with narcolepsy, obstructive sleep apnea/hypopnea syndrome, and "shift work sleep disorder." The mechanisms of its action is still unclear (Müller et al. 2013), but are believed to be different from those of methylphenidate and amphetamine. Effects on dopamine and norepinephrine presumably play an important role in its cognition-enhancing effects, although GABA, histamine, glutamate, and orexin/hypocretin may also be involved (Dresler et al. 2013; de Jongh et al. 2008).

In 2008, we concluded that modafinil can sustain alertness and cognitive performance during prolonged periods of wakefulness. In addition to its wake-promoting properties, modafinil also enhanced working memory, stopped signal reaction time, and improved spatial planning (Turner et al. 2003a), visual pattern recognition memory, speed of responding, and sustained attention (Randall et al. 2005a). Some authors, however, concluded that the effects of modafinil were limited, enhancing performance in only 6 out of 29 tests (Randall et al. 2004), and that the enhancement was limited to very specific, simple tasks (Randall et al. 2005a). We also reported on findings that modafinil was most effective in below-average performing subjects (Müller et al. 2004) and subjects with a lower IQ (Randall et al. 2005b).

After reviewing the findings of 31 randomized controlled trials, Repantis et al. (2010a) concluded that in subjects who were not sleep-deprived, modafinil had positive, though moderate enhancing effects on attention, while no effect could be detected on memory, mood, or motivation. Effects on executive functions could not be analyzed. In sleep-deprived individuals, a single dose of modafinil had strong positive effects on executive functioning, memory, and wakefulness. During sustained sleep deprivation over several days, however, repeated administration of modafinil maintained wakefulness for up to 4 days, but did not sustain attention and executive functioning, a potentially dangerous combination which we will come back to in section 3.5.4.

More recently, Müller et al. (2013) found that modafinil enhanced spatial working memory, planning, and decision-making at the most difficult levels of their cognitive tests in non-sleep-deprived participants. They also found an improved delayed visual pattern recognition memory and a large increase in task motivation. Modafinil did not improve paired associates learning and digit span, however. It had an inconsistent effect on creativity.

3.2.4 Side effects

So far, we have considered the benefits of CEs, but what about the harms? For modafinil, Ragan et al. (2013) argue that the European Medicines Agency would not approve it for cognitive enhancement purposes as the harms, such as the risk of serious skin reactions, suicidality, depression, psychosis, and adverse cardiovascular events, outweigh the benefits. While this sounds very alarming, Repantis et al. (2010a) conclude in their review on methylphenidate and modafinil that the drugs were well tolerated, although some side effects, normally benign, were reported. They say, "In total, these drugs seem to be well tolerated even by this population where the trade-off between side effects and improvement may be less clear" (p. 203). Donepezil also appears to be well tolerated, with no or benign side effects (Repantis et al. 2010b).

3.2.5 Summary

In summary, donepezil appears to enhance different types of memory, with both acute and repeated administration, although it is difficult to draw firm conclusions based on the small number of studies.

The cognition-enhancing effects of methylphenidate are somewhat disappointing, being limited to memory, specifically spatial working memory, and perhaps enhanced recall and recognition of verbal materials at longer test intervals. For amphetamine, there is stronger evidence for the enhancement of the consolidation of declarative memory, especially when longer periods intervene between learning and testing. For both methylphenidate and amphetamine, evidence for enhancement of executive functions (working memory and cognitive control) is mixed, if not to say contradictory, with reports of enhancement, null results, and even impairment, depending on the task and the individual.

And finallly, with modafinil, a clear enhancing effect is found on attention in non-sleep-deprived subjects, while in sleep-deprived participants, a single dose of modafinil had strong positive effects on executive functioning, memory, and wakefulness.

3.3 Explaining the equivocal findings

When studying the reviews by Repantis et al. (2010a,b) and Smith and Farah (2011), it becomes obvious that the evidence for cognition-enhancing effects of donepezil, stimulants, and modafinil is mixed and sometimes contradictory, at least for some cognitive domains. The situation might even be worse as a result of the publication bias against null results, otherwise known as the "file drawer effect" (Smith and Farah 2011; Ilieva et al. 2013). For example, Smith and Farah (2011) found that null results were invariably reported in studies using multiple tasks, while published single-task studies all found enhancement, suggesting that single-task studies with null results end up in file drawers. A number of factors might explain the equivocal findings.

3.3.1 Small effects and insufficient statistical power

First, the effects of CEs might simply be small, which has been frequently suggested, for example by Hall and Lucke (2010) and Ilieva et al. (2013). Combined with the fact that studies have frequently been underpowered, with fewer than 40 participants in between-subject designs, this could explain the mix of positive and null results (Ilieva et al. 2013).

In support of this hypothesis, Smith and Farah (2011) mention that several imaging studies have shown effects of methylphenidate and amphetamine on task-related activation in the absence of effects on cognitive performance, suggesting that the effects are so subtle that they can sometimes only be detected by the most sensitive measures. In section 3.4, we will focus on whether potentially small effects are of any practical value.

3.3.2 Insufficient attention to individual differences

Another explanation for null results reported in the literature is insufficient attention to individual differences, such as differences in baseline cognitive ability and COMT genotype of participants (Ilieva et al. 2013). We will take a detailed look at the role of baseline differences in section 3.5.2. Null results might well be explained by the fact that some individuals clearly benefit from the drug and show enhancement, while others do not, or even show impairment. There is plenty of evidence that individual differences modulate the effect of CEs, but this factor has generally been neglected in most studies (Smith and Farah 2011). As Hills and Hertwig (2011) summarize, "such investigations should report expectations relative to individual differences in baseline performance in the task, as well as more general measures of intelligence. Between-subjects designs that overlook such differences are almost guaranteed to over- or underestimate actual effects and invite improper generalizations of the usefulness to people of different abilities" (p. 376).

3.3.3 Task characteristics: difficulty and novelty

Some of the negative findings might also simply reflect ceiling effects. As pointed out by Müller et al. (2004), "in relatively high performing subjects without brain pathology or experimentally induced impairment it is difficult to improve cognitive performance with any given drug (ceiling effect)" (p. 166). High-performing subjects might therefore also benefit from dopamine agonists, methylphenidate, and modafinil, as long as the task is difficult enough to allow for an improvement of their performance.

In support of this view, Müller et al. (2013) found modafinil to enhance spatial working memory, planning, and decision-making at the most difficult levels of their cognitive tests in non-sleep-deprived participants. Interestingly, the effect on spatial working memory was not found by Turner et al. (2003a), who used an easier version of the task at which participants might not have been able to show further improvement.

Another important task characteristic might be novelty. Elliott et al. (1997) found methylphenidate to either enhance or impair cognitive performance on tests of spatial working memory and planning, depending on the familiarity and requirements of the task. The authors suggest that methylphenidate enhances executive function on novel tasks, but impairs previously established performance.

3.3.4 Single-dose studies

A fourth and final factor is the abundance of single-dose studies. Null results and detrimental effects of CEs may result from a dose which is too low or too high, as some CEs, especially stimulants, appear to follow U-shaped dose-response curves (Ilieva et al. 2013; Repantis et al. 2010a; Smith and Farah 2011). As pointed out by Smith and Farah (2011), the optimal dose depends on individual characteristics and the task at hand, and therefore cannot be easily predicted on the basis of data from previous experiments. In animal research, it is common practice to use a dose range, typically a placebo and three doses. As we will see in section 3.5.1, the use of such a dose range, as was done by Tannock et al. (1995), who studied the effects of methylphenidate (0, 0.3, 0.6, and 0.9 mg/kg) in children with ADHD, even allows for the detection of simultaneous U-shaped and linear dose–response curves, depending on the outcome measure.

3.3.5 Summary

Smith and Farah (2011) summarize the problem succinctly: “Optimizing the cognitive effects of a stimulant would therefore require, in effect, a search through a high-dimensional space whose dimensions are dose; individual characteristics such as genetic, personality, and ability levels; and task characteristics. The mixed results in the current literature may be due to the lack of systematic optimization” (p. 19).

Future research should take all these considerations into account, although this proves to be a tremendous challenge. For example, in studying the effects of modafinil, Müller et al. (2013) recently controlled for task difficulty, but used a single fixed dose (200 mg) and did not control for baseline differences in ability. In studying the effects of MAS, Ilieva et al. (2013) did control for baseline performance, COMT genotype, and task novelty. They also ensured sufficient statistical power to detect a medium-size effect. On the other hand, they used a single fixed dose (20 mg) and did not adequately study the role of task difficulty.

3.4 Practical use in real-life scenarios

How effective are these drugs in enhancing real-world cognitive performance? According to Hall and Lucke (2010) the effects of CEs are small under laboratory conditions and we should be skeptical about neuroenhancement claims. They point out that it is generally unclear whether enhancement effects observed in the lab translate into benefits in everyday life (Hall and Lucke 2010).

Smith and Farah (2011), however, argue that “success in academic and occupational competitions often hinges on the difference between being at the top or merely near the top” (p. 19). Moreover, they suggest that stimulants may be more helpful than laboratory tasks indicate. Subjects in the lab are usually tested soon after learning, but the enhancing effects of amphetamine on episodic memory were only detected when longer retention intervals were used. Obviously, it is exactly this long-term effect on learning that is of practical value.

Ilieva et al. (2013) also wonder whether effects might be larger in real-world scenarios, when there are more distractions present than in a laboratory setting, or tasks take longer and therefore become more tedious. They also suggest that MAS might have a larger effect on motivation to work, which presumably plays a much bigger role in daily life than in the laboratory. Such effects on motivation have indeed been found. Volkow et al. (2004) already reported that methylphenidate increased a participant's interest in a dull mathematical task, and recently, Müller et al. (2013) found a large increase in task motivation after modafinil administration.

Another reason why benefits in real-world scenarios may be underestimated is that while most studies focus on non-sleep-deprived, well-rested healthy individuals, effects of CEs, for example modafinil, are much more pronounced in subjects that are sleep-deprived. On the effects of MAS on cognition Ilieva et al. (2013) write, "If it does enhance cognition in healthy and adequately-rested young adults, the effects are likely to be small." However, although healthy, a lot of people are far from adequately rested. For example, 13–33% of the Australian population have regular difficulty either getting to sleep or staying asleep (Cunnington et al. 2013). For a substantial part of the population, effects of CEs on sleep-deprived subjects might therefore approximate practical benefits in daily life better than studies on well-rested individuals.

Further support for the viewpoint that CEs may be more helpful than laboratory tasks indicate can be found if we look at the effects of caffeine. In the laboratory, the effects of caffeine on cognition appear to be small and often contradictory (Harrell and Julaino 2009; Nehlig 2010), as is the case with the CEs which are the topic of this chapter. If we look at the effects of caffeine in the workplace, however, the picture changes dramatically. Smith (2005) studied the association between caffeine consumption and cognitive failures and accidents in 1555 workers at high risk of having an accident at work (because of exposure to physical hazards and irregular or long working hours). After controlling for confounds, he found that higher caffeine consumption (>220 mg) was associated with half the risk of cognitive failures and half the risk for accidents at work, which were defined as injuries that required medical attention.

For the CEs discussed in this chapter, studies in the workplace or at least approximating real-world scenarios are scarce. One example is a study by Sugden et al. (2012) on the effects of modafinil on sleep-deprived doctors. The authors looked at the effects of modafinil on several neuropsychological tests and on clinical psychomotor performance in a surgical simulator, both after one night of sleep deprivation. Results showed that sleep-deprived doctors worked more efficiently when solving working memory and planning problems, and displayed less impulsivity and more flexibilty to redirect attention. Given the effects of modafinil on cognition in sleep-deprived subjects, which we already discussed, these results are not very new and surprising. Unfortunately, modafinil did not enhance performance in the surgical simulator, which would have been a new finding approaching real-life scenarios, even though the control group performed suboptimally.

The difficulty, of course, is that it is indeed generally unclear whether enhancement effects observed in the lab translate into benefits in everyday life, as Hall and Lucke (2010) write. While studies that approximate real-world cognitive performance, by using a simulator for example, might have a high face validity they also tap into a wide variety of cognitive functions, making it very difficult to determine which specific cognitive functions improve after drug administration.

Take for example (and also as another example of studies in the workplace) the finding that donepezil improved the retention of training in healthy pilots tested in a flight simulator (Yesavage et al. 2002), widely cited in both ethical discussions and popular scientific articles on cognitive enhancement. Yesavage et al. (2002) trained 18 pilots in a flight simulator. Afterwards, half of the subjects were instructed to take donepezil (5 mg) for 30 days and the other half were given capsules containing a placebo. Both subjects and experimenters were blind to the treatment condition. On day 30, subjects returned to the laboratory to perform two test flights.

In the donepezil group, the flight performance on day 30 was found to be similar to performance after initial training, whereas in the placebo group, flight performance declined. The authors interpret these results as an improvement of the ability to retain a practiced skill. It seems equally feasible, however, that the increased performance in the donepezil group, relative to the placebo group, can be explained not by increased retention of the learned skills (long-term memory), but by improving attention or working memory during the test flights on day 30.

Of course, most researchers prefer the controlled environment of the laboratory, maintaining the uncertainty of whether and how effects observed in the lab translate into benefits in daily life.

3.5 Factors that limit the usefulness of cognitive enhancing drugs

Discussions of ethical issues raised by the use of cognition-enhancing drugs, such as safety, coercion, fairness, equality, and authenticity, can benefit both from realistic expectations of their efficacy, as already discussed, and from an understanding of the limitations of CEs, limitations that might prove to be fundamental and insuperable.

3.5.1 Simultaneous linear and U-shaped dose–response curves

Depending on the task, some CEs, specifically psychostimulants, can simultaneously exert both linear and quadratic (U-shaped) effects. Tannock et al. (1995), for example, studied the effects of methylphenidate (0, 0.3, 0.6, and 0.9 mg/kg) on cognitive flexibility and overt behavior in children with ADHD. They found that while the effects on behavior showed a linear dose–response curve, the function for response inhibition was U-shaped, in that the high dose was less effective in enhancing response inhibition than lower doses.

A similar finding has been reported by Konrad et al. (2004), also in children with ADHD. The authors found that methylphenidate (0, 0.25, and 0.5 mg/kg) improved alertness and focused and sustained attention in a linear fashion. But on two executive tasks, response inhibition and set-shifting, the high dose was less effective than the low dose in enhancing performance. Alertness and sustained attention appear to be localized in the frontosubcortical network of the right hemisphere, while executive functions are associated with activity in the prefrontal cortex. The authors therefore speculate that these different brain areas display different dose–response curves.

These findings are potentially problematic for the practical use of CEs in healthy individuals, as doses most effective in facilitating one behavior could simultaneously exert null or even detrimental effects on other cognitive domains.

3.5.2 Baseline dependency

The effects of a number of different cognitive enhancers also seem to depend on the subjects' baseline working memory capacity. Individuals with a low working memory capacity improve on dopamine receptor agonists, while high-performing subjects are either not affected or get worse. This pattern was found for the D2 dopamine receptor agonist bromocriptine (Gibbs and D'Esposito 2005a,b; Kimberg et al. 1997; Mehta et al. 2001), and for amphetamine (Mattay et al. 2000, 2003). Two notable exceptions are Kimberg et al. (2001) and Kimberg and D'Esposito (2003), where bromocriptine and the dopamine agonist pergolide, respectively, improved performance for high-span subjects, but were harmful for low-span subjects.

More recently, Ilieva et al. (2013) took baseline cognitive performance into account when studying the effects of MAS (Adderall) on 13 measures of cognitive ability. They found a significant interaction between drug and baseline performance for word recall and embedded figures, a measure of convergent creativity, and a trend toward significance for Raven's Progressive Matrices, a test of nonverbal intelligence. That is, MAS tended to enhance performance for below-median baseline performers, while somewhat reducing performance for above-median performers.

Similar findings have been reported for methylphenidate and modafinil. Subjects with lower baseline working memory capacity showed the greatest improvement in spatial working memory after methylphenidate administration (Mehta et al. 2000). And while Müller et al. (2004) found modafinil to improve both maintenance and manipulation processes in difficult and monotonous working memory tasks, the effects were most pronounced in lower performing subjects. Similarly, modafinil improved accuracy in a sustained attention task, but only in subjects with a lower (although still above-average) IQ (Randall et al. 2005a,b). To the best of my knowledge, no studies have found impaired performance in high span subjects after methylphenidate administration. In the case of modafinil, such an effect has only been reported when studying creativity (Mohamed 2012). Therefore, the baseline-dependent effects of these drugs might perhaps be better explained by ceiling effects, which we discussed in section 3.3.3.

The effect of MAS (Adderall) and modafinil on creativity also appear to be dependent on baseline performance. Two tests of creativity are typically used in the laboratory: measures of convergent creativity, where the aim is to produce a set answer or arrive at a particular goal. Measures of divergent creativity, on the other hand, require spontaneous, free-flowing thinking, where the aim is to generate many different ideas (Mohamed

2014). On the remote associates test, a measure of convergent creativity, MAS enhanced creativity for low-performing individuals, while impairing it for high-performing subjects (Farah et al. 2009). Modafinil was also found to reduce convergent thinking for highly creative subjects, and the impairing effect of modafinil on divergent thinking was more pronounced in creative participants (Mohamed 2012).

All of these findings seem to point to an inverted U-model, which predicts optimal performance at intermediate (prefrontal) catecholamine levels, and impairment at levels that are either too low or too high. Presumably, low-performing individuals find themselves on the up slope of the inverted U, and therefore benefit from administration of dopamine agonists. By contrast, high-performing subjects who are located at or near the peak of the inverted U do not benefit. Administration of dopamine agonists therefore leads to an "overdose" of prefrontal dopamine and a consequent deterioration in performance.

Results from Egan et al. (2001) and Mattay et al. (2003) support such a view. Both groups made use of the common polymorphism in the COMT gene. The enzyme COMT breaks down dopamine and norepinephrine. Whether subjects have *val/val, val/met*, or *met/met* alleles of COMT determines an individual's endogenous dopamine activity. First, Egan et al. (2001) reported that *met/met* subjects, with high levels of prefrontal cortical dopamine, show better baseline working memory ability than *val/val* subjects, with relatively less prefrontal dopamine. And Mattay et al. (2003) found that increasing prefrontal dopamine levels with amphetamine improved working memory performance and enhanced efficiency of prefrontal cortex function in *val/val* subjects, while it decreased working memory performance and efficiency of prefrontal cortex information processing in *met/met* subjects.

These findings are particularly relevant to an observation made by Bostrom (2008):

> it seems to me (based on anecdotal evidence and personal observations) that people who are already endowed with above-average cognitive capacities are at least as eager, and, from what I can tell, actually more eager to obtain further improvements in these capacities than are people who are less talented in these regards (p. 117).

Ironically, then, it would seem that the people who are most eager to enhance themselves might not benefit from cognitive enhancers, while people who would benefit most might be less motivated to seek enhancement. Although little is known about the motivations for taking CEs, a study by McCabe et al. (2005) seems to refute Bostrom's claim. Among US college students, illicit use was higher among students who earned lower grade point averages.

In terms of equality, one could argue that the phenomenon of baseline dependency leads to a leveling of the playing field, instead of cognitive enhancers promoting inequality. Nonetheless, wealthy low-performing individuals may be able to afford these drugs, while poor low performers might not (de Jongh et al. 2008).

3.5.3 **Trade-offs**

Third, evidence suggests that there could be important and perhaps insurmountable trade-offs. That is, pharmacological enhancement of one task might necessarily be associated with impairment in another area, because different cognitive abilities are interdependent. Hills and Hertwig (2011) describe this as a *gain/loss asymmetry*: for any trait that had reached an evolutionary stable state, or in other words, an optimal trade-off, shifting the value of trait A upward through the use of CEs leads to a simultaneous loss in performance on trait B of a larger magnitude than the gain on trait A.

A famous example of a less than optimal trade-off in the evolution of the human body is the size of the female pelvis. Hills and Hertwig (2011) ask themselves: "Why hasn't evolution improved the survival chances of both mother and baby by selecting for a larger female pelvis?" (p. 374). The answer is that expanding the birth canal would come at the expense of efficient bipedal locomotion. These trade-offs apply to cognition as well. One example is the famous case of Shereshevskii, or "S," a man with an unlimited memory, but poor face recognition. The price that he paid for his phenomenal memory, which did not select important details and forget the rest, was that he was impaired in important cognitive processes such as abstraction, generalization, and trend detection, which are important in recognizing faces (Hills and Hertwig 2011).

Here, we will look at four potential trade-offs between and within different cognitive domains that have emerged out of research on animals and human subjects.

3.5.3.1 Long-term memory versus working memory

The first potential trade-off involves the opposite "chemical needs" of the prefrontal cortex, involved in working memory, and the hippocampus, critical for long-term memory (LTM). Animal research shows this trade-off is a real possibility: Rolipram is a drug which indirectly increases protein kinase A (PKA) activity by inhibiting phosphodiesterase (PDE) activity, which in turn prevents the breakdown of cyclic adenosine monophosphate (cAMP). It has been shown to improve LTM consolidation and to facilitate long-term potentiation in aged mice. This has prompted pharmaceutical companies to develop drugs that enhance PKA activity as a treatment for age-related cognitive decline in humans (Ramos et al. 2003).

While PKA activation in the hippocampus appears to keep memory fixed and enhance long-term storage of information, updating of working memory in the prefrontal cortex rather calls for erasure of information. Indeed, Ramos et al. (2003) showed that in aged rats, activation of PKA in the prefrontal cortex exacerbates (age-related) cognitive deficits. Similarly, they found that rolipram impaired prefrontal cortical cognitive performance in aged monkeys.

3.5.3.2 Stability versus flexibility of long-term memory

A second trade-off could occur between stability versus flexibility in updating LTM. While stability is an important feature of LTM, memory traces still need to be updated or replaced from time to time to incorporate new information, such as the name of the new president or your new partner, or when you move, your new address and postal code. The fact that memories can become "too stable" was demonstrated in animal research on the effects of the drug Rimonabant (SR141716A), a CB_1 cannabinoid receptor antagonist. Animal data suggest it also has cognition-enhancing effects: In mice, Rimonabant improved memory acquisition and consolidation (Takahashi et al., 2005), and in food-storing birds, it enhanced LTM for the location of a hidden food supply (Shiflett et al. 2004).

However, while blocking the CB_1 receptor led to more robust LTMs, it also disrupted the ability of new information to modify those memories (Shiflett et al. 2004). Similarly, when mice lacking the CB_1 receptor were tested in a water maze, they kept returning to the original location of the hidden platform, despite being repeatedly shown the new location (Varvel and Lichtman 2002).

3.5.3.3 Stability versus flexibility of working memory

A similar trade-off between stability and flexibility could arise within the domain of working memory. As Bilder et al. (2004) write:

> Increased cognitive stability benefits certain functions, such as working memory maintenance tasks (i.e. keeping representations "on line"), sustained attention tasks, and tasks demanding freedom from distraction. On the other hand, excessive stability yields inflexibility and difficulty in responding appropriately to external change by modifying ongoing behavioural programs or shifting attention to new foci. This may result in excessive repetition of maladaptive behaviours, perseveration, stereotypy, and a failure to detect novelty. It may also yield difficulty updating the contents of working memory representations (p. 1947).

According to the tonic/phasic dopamine theory, high amplitude transient phasic dopamine release, mediated by dopamine neuron burst firing, may be important for updating or resetting working memory traces. On the other hand, constant low-level "background" tonic dopamine may enhance the stability of memory traces. While behaviorally relevant stimuli trigger the phasic component of dopamine release, tonic dopamine levels regulate the amplitude of the phasic response by acting on autoreceptors on dopamine terminals. That is, increases in tonic dopamine levels suppress the phasic response (Bilder et al. 2004; Nolan et al. 2004). This suggests that manipulations which increase cognitive stability might lead to a decreased capacity to flexibly alter behavior.

The COMT polymorphism, which we described in section 3.5.2, provides a test of the tonic/phasic dopamine theory. Presumably, *val/val* subjects have increased phasic and reduced tonic dopamine transmission subcortically and decreased dopamine concentrations cortically. As a result, one would expect decreased cognitive stability, but increased flexibility. In contrast, *met/met* subjects have decreased phasic and increased tonic dopamine transmission subcortically, and increased dopamine concentrations cortically. This presumably leads to an increased cognitive stability, but decreased flexibility (Bilder et al. 2004).

To adequately distinguish between stability and flexibility, Nolan et al. (2004) used a competing programs task, which required subjects to alternate between two rules of responding: imitation and reversal. Learning and maintenance of the imitation rule required cognitive stability, while flexibility was needed to switch rules in the reversal condition and to inhibit the previously learned response. Do we see a trade-off between the stability and flexibility of working memory processes on this task? Compared to *val/val* subjects, *met/met* subjects indeed showed better acquisition of the imitation rule (increased stability), but performed worse in the reversal condition (decreased flexibility).

Although the trade-off between stability and flexibility of working memory becomes apparent as a result of natural differences in genotype in this case, it might also be induced by CEs. For example, Bilder et al. (2004) suggest that at low doses, psychostimulants, such as amphetamine, enhance phasic dopamine transmission by blocking reuptake of released dopamine, thereby increasing flexibility at the cost of reduced stability. In contrast, higher doses may preferentially increase tonic dopamine, increasing stability at the cost of reduced flexibility.

3.5.3.4 Attentional focus versus creativity

A fourth and final trade-off could occur between focus versus creativity. Drugs that heighten focus might dampen creativity, as a certain degree of distractibility may be required for creative thought. Charles Darwin is a famous example of a researcher who attributed his insights to his ability to notice irrelevant stimuli and his tendency to get easily distracted. The association between creativity and distractibility is supported by links between creativity and ADHD, and by the finding that highly creative subjects showed more intrusion errors on a dichotic listening task (Farah et al. 2009).

Does enhancing our attention with CEs impair our creativity? Not necessarily. Farah et al. (2009) found that the amphetamine Adderall actually enhanced two measures of creativity, instead of dampening it. It only impaired the creativity of highly creative individuals, which is another example of baseline dependency. Similarly, modafinil, which has positive, though moderate enhancing effects on attention (Repantis et al. 2010a), was found to reduce convergent thinking for highly creative subjects, and the impairing effect of modafinil on divergent thinking were more pronounced in creative participants (Mohamed 2012). Müller et al. (2013) found inconsistent effects on creativity, which did not reach significance.

Therefore, it remains to be seen whether there is a fourth trade-off between attention and creativity.

3.5.4 Overconfidence

The fourth factor which may limit the practical use of CEs is overconfidence, or impaired self-monitoring, the ability to accurately assess one's cognitive performance. Baranski and Pigeau (1997) found modafinil to induce overconfidence in sleep-deprived individuals; that is, their subjective estimates of performance exceeded their actual cognitive performance (although only in retrospective and not in prospective estimations). In a more recent study in non-sleep-deprived individuals, however, Baranski et al. (2004) did not find modafinil to induce overconfidence, leading them to conclude that modafinil per se does not generally induce overconfidence, but in individuals who are sleep deprived "the marked improvement in subjective sleepiness and increased vigor following a relatively large single administration of modafinil may be misinterpreted to imply a concomitant improvement in cognitive performance" (p. 330).

Repantis et al. (2010a) expresses a similar concern. They conclude that during sustained sleep deprivation over several days, repeated administration of modafinil maintained wakefulness for up to 4 days, but did not sustain attention and executive functioning, effects which might induce overconfidence. For MAS (Adderall), the results reported by Ilieva et al. (2013) also point to overconfidence. While the drug did not enhance cognition on most tasks and for most participants, they nevertheless perceived the drug as enhancing their cognition.

Overconfidence could potentially result in unnecessary risk-taking and dangerous situations if we require soldiers or surgeons to use these drugs during long missions and night shifts, respectively. We should be especially careful, as Baranski et al. (2004) warn, when it concerns occasional users who are sleep-deprived, as they are most likely to misinterpret the vigilance-enhancing properties of the drug as a sign of general cognitive enhancement.

3.6 Conclusion and future directions

In summary, in human laboratory studies CEs do appear to enhance different types of memory, including declarative memory when longer periods intervene between learning and testing. And while attention can be reliably enhanced, evidence for enhancement of executive functions (working memory and cognitive control) is mixed and sometimes contradictory, with reports of enhancement, null results, and even impairment, depending on the task and the individual. In sleep-deprived participants, strong positive effects of CEs on executive functioning, memory, and wakefulness can be obtained. To make sense of the mixed and contradictory findings, future studies need to pay closer attention to baseline difference, genotype, dose, task difficulty, and task novelty.

As for practical benefits of CEs in real-world scenarios, there is simply insufficient evidence to draw firm conclusions. While the practical benefits may indeed be small, I have listed several reasons why they may be underestimated in laboratory studies. More researchers need to follow the example of Yesavage et al. (2002), Smith (2005), and Sugden et al. (2012) and study the effects of CEs in the workplace or, when patient or passenger safety is involved, in simulators.

A number of factors, however, appear to limit the practical use of CEs, namely, baseline dependency, trade-offs, and overconfidence. As Hills and Hertwig (2011) state, "Enhancement is only meaningful with respect to specific individuals in specific environments. Although the empirical possibility of a domain-general, cognitively enhanced 'supermind' remains, evolutionary theory would suggest it is extremely unlikely" (p. 376).

What about the possibility of more effective future CEs? Although considerable effort has been directed at the development of memory-enhancing drugs, drugs that target the specific mechanisms that underlie memory formation, the outlook is rather bleak. As pointed out by Ragan et al. (2013), the problem is that pharmaceutical companies cannot develop and test drugs for non-medical use, only for defined and recognized medical conditions. Moreover, of 34 possible drug targets for improving cognitive function, only 8 are currently in clinical development by the top ten pharmaceutical companies. Of these 8, only a minority will prove effective and safe, fewer will prove to be more effective and safer than current treatments, and only a very small subset of these might prove to be useful for CE purposes (Ragan et al. 2013).

References

Baranski, J., and R. Pigeau. 1997. Self-monitoring cognitive performance during sleep deprivation: effects of modafinil, d-amphetamine and placebo. *Journal of Sleep Research* **6**:84–91.

Baranski, J. V., R. Pigeau, P. Dinich, and I. Jacobs. 2004. Effects of modafinil on cognitive and meta-cognitive performance. *Human Psychopharmacology* **19**(5):323–32.

Barch, D. M., and C. S. Carter. 2005. Amphetamine improves cognitive function in medicated individuals with schizophrenia and in healthy volunteers. *Schizophrenia Research* **77**:43–58.

Bilder, R. M., J. Volavka, H. M. Lachman, and A. A. Grace. 2004. The catechol-O-methyltransferase polymorphism: relations to the tonic-phasic dopamine hypothesis and neuropsychiatric phenotypes. *Neuropsychopharmacology* **29**:1943–61.

Bostrom, N. 2008. Why I want to be a posthuman when I grow up. In *Medical enhancement and posthumanity*, ed. B. Gordijn and R. Chadwick, 107–37. Berlin: Springer.

Chuah, L. Y. M., and M. W. L. Chee. 2008. Cholinergic augmentation modulates visual task performance in sleep-deprived young adults. *Journal of Neuroscience* **28**:11369–77.

Chuah, L. Y. M., D. L. Chong, A. K. Chen, et al. 2009. Donepezil improves episodic memory in young individuals vulnerable to the effects of sleep deprivation. *Sleep* **32**:999.

Cunnington, D., M. F. Junge, and A. T. Fernando. 2013. Insomnia: prevalence, consequences and effective treatment. *Medical Journal of Australia* **199**:S36–40.

Dresler, M., A. Sandberg, K. Ohla, C. Bublitz, C. Trenado, A. Mroczko-Wasowicz, S. Kühn, and D. Repantis. 2013. Non-pharmacological cognitive enhancement. *Neuropharmacology* **64**:529–43.

Egan, M. F., T. E. Goldberg, B. S. Kolachana, J. H. Callicott, C. M. Mazzanti, R. E. Straub, D. Goldman, and D. R. Weinberger. 2001. Effect of COMT Val108/158 Met genotype on frontal lobe function and risk for schizophrenia. *Proceedings of the National Academy of Science USA* **98**:6917–22.

Elliott, R., B. J. Sahakian, K. Matthews, A. Bannerjea, J. Rimmer, and T. W. Robbins. 1997. Effects of methylphenidate on spatial working memory and planning in healthy young adults. *Psychopharmacology (Berlin)* **131**:196–206.

Farah, M. J., C. Haimm, G. Sankoorikal, M. E. Smith, and A. Chatterjee. 2009. When we enhance cognition with Adderall, do we sacrifice creativity? A preliminary study. *Psychopharmacology* **202**(1–3):541–7.

FitzGerald, D. B., G. P. Crucian, J. B. Mielke, B. V. Shenal, D. Burks, K. B. Womack, G. Ghacibeh, V. Drago, P. S. Foster, E. Valenstein, and K. M. Heilman. 2008. Effects of donepezil on verbal memory after semantic processing in healthy older adults. *Cognitive and Behavioral Neurology* **21**(2):57–64.

Gibbs, S. E., and M. d'Esposito. 2005a. A functional MRI study of the effects of bromocriptine, a dopamine receptor agonist, on component processes of working memory. *Psychopharmacology (Berlin)* **180**:644–53.

Gibbs, S. E., and M. d'Esposito. 2005b. Individual capacity differences predict working memory performance and prefrontal activity following dopamine receptor stimulation. *Cognitive, Affective, and Behavioral Neuroscience* **5**:212–21.

Greely, H., B. Sahakian, J. Harris, R. C. Kessler, M. Gazzaniga, P. Campbell, and M. J. Farah. 2008. Towards responsible use of cognitive-enhancing drugs by the healthy. *Nature* **456**(7223):702–5.

Hall, W. D., and J. C. Lucke. 2010. The enhancement use of neuropharmaceuticals: more scepticism and caution needed. *Addiction* **105**(12):2041–3.

Harrell, P. T., and L. M. Juliano. 2009. Caffeine expectancies influence the subjective and behavioral effects of caffeine. *Psychopharmacology (Berlin)* **207**:335–42.

Hills, T., and R. Hertwig. 2011. Why aren't we smarter already: Evolutionary trade-offs and cognitive enhancements. *Current Directions in Psychological Science* **20**:373-7.

Hornung, O. P., F. Regen, H. Danker-Hopfe, M. Schredl, and I. Heuser. 2007. The relationship between REM sleep and memory consolidation in old age and effects of cholinergic medication. *Biological Psychiatry* **61**:750–7.

Hutchison, C. W., P. J. Nathan, L. Mrazek, and C. Stough. 2001. Cholinergic modulation of speed of early information processing: the effect of donepezil on inspection time. *Psychopharmacology (Berlin)* **155**:440–2.

Ilieva, I., J. Boland, and M. J. Farah. 2013. Objective and subjective cognitive enhancing effects of mixed amphetamine salts in healthy people. *Neuropharmacology* **64**:496–505.

de Jongh, R., I. Bolt, M. Schermer, and B. Olivier. 2008. Botox for the brain: enhancement of cognition, mood and pro-social behaviour and blunting of unwanted memories. *Neuroscience and Biobehavioral Reviews* **32**:760–76.

Juengst, E. 1998. What does enhancement mean? In *Enhancing human traits: Ethical and social implications*, ed. E. Parens. Washington, DC: Georgetown University Press.

Kimberg, D. Y., G. K. Aguirre, J. Lease, and M. d'Esposito. 2001. Cortical effects of bromocriptine, a D-2 dopamine receptor agonist, in human subjects, revealed by fMRI. *Human Brain Mapping* **12**:246–57.

Kimberg, D. Y., and M. d'Esposito. 2003. Cognitive effects of the dopamine receptor agonist pergolide. *Neuropsychologia* **41**:1020–7.

Kimberg, D. Y., M. D'Esposito, and M. J. Farah. 1997. Effects of bromocriptine on human subjects depends on working memory capacity. *NeuroReport* **8**:3581–5.

Konrad, K., T. Gunther, C. Hanisch, and B. Herpertz-Dahlmann. 2004. Differential effects of methylphenidate on attentional functions in children with attention-deficit/hyperactivity disorder. *Journal of the American Academy of Child and Adolescent Psychiatry* **43**:191–8.

McCabe, S. E., J. R. Knight, C. J. Teter, and H. Wechsler. 2005. Non-medical use of prescription stimulants among US college students: prevalence and correlates from a national survey. *Addiction* **100**:96–106.

Mattay, V. S., J. H. Callicott, A. Bertolino, I. Heaton, J. A. Frank, R. Coppola, K. F. Berman, T. E. Goldberg, and D. R. Weinberger. 2000. Effects of dextroamphetamine on cognitive performance and cortical activation. *Neuroimage* **12**:268–75.

Mattay, V. S., T. E. Goldberg, F. Fera, A. R. Hariri, A. Tessitore, M. F. Egan, B. Kolachana, J. H. Callicott, and D. R. Weinberger. 2003. Catechol O-methyltransferase val158-met genotype and individual variation in the brain response to amphetamine. *Proceedings of the National Academy of Science USA* **100**:6186–91.

Mehlman, M. J. 2004. Cognition-enhancing drugs. *Milbank Quarterly* **82**(3):483–506.

Mehta, M., A. Owen, B. Sahakian, N. Mavaddat, J. Pickard, and T. Robbins. 2000. Methylphenidate enhances working memory by modulating discrete frontal and parietal lobe regions in the human brain. *Journal of Neuroscience* **20**(RC65):1–6.

Mehta, M. A., R. Swainson, A. D. Ogilvie, J. Sahakian, and T. W. Robbins. 2001. Improved short-term spatial memory but impaired reversal learning following the dopamine D(2) agonist bromocriptine in human volunteers. *Psychopharmacology (Berlin)* **159**:10–20.

Mohamed, A. D. 2014. The effects of modafinil on convergent and divergent thinking of creativity: a randomized controlled trial. *Journal of Creative Behavior* **0**:1–21.

Müller, U., J. B. Rowe, T. Rittman, C. Lewis, T. W. Robbins, and B. J. Sahakian. 2013. Effects of modafinil on non-verbal cognition, task enjoyment and creative thinking in healthy volunteers. *Neuropharmacology* **64**:490–5.

Müller, U., N. Steffenhagen, R. Regenthal, and P. Bublak. 2004. Effects of modafinil on working memory processes in humans. *Psychopharmacology* **177**:161–9.

Mumenthaler, M. S., J. A. Yesavage, J. L. Taylor, R. O'Hara, L. Friedman, H. Lee, and H. C. Kraemer. 2003. Psychoactive drugs and pilot performance: a comparison of nicotine, donepezil, and alcohol effects. *Neuropsychopharmacology* **28**:1366–73.

Nathan, P. J., A. Baker, E. Carr, J. Earle, M. Jones, M. Nieciecki, C. Hutchison, and C. Stough. 2001. Cholinergic modulation of cognitive function in healthy subjects: acute effects of donepezil, a cholinesterase inhibitor. *Human Psychopharmacology* **16**:481–3.

Nehlig, A. 2010. Is caffeine a cognitive enhancer? *Journal of Alzheimer's Disease* **20**(Suppl 1):S85–94.

Nolan, K. A., R. M. Bilder, H. M. Lachman, and J. Volavka. 2004. Catechol O-methyltransferase Val158Met polymorphism in schizophrenia: differential effects of Val and Met alleles on cognitive stability and flexibility. *American Journal of Psychiatry* **161**:359–61.

Ragan, C. I., I. Bard, I. Singh, and Independent Scientific Committee on Drugs. 2013. What should we do about student use of cognitive enhancers? An analysis of current evidence. *Neuropharmacology* **64**:588–95.

Ramos, B. P., S. G. Birnbaum, I. Lindenmayer, S. S. Newton, R. S. Duman, and A. F. Arnsten. 2003. Dysregulation of protein kinase a signaling in the aged prefrontal cortex: new strategy for treating age-related cognitive decline. *Neuron* **40**:835–45.

Randall, D. C., N. L. Fleck, J. M. Shneerson, and S. E. File. 2004. The cognitive-enhancing properties of modafinil are limited in non-sleep-deprived middle-aged volunteers. *Pharmacology Biochemistry and Behavior* 77:547–55.

Randall, D. C., J. M. Shneerson, and S. E. File. 2005b. Cognitive effects of modafinil in student volunteers may depend on IQ. *Pharmacology Biochemistry and Behavior* **82**:133–9.

Randall, D. C., A. Viswanath, P. Bharania, S. M. Elsabagh, D. E. Hartley, J. M. Shneerson, and S. E. File. 2005a. Does modafinil enhance cognitive performance in young volunteers who are not sleep-deprived? *Journal of Clinical Psychopharmacology* **25**:175–9.

Repantis, D., O. Laisney, and I. Heuser. 2010a. Acetylcholinesterase inhibitors and memantine for neuroenhancement in healthy individuals: a systematic review. *Pharmacological Res*earch **61**:473–81.

Repantis, D., P. Schlattmann, O. Laisney, and I. Heuser. 2010b. Modafinil and methylphenidate for neuroenhancement in healthy individuals: a systematic review. *Pharmacological Research* **62**:187–206.

Riis, J., J. P. Simmons, and G. P. Goodwin. 2008. Preferences for enhancement pharmaceuticals: the reluctance to enhance fundamental traits. *Journal of Consum Research* **35**:495–508.

Shiflett, M. W., A. Z. Rankin, M. L. Tomaszycki, and T. J. Devoogd. 2004. Cannabinoid inhibition improves memory in food-storing birds, but with a cost. *Proceedings: Biological Sciences* **271**:2043–8.

Smith, A. P. 2005. Caffeine at work. *Human Psychopharmacology* **20**:441–5.

Smith, M. E., and M. J. Farah. 2011. Are prescription stimulants "smart pills"? The epidemiology and cognitive neuroscience of prescription stimulant use by normal healthy individuals. *Psychological Bulletin* **137**(5):717–41.

Snyder, P. J., M. M. Bednar, J. R. Cromer, and P. Maruff. 2005. Reversal of scopolamine-induced deficits with a single dose of donepezil, an acetylcholinesterase inhibitor. *Alzheimer's and Dementia* **1**:126–35.

Sugden, C., C. R. Housden, R. Aggarwal, B. J. Sahakian, and A. Darzi. 2011. Effect of pharmacological enhancement on the cognitive and clinical psychomotor performance of sleep-deprived doctors: a randomized controlled trial. *Annals of Surgery* **255**:222–7.

Takahashi, R. N., F. A. Pamplona, and M. S. Fernandes. 2005. The cannabinoid antagonist SR141716A facilitates memory acquisition and consolidation in the mouse elevated T-maze. *Neuroscience Letters* **380**:270–5.

Tannock, R., R. Schachar, and G. Logan. 1995. Methylphenidate and cognitive flexibility: dissociated dose effects in hyperactive children. *Journal of Abnormal Child Psychology* **23**:235–66.

Teter, C. J., S. E. McCabe, K. Lagrange, J. A. Cranford, and C. J. Boyd. 2006. Illicit use of specific prescription stimulants among college students: prevalence, motives, and routes of administration. *Pharmacotherapy* **26**:1501–10.

Turner, D. C., T. W. Robbins, L. Clark, A. R. Aron, J. Dowson, and B. J. Sahakian. 2003a. Cognitive enhancing effects of modafinil in healthy volunteers. *Psychopharmacology* **165**(3):260–9.

Turner, D. C., T. W. Robbins, L. Clark, A. R. Aron, J. Dowson, and B. J. Sahakian. 2003b. Relative lack of cognitive effects of methylphenidate in elderly male volunteers. *Psychopharmacology* **168**(4):455–64.

Varvel, S. A., and A. H. Lichtman. 2002. Evaluation of CB1 receptor knockout mice in the Morris water maze. *Journal of Pharmacology and Experimental Therapeutics* **301**:915–24.

Volkow, N.D., G. J. Wang, J. S. Fowler, F. Telang, L. Maynard, J. Logan, S. J. Gatley, N. Pappas, C. Wong, P. Vaska, W. Zhu, and J. M. Swanson. 2004. Evidence that methylphenidate enhances the saliency of a mathematical task by increasing dopamine in the human brain. *American Journal of Psychiatry* **161**(7):1173–80.

Yesavage, J. A., M. S. Mumenthaler, J. L. Taylor, L. Friedman, R. O'Hara, J. Sheikh, J. Tinklenberg, and P. J. Whitehouse. 2002. Donepezil and flight simulator performance: effects on retention of complex skills. *Neurology* **59**(1):123–5.

Zaninotto, A. L., O. F. Bueno, M. Pradella-Hallinan, S. Tufik, J. Rusted, C. Stough, and S. Pompéia. 2009. Acute cognitive effects of donepezil in young, healthy volunteers. *Human Psychopharmacology* **24**(6):453–64.

Chapter 4

Neuroenhancement: a call for better evidence on safety and efficacy

Charles F. Massie, Eric M. Yamga, and Brendon P. Boot[1]

4.1 Current evidence on the safety and efficacy of neuroenhancers

Interest in neuroenhancement is rising in parallel with the rapidly rising rate of attention deficit hyperactivity disorder (ADHD) diagnosis (Graf et al. 2013): between 1990 and 2013, the number of children diagnosed with ADHD rose from 600,000 to 3.5 million (Schwarz 2013). The number of adolescents abusing ADHD medications also rose (Setlik et al. 2009). Methylphenidate (Ritalin), mixed amphetamine salts (Adderall), and modafinil are three of the most commonly purported drugs used for neuroenhancement among college students (Partridge et al. 2013; Ragan et al. 2013). Clinical studies with healthy individuals are yet to show that these drugs are safe in this context, but their safety profile in ADHD patients provides general reassurance.

In parallel, referring to prescription drugs as "neuroenhancers" begs the question of their efficacy. The word suggests that these drugs that aim at cognitive enhancement do in fact succeed in enhancing our cognition. Reports studying the efficacy of methylphenidate, mixed amphetamine salts, and modafinil as well as drugs for dementia (memantine and cholinesterase inhibitors) suggest that the benefits of cognitive enhancers in healthy individuals, if they exist, are modest at best, and that expectations exceed actual effects (Ilieva et al. 2013; Outram 2010; Repantis et al. 2010a,b; Smith and Farah 2011).

4.1.1 Methylphenidate (Ritalin)

Methylphenidate use can cause dry mouth, anorexia, weight loss, depression, vertigo, and tachycardia, the latter being the most commonly reported side effect (Outram 2010; Repantis et al. 2010b). Less commonly it is associated with hepatomegaly, suicidal ideation, and short-term pubertal growth suppression (Thomas et al. 2013). While the drug has a relatively safe profile when used as directed, the risk for addiction is substantial. Patients who take the medication in accordance with the guidelines do not report euphoria, but regular use of larger doses can produce dependence (Outram 2010; Repantis et al. 2010b). The reinforcing effects of large doses and/or intravenous administration of methylphenidate are similar to those of cocaine and D-amphetamine (Outram 2010). More than 1000 cases of methylphenidate intoxication are reported each year in the United States (Iversen 2008).

In terms of efficacy, once again the evidence is limited. In single-dose studies of healthy individuals, methylphenidate improved memory, but had no effect on attention, mood, or executive function (Repantis et al. 2010b). There are two studies of repeated drug administration (Gilbert et al. 1973; Gobbi et al. 2003), but cognitive effects cannot be determined because pertinent cognitive variables were not studied (Repantis et al. 2010b). Methylphenidate shows no efficacy in enhancing cognitive functions in sleep-deprived individuals whether it is a single dose after a sleepless night (Bray et al. 2004) or several doses during a sustained sleep-deprivation period of 64 hours (Babkoff et al. 1992). Instead, people overestimated their own performance (Bray et al. 2004).

4.1.2 Mixed amphetamine salts (Adderall)

The number of adolescents abusing ADHD medications is rising alongside the increasing number of ADHD diagnoses (Setlik et al. 2009). In McCabe et al.'s (2005) survey done among students across 119 American colleges, a median of 3% of students admitted to nonmedical stimulant use for that academic year. In another study, Prudhomme White et al. (2006) found that, among those who abused the medication, 65% reported abusing the drug for its euphoric effects, and 69% to improve attention. A significant number (40.3%) reported snorting the drug to induce a rapid peak of the stimulant in the blood to promote euphoria. The same subjects report that they use these prescription drugs because of their lower cost and ready availability compared with street amphetamines. It is sobering to remind ourselves that this medication is the same substance that was prohibited by the American government in 1965 (Public Law 1965), and the United Nations at the 1971 Convention on Psychotropic Substances (UN General Assembly 1975). Their widespread prescription for neuroenhancement would further facilitate the abuse of and dependence on this drug. We have a moral and legal contradiction: street amphetamines are prohibited, yet the same substance is prescribed widely despite evidence that it is encouraging abuse and addiction.

The evidence for cognitive enhancement with mixed amphetamines is inconclusive. A double-blind, placebo-controlled trial studying the effects of mixed amphetamine salts in healthy people across cognitive functions including memory, creativity, intelligence, and scholastic achievements found little evidence that a standard clinical dose of Adderall enhanced performance (Ilieva et al. 2013). Smith and Farah's (2011) review of more than 50 experiments found mixed results. Over a third of the experiments found no effect on cognitive function, while some reported evidence for consolidation of long-term declarative memory. Ilieva et al. (2013) argue that inadequate power and publication bias make it very "difficult to estimate the true robustness and effect size of cognitive enhancement with mixed amphetamine salts and other stimulant medications by surveying the published literature" (p. 504). They also noted that mixed amphetamine salts had a small but reliable effect on the judgment of one's own enhancement compared to placebo, which may help to explain their popularity as cognitive enhancers.

4.1.3 **Modafinil (Provigil)**

Modafinil is a wakefulness-promoting agent that is licensed to treat patients suffering from daytime sleepiness caused by narcolepsy, sleep apnea, or shift-work sleep disorder. It is generally well tolerated in short-term studies. Most commonly reported adverse effects include headache, dizziness, gastrointestinal complaints, and sleep disturbances (insomnia in non-sleep-deprived patients) (Repantis et al. 2010b). Even though most studies have proven safety in modafinil's off-label use in the short term (Westcott 2005), none have been able to claim the same in the long term. On the contrary, emerging neurobiological evidence shows that this drug has a non-negligible potential for addiction on the basis of molecular, imaging, and behavioral studies (Mohamed 2012) and that it increases dopamine concentration in the nucleus accumbens (Volkow et al. 2009). There have also been reports of psychotic events induced by modafinil (Wu et al. 2008). The absence of precise neurobiological mechanism of action, inadequate clinical experience, and a dearth of empirical data addressing its long-term use make it difficult to assume safety of modafinil in enhancing cognitive function (Minzenberg et al. 2008).

Modafinil is already used by military personnel to battle fatigue that can result in diminished alertness and performance in critical situations. A memorandum from the US Air Force (Taylor and Keys 2003) approves the use of modafinil for well-defined training and combat conditions considered extremely intense and lengthy operations that require extended wakefulness and alertness (bombing missions can last 20–50 hours in duration). Modafinil, referred to in the military community as the "go-pill," shows attractive effects, especially for helicopter pilots: it mimics the stimulating effects of amphetamines (namely, dextroamphetamine) by promoting wakefulness in sleep-deprived individuals, without the adverse effects associated with amphetamine use (behavioral modification, addictive potential, sleep rebound effects) (Estrada et al. 2012; Westcott 2005). In a recent survey investigating stimulant use in F-15 aircrew, 58.6% of study participants (n = 29) reported using stimulants and their use was observed in 35% of surveyed sorties. Again in this study, in-flight stimulants were effective in decreasing fatigue without being associated with post-flight side effects (Gore et al. 2010).

A single dose of modafinil in healthy individuals has been reported to have a cognitive-enhancing effect in cases of moderate sleep deprivation, with an important impact on executive function that persisted over time (Cohen's $d = 3.3$, $p < 0.0001$). A positive effect on memory was found, which did not persist over time, and decreased the longer the sleep deprivation lasted. Finally, wakefulness was significantly and strongly improved by a single dose of modafinil with a stronger effect with each extra hour of sleep deprivation (Repantis et al. 2010b). Usefulness of repeated doses in long-lasting sleep deprivation, however, is questionable: only wakefulness (and not executive functions, nor attention) was significantly affected by modafinil (Repantis et al. 2010b).

4.1.4 Cholinesterase inhibitors and memantine

Repantis et al. (2010a) conducted a systematic review of the use of cholinesterase inhibitors (donepezil, galantamine, and rivastigmine) and memantine as neuroenhancers in healthy individuals. Twelve of the 20 included trials studied donepezil, seven memantine, and one studied rivastigmine. No trials of galantamine met the inclusion criteria. In six small ($n \leq 48$) trials lasting 14–42 days, the following results were found:

a) Episodic memory was adversely affected in two trials (Beglinger et al. 2004, 2005) and improved in one study (Gron et al. 2005).

b) In a sleep deprivation trial, donepezil reduced memory and attention deficits resulting from 24 h of sleep deprivation (Chuah et al. 2009). Under conditions of wakefulness, however, the drug had no significant effect on performance on neuropsychological tests (Chuah and Chee 2008).

c) Rivastigmine improved working visuospatial memory but impaired episodic memory in a small trial of 16 elderly patients (Wezenberg et al. 2005).

d) Donepezil improved the retention of "practical skills" on complex aviation tasks and verbal memory for semantically processed words.

e) No trials of repeated doses of memantine met the review inclusion criteria.

Thus, the evidence to date suggests that the effects of donepezil and rivastigmine on cognition in healthy adults are modest and inconsistent. The authors of the review conclude that the available data are underpowered to draw definitive conclusions on cholinesterase inhibitors and memantine as cognitive enhancers because of missing large longitudinal studies (Repantis et al. 2010a). Publication bias and demand characteristics suggest that these modest results may be overestimates of the true effects.

4.2 Ethical implications of the data for neuroenhancement

4.2.1 Prescribing drugs to healthy people demands that we re-evaluate the traditional risk–benefit ratio

When prescribing drugs to patients, a physician must assess a risk–benefit ratio. The ratio is evaluated by comparing the risk of harm associated with the treatment to its potential benefits in terms of reduction in morbidity or mortality. When treating a disease, some adverse pharmacological effects can be deemed acceptable if they are not worse than the natural course of the disease, and if the benefits of the drug outweigh these adverse consequences. In the case of neuroenhancement, however, the benefits cannot be measured in terms of reductions in morbidity or mortality, since none exist. Instead, these benefits must arise from a sought-after goal. Considering that this goal has less teleological importance than a reduction in morbidity or mortality, the calculus of risk–benefit analysis must change accordingly (Drabiak-Syed 2011; Wasson 2011). When giving drugs to healthy people, the tolerance threshold for what is "acceptable" risk should be significantly reduced. We should tolerate low risks for minor harm only.

4.2.1.1 Elements of risk–benefit calculations

Evaluating the risk–benefit ratio in clinical practice requires three key elements: first, knowledge of the benefits; second, knowledge of the risks; and finally, a judgment of whether the benefits are worth the risks. LASIK eye surgery illustrates these points. Even though multiple risks exist, patients are able to make a rational decision based on well-established facts of long-term and short-term undesirable effects. The procedure is very well studied, so physicians are able to objectively counsel their patients with specific information to help them in their decision-making. Some ethicists argue that neuroenhancement should be limited to drugs in which there are no "major risks" (Bostrom and Sandberg 2009). When it comes to minor risks, they argue that there is consensus amongst ethicists that patient autonomy prevails, even when the procedure does not reduce mortality (Bostrom and Sandberg 2009). Of course, what constitutes "major and minor risks" and what constitutes "acceptable" adverse events will undoubtedly be subject to controversy and is, to some extent, individual-specific.

4.2.1.2 No consensus on long-term risks and benefits for neuroenhancement

The evaluation of the risk–benefit ratio in the context of neuroenhancement is difficult because important information about long-term risks and benefits is lacking (Drabiak-Syed 2011; Lucke et al. 2011). Whilst the American Academy of Neurology (AAN) guidance suggested short-term use of these medications, clinical experience suggests that patients will request long-term prescriptions (Boot et al. 2012). Sustained use may produce cumulative effects such as tolerance and addition that make short-term assessment insufficient: long-term studies need to be conducted. (Drabiak-Syed 2011; Hall 2004). Similarly, the benefits of neuroenhancing drugs have not been adequately characterized. Benefits are likely to vary from one individual to another, and the cost–benefit ratio will similarly come down to a personal choice (Thaler 2009). The principle of nonmaleficence dictates that an unproven benefit for a healthy individual is not a sufficient reason for prescribers to ignore possible harms. We should therefore wait for the evidence on the benefits and harms before prescribing drugs for enhancement.

While recognizing the lack of evidence on risks of healthy individuals taking medications for neuroenhancement, the AAN Ethics Committee (Larriviere et al. 2009) presupposed that evidence will eventually emerge that demonstrates that benefits outweigh the risks of harm. This claim is based on clinical experience with the same drugs given to patients and on the belief that their safety and efficacy profiles from this clinical use can be applied to young and healthy individuals. There are several problems with this reasoning. For the most part, there is a clear molecular rationale for the treatment of patients, but not for their use in healthy individuals. For example, positron emission tomograph (Nordberg et al. 1990; Zubieta et al. 2001) and pathology studies (Bird et al. 1983; Reinikainen et al. 1988; Rossor et al. 1984) confirm that Alzheimer's patients have marked cholinergic dysfunction. Therefore, the use of cholinesterase inhibitor medication is coherent at the molecular level, as it is designed to increase acetylcholine concentration in the brain. In young, healthy subjects with normal levels of acetylcholine, it's not at all clear that inhibiting cholinesterase will be beneficial.

4.2.2 Unintended consequences of neuroehancement prescriptions

The most adventurous young adults are likely to use drugs for unapproved purposes and consume them more than the recommended therapeutic dose and for as long as necessary to achieve desired results: we know very little about the possible risks associated with such a regimen (Hall et al. 2004). Experience with performance-enhancing drugs amongst athletes demonstrates that even if the safety profile for neuroenhancers becomes available and risks become apparent, some users may disregard these warnings. Anabolic steroids are widely used, for example, despite their links to myocardial infarction and liver malignancy (NIDA 2006).

Neuroenhancement medications are frequently diverted to people without prescriptions. For example, 2.8% of US 12th-graders report use of diverted mixed amphetamine salts (Adderall) in the past month (Johnston et al. 2010). It is therefore important to document a safety profile that includes not only the target age population but also other age groups that may put themselves at risk by taking medications that were not prescribed for them. For example, what are the developmental consequences for such long-term use in children and teenagers?

4.2.3 Media portrayal of neuroenhancement

Thaler (2009) argues that risk–benefit calculations should be left up to the patient. Respect for patient autonomy dictates that he or she should be allowed to choose the risks he or she is willing to endure for a given benefit (Thaler 2009). This includes an important requirement: that patients understand the facts. Media and web-based coverage of cognitive enhancement is often uncritically enthusiastic about its benefits (Forlini and Racine 2009). Reports emphasize benefits while ignoring or under-reporting adverse effects of their use (Partridge et al. 2011). The role of the prescriber is to educate the patient with an objective, nuanced portrait of what is achievable with these medications that counterbalances these media-generated expectations: only then will a patient's decision be truly autonomous. The lack of consensus on neuroenhancement's risks and benefits renders this exchange of knowledge between the prescriber and patient more difficult.

4.2.4 The limitations of the regulatory framework for prescription medicines

The AAN guidance on neuroenhancement (Larriviere et al. 2009) argues that "pharmaceutical companies have financial incentives to develop medications that are effective and safe within an acceptable risk-benefit ratio" (ibid. 1408), yet experience shows that this is not always the case. The toxic effects of fen-phen (Connolly et al. 1997; Kushner 1997), and COX2 inhibitors, are examples in which corporate financial incentives did not preclude the widespread use of harmful drugs. Pharmaceutical companies can act in ways that are contrary to public interest, such as the inappropriate promotion of slow release oxycodone for chronic pain (Hall and Degenhardt 2007) and the suppression of evidence from clinical trials that showed only equivocal benefits of selective serotonin reuptake inhibitor (SSRI) drugs in depression (e.g. Kirsch et al. 2008). Other controversial events also occurred in relation to neuroenhancers. In 2006 and 2007, Cephalon (the maker of modafinil) changed its marketing campaign in order to expand modafinil indications and thereby increase its sales. Cephalon also claimed that modafinil was an addiction-free drug; post-marketing surveillance revealed that this statement was false (Volkow et al. 2009).

The AAN Ethics Committee also suggests that the U.S. Food and Drug Administration's (FDA) oversight of medical prescriptions will minimize harm in the use of neuroenhancement medications. The FDA's oversight is restricted to licensing and post-market surveillance of medications for *the treatment of disease*. When used for the purpose of cognitive enhancement, these drugs are used off-label (Boot et al. 2012; Hall et al. 2010). Although regulations governing medicinal drugs ensure that they are safe and effective for their therapeutic indications, there is no equivalent procedure for unregulated "off-label" uses (Greely et al. 2008). Thus, in suggesting that it is ethically legitimate to use the drugs off-label, the AAN Ethics Committee circumvents one of the safeguards that they are using to support this legitimization. The evidence on the risk versus benefit of neuroenhancement use would then come from epidemiological research on health outcomes in people taking these drugs. If these drugs turn out to have serious adverse effects like the historical examples of cocaine and amphetamine, their premature use could harm patients and (rightly) undermine trust in the medical profession.

4.2.5 **Lessons for neuroenhancement from history**

Excitement over neuroenhancement is not a new phenomenon. The discovery of several substances with purported neuroenhancing qualities have been followed by predictably a cycle of excitement, disillusionment, and prohibition. Cocaine, and later amphetamines, were used for their purported neuroenhancing capacities at the start of the twentieth century. Among other things, cocaine was reported to improve mental performance and allow users to sustain high levels of attention for prolonged periods of time, not unlike the current claims made for methylphenidate and modafinil (de Jongh et al. 2008; Sahakian et al. 2007). The success of cocaine led to it being prescribed more and to its nonmedical use in an increasingly high number of people. Eventual reports of harm and disillusionment by the physicians led to a steep decline in prescribing, followed by a ban of the substance. The story of amphetamines is a similar one. Like cocaine before them, amphetamines were seen as "essential for maintaining optimal performance in an increasingly fast-paced modern life" (Iversen 2008). Drug-induced psychosis and dependence emerged from the widespread use of these drugs before their eventual prohibition (Bell et al. 2012). There are more recent examples of this cycle. Some clinicians have advocated the use of 3,4-methylenedioxy-*N*-methylamphetamine (MDMA or ecstasy) to facilitate emotional communication in psychotherapy and relief of anxiety (Johansen and Krebs

2009; Savulescu and Sandberg 2008)—emotional enhancement in this case. With these perspectives in mind, we need to remember the neurotoxic effects of regular MDMA use on cognition (dysfunction in working, short- and longer-term memory), behavior (impulsivity, suicidality, psychosis), and mood (depression) brought on by damage to serotonergic nerve terminals (Boot et al. 2000; Halpin et al. 2014).

4.2.6 Limitations of assessment of neuroenhancers

The AAN Ethics Committee suggests that one way to deal with our ignorance of potential long-term risks of neuroenhancers is to establish short-term goals and to discontinue any medication as soon as the goals have been reached or deemed unreachable (Larriviere et al. 2009). However, this is unlikely to be successful because of the following:

a) It is difficult to specify discrete goals that would enable assessment of whether the intended effects have been achieved (Boot et al. 2012).
b) Success in reaching a first goal may simply lead to the desire to use enhancement drugs to achieve an ever-expanding set of goals—this increases the risk of chronic use and possibly dependence.
c) The most common motivation for those seeking neuroenhancers is to improve concentration at work (Maher 2008). Since this task is ongoing, the desire to use neuroenhancers is likely to be ongoing.
d) Assessment of goal success relies on the patient's assessment. The medications may impair the patient's judgment, for example, by inducing tolerance and addiction (Drabiak-Syed 2011).

There is no evidence that small improvements in cognitive function under laboratory conditions will translate into significant clinical enhancements in "real-world" settings (Ilieva and Farah 2013; Smith and Farah 2011), or that those acute effects will persist with long-term use (Outram 2010). Some recent studies have noted that improving cognitive abilities may come at the expense of other brain functions such as creativity or divergent thinking (Farah et al. 2009), because of the interplay between cognitive and emotional areas that happens via the limbic system: cognition improvement was associated with changes in affect and development of anxiety (Drabiak-Syed 2011).

4.3 Conclusion

There is insufficient evidence of long-term, real-world improvement in performance in subjects taking the drugs purported to be neuroenhancers to conclude that they "enhance" the lives of those who take them. Nor is there convincing evidence that these drugs enhance cognition in the short term, since what is regarded as "enhancement" generally involves a trade-off of some kind, such as a possible trade-off, for example, between cognitive processing and creativity (Farah et al. 2009). Despite the claim that the cultural acceptance of neuroenhancers is rapidly increasing (Maher 2008), for now, and until we have the appropriate data for safety and efficacy, we should establish strict boundaries regarding the use and promotion of neuroenhancers. At the same time, the medical research community needs to rapidly expand our knowledge of the risk and benefits of these medications in healthy people. Only in these ways can we avoid a repetition of the costly missteps so common in the history of "neuroenhancement."

Note

1. Regarding author funding and conflict of interest specifc to this work, Messrs Massie and Yamga have nothing to disclose. Dr. Boot receives funding from the Michael J. Fox Foundation and the Parkinson's disease Research Foundation, and has served as an investigator for clinical trials sponsored by Pfizer, Janssen, Avid, and Bristol-Myers Squibb.

References

Babkoff, H., T. L. Kelly, L.T. Matteson, et al. 1992. Pemoline and methylphenidate: interaction with mood, sleepiness, and cognitive performance during 64 hours of sleep deprivation. *Military Psychology* **4**:235–65.

Beglinger, L. J., B. L. Gaydos, D. A. Kareken, O. Tangphao-Daniels, E. R. Siemers, and R. C. Mohs. 2004. Neuropsychological test performance in healthy volunteers before and after donepezil administration. *Journal of Psychopharmacology* **18**:102–8.

Beglinger, L. J., O. Tangphao-Daniels, D. A. Kareken, L. Zhang, R. Mohs, and E. R. Siemers. 2005. Neuropsychological test performance in healthy elderly volunteers before and after donepezil administration: a randomized, controlled study. *Journal of Clinical Psychopharmacology* **25**:159–65.

Bell, S. K., J. C. Lucke, and W. D. Hall. 2012. Lessons for enhancement from the history of cocaine and amphetamine use. *AJOB Neuroscience* **3**:24–9.

Bird, T. D., S. Stranahan, S. M. Sumi, and M. Raskind. 1983. Alzheimer's disease: choline acetyltransferase activity in brain tissue from clinical and pathological subgroups. *Annals of Neurology* **14**:284–93.

Boot, B. P., I. S. McGregor, and W. Hall. 2000. MDMA (Ecstasy) neurotoxicity: assessing and communicating the risks. *The Lancet* **355**:1818–21.

Boot, B. P., B. Partridge, and W. Hall. 2012. Letter to the Editor: Better evidence for safety and efficacy is needed before neurologists prescribe drugs for neuroenhancement to healthy people. *Neurocase* **18**:181–4.

Bostrom, N., and A. Sandberg. 2009. Cognitive enhancement: Methods, ethics, regulatory challenges. *Science and Engineering Ethics* **15**:311–41.

Bray, C. L., K. S. Cahill, J. T. Oshier, et al. 2004. Methylphenidate does not improve cognitive function in healthy sleep-deprived young adults. *Journal of Investigative Medicine* **52**:192–201.

Chuah, L. Y. M., and M. W. L. Chee. 2008. Cholinergic augmentation modulates visual task performance in sleep-deprived young adults. *Journal of Neuroscience* **28**:11369–77.

Chuah, L. Y. M., D. L. Chong, A. K. Chen, et al. 2009. Donepezil improves episodic memory in young individuals vulnerable to the effects of sleep deprivation. *Sleep* **32**:999.

Connolly, H. M., J. L. Crary, M. D. McGoon, et al. 1997. Valvular heart disease associated with fenfluramine–phentermine. *New England Journal of Medicine* **337**:581–8.

Drabiak-Syed, K. 2011. Physicians prescribing "medicine" for enhancement: Why we should not and cannot overlook safety concerns. *American Journal of Bioethics* **11**:17–19.

Estrada, A., A. M. Kelley, C. M. Webb, J. R. Athy, and J. S. Crowley. 2012. Modafinil as a replacement for dextroamphetamine for sustaining alertness in military helicopter pilots. *Aviation, Space, and Environmental Medicine* **83**:556–67.

Farah, M. J., C. Haimm, G. Sankoorikal, and A. Chatterjee. 2009. When we enhance cognition with Adderall, do we sacrifice creativity? A preliminary study. *Psychopharmacology* **202**:541–7.

Forlini, C., and E. Racine. 2009. Disagreements with implications: diverging discourses on the ethics of non-medical use of methylphenidate for performance enhancement. *BMC Medical Ethics* **10**:9.

Gilbert, J. G., K. J. Donnelly, L. E. Zimmer, and J. F. Kubis. 1973. Effect of magnesium pemoline and methylphenidate on memory improvement and mood in normal aging subjects. *International Journal of Aging and Human Development* **4**:35–51.

Gobbi, G., S. Slater, N. Boucher, G. Debonnel, and P. Blier. 2003. Neurochemical and psychotropic effects of bupropion in healthy male subjects. *Journal of Clinical Psychopharmacology* **23**:233–9.

Gore, R. K., T. S. Webb, and E. D. A. Hermes. 2010. Fatigue and stimulant use in military fighter aircrew during combat operations. *Aviation, Space, and Environmental Medicine* **81**:719–27.

Graf, W. D., S. K. Nagel, L. G. Epstein, G. Miller, R. Nass, and D. Larriviere. 2013. Pediatric neuroenhancement: ethical, legal, social, and neurodevelopmental implications. *Neurology* **80**:1251–60.

Greely, H., B. Sahakian, J. Harris, et al. 2008. Towards responsible use of cognitive-enhancing drugs by the healthy. *Nature* **456**:702–5.

Grön, G., M. Kirstein, A. Thielscher, M. W. Riepe, and M. Spitzer. 2005. Cholinergic enhancement of episodic memory in healthy young adults. *Psychopharmacology* **182**:170–9.

Hall, W. 2004. Feeling "better than well." *EMBO Reports* **5**:1105–9.

Hall, W., and L. Degenhardt. 2007. Regulating opioid prescribing to provide access to effective treatment while minimizing diversion: an overdue topic for research. *Addiction* **102**:1685–8.

Hall, W. D., and J. C. Lucke. 2010. The enhancement use of neuropharmaceuticals: More scepticism and caution needed. *Addiction* **105**:2041–3.

Halpin, L. E., S. A. Collins, and B. K. Yamamot. 2014. Neurotoxicity of methamphetamine and 3, 4-methylenedioxymethamphetamine. *Life Sciences* **97**:37–44.

Ilieva, I., J. Boland, and M. J. Farah. 2013. Objective and subjective cognitive enhancing effects of mixed amphetamine salts in healthy people. *Neuropharmacology* **64**:496–505.

Ilieva, I. P., and M. J. Farah. 2013. Enhancement stimulants: perceived motivational and cognitive advantages. *Frontiers in Neuroscience* 7:198.

Iversen, L. 2008. *Speed, Ecstasy, Ritalin: The Science of Amphetamines*. Oxford: Oxford University Press.

Johansen, P., and T. S. Krebs. 2009. How could MDMA (ecstasy) help anxiety disorders? A neurobiological rationale. *Journal of Psychopharmacology* **23**:389–91.

Johnston, L. D., P. M. O'Malley, J. G. Bachman, and J. E. Schulenberg. 2010. Monitoring the future: National survey results on drug use, 1975–2009. Volume I: Secondary school students. NIH Publication 10-7584. *National Institute on Drug Abuse (NIDA).*

de Jongh, R., I. Bolt, M. Schermer, and B. Olivier. 2008. Botox for the brain: enhancement of cognition, mood and pro-social behavior and blunting of unwanted memories. *Neuroscience and Biobehavioral Reviews* 32:760–76.

Kirsch, I., B. J. Deacon, T. B. Huedo-Medina, A. Scoboria, T. J. Moore, and B. T. Johnson. 2008. Initial severity and antidepressant benefits: a meta-analysis of data submitted to the Food and Drug Administration. *PLoS Medicine* **5**:e45.

Kushner, R. 1997. The treatment of obesity: a call for prudence and professionalism. *Archives of Internal Medicine* **157**:602–4.

Larriviere, D., M. A. Williams, M. Rizzo, and R. J. Bonnie. 2009. Responding to requests from adult patients for neuroenhancements. Guidance of the Ethics, Law and Humanities Committee. *Neurology* 73:1406–12.

Lucke, J. C., S. Bell, B. Partridge, and W. D. Hall. 2011. Deflating the neuroenhancement bubble. *AJOB Neuroscience* **2**:38–43.

McCabe, S. E., J. R. Knight, C. J. Teter, and H. Wechsler. 2005. Non-medical use of prescription stimulants among US college students: prevalence and correlates from a national survey. *Addiction* **100**:96–106.

Maher, B. 2008. Poll results: look who's doping. *Nature* **452**:674–5.

Minzenberg, M. J., A. J. Watrous, J. H. Yoon, S. Ursu, and C. S. Carter. 2008. Modafinil shifts human locus coeruleus to low-tonic, high-phasic activity during functional MRI. *Science* **322**:1700–2.

Mohamed, A. D. 2012. Modafinil has the potential for addiction. *AJOB Neuroscience* **3**:36–8.

National Institute on Drug Abuse (NIDA). 2001; 2006. Report series: anabolic steroid abuse. Available at http://www.drugabuse.gov/sites/default/files/rrsteroids_0.pdf

Nordberg, A., A. Adem, L. Nilsson-Håkansson, et al. 1990. New approaches to clinical and postmortem investigations of cholinergic mechanisms. *Progress in Brain Research* **84**:313.

Outram, S. M. 2010. The use of methylphenidate among students: the future of enhancement? *Journal of Medical Ethics* **36**:198–202.

Partridge, B., S. Bell, J. Lucke, and W. Hall. 2013. Australian university students' attitudes towards the use of prescription stimulants as cognitive enhancers: perceived patterns of use, efficacy and safety. *Drug and Alcohol Review* **32**:295–302.

Partridge, B. J., S. K. Bell, J. C. Lucke, S. Yeates, and W. D. Hall. 2011. Smart drugs "as common as coffee": media hype about neuroenhancement. *PLoS One* **6**:e28416.

Public Law 89–74, 1965. "Drug Abuse Control Amendments of 1965". Available at http://www.gpo.gov/fdsys/pkg/STATUTE-79/pdf/STATUTE-79-Pg226.pdf.

Ragan, C. I., I. Bard, and I. Singh. 2013. What should we do about student use of cognitive enhancers? An analysis of current evidence. *Neuropharmacology* **64**:588–95.

Reinikainen, K. J., P. J. Riekkinen, L. Paljärvi, et al. 1988. Cholinergic deficit in Alzheimer's disease: a study based on CSF and autopsy data. *Neurochemical Research* **13**:135–46.

Repantis, D., O. Laisney, and I. Heuser. 2010a. Acetylcholinesterase inhibitors and memantine for neuroenhancement in healthy individuals: a systematic review. *Pharmacological Res*earch **61**:473–81.

Repantis, D., P. Schlattmann, O. Laisney, and I. Heuser. 2010b. Modafinil and methylphenidate for neuroenhancement in healthy individuals: a systematic review. *Pharmacological Research* **62**:187–206.

Rossor, M. N., L. L. Iversen, G. P. Reynolds, C. Q. Mountjoy, and M. Roth. 1984. Neurochemical characteristics of early and late onset types of Alzheimer's disease. *British Medical* Journal (Clinical Research Ed.) **288**:961.

Sahakian, B., and S. Morein-Zamir. 2007. Professor's little helper. *Nature* **450**:1157–59.

Savulescu, J., and A. Sandberg. 2008. Neuroenhancement of love and marriage: the chemicals between us. *Neuroethics* **1**:31–44.

Schwarz, A. 2013. The selling of attention deficit disorder. *New York Times*.

Setlik, J., G. R. Bond, and M. Ho. 2009. Adolescent prescription ADHD medication abuse is rising along with prescriptions for these medications. *Pediatrics* **124**:875–80.

Smith, M. E., and M. J. Farah. 2011. Are prescription stimulants "smart pills"? The epidemiology and cognitive neuroscience of prescription stimulant use by normal healthy individuals. *Psychological Bulletin* **137**:717.

Taylor, G. P., Jr., and R. E. Keys. 2003. *Modafinil and management of aircrew fatigue*. Washington, DC: United States Department of the Air Force.

Thaler, D. S. 2009. Improving introspection to inform free will regarding the choice by healthy individuals to use or not use cognitive enhancing drugs. *Harm Reduction Journal* **6**:10.

Thomas, R., G. K. Mitchell, and L. Batstra. 2013. Attention-deficit/hyperactivity disorder: are we helping or harming? *BMJ* November 5:347.

UN General Assembly. 1975. *1971 Convention on Psychotropic Substances.* [1975] UNGA 94; A/RES/3443 (XXX). Available at http://www.worldlii.org/int/other/UNGA/1975/94.pdf.

Volkow, N. D., J. S. Fowler, J. Logan, et al. 2009. Effects of modafinil on dopamine and dopamine transporters in the male human brain: clinical implications. *JAMA* **301**:1148–54.

Wasson, K. 2011. Medical and genetic enhancements: Ethical issues that will not go away. *American Journal of Bioethics* **11**:21–2.

Westcott, K. J. 2005. Modafinil, sleep deprivation, and cognitive function in military and medical settings. *Military Medicine* **170**(4):333–5.

Wezenberg, E., R. J. Verkes, B. G. C. Sabbe, G. S. F. Ruigt, and W. Hulstijn. 2005. Modulation of memory and visuospatial processes by biperiden and rivastigmine in elderly healthy subjects. *Psychopharmacology* **181**:582–94.

White, B. P., K. A. Becker-Blease, and K. Grace-Bishop. 2006. Stimulant medication use, misuse, and abuse in an undergraduate and graduate student sample. *Journal of American College Health* **54**:261–8.

Wu, P., S. Jones, C. J. Ryan, D. Michail, and T. D. Robinson. 2008. Modafinil-induced psychosis. *Internal Medicine Journal* **38**:677–8.

Zubieta, J.-K., R. A. Koeppe, K. A. Frey, et al. 2001. Assessment of muscarinic receptor concentrations in aging and Alzheimer disease with [11C]NMPB and PET. *Synapse* **39**(4):275–87.

Chapter 5

Exaggerating the benefits and downplaying the risks in the bioethical debate on cognitive neuroenhancement

Andreas Heinz and Sabine Müller

5.1 Introduction

Cognitive neuroenhancement, i.e., the improvement of intellectual capabilities through medical means without therapeutic intentions, has been debated in universities and in the media for more than a decade. Although sound empirical data about the prevalence rates of drug use for neuroenhancement are scarce, alarming estimations have been published: often cited is a *Nature* paper of neuroethicist Henry Greely and colleagues which quotes a survey according to which 7% of students in US universities had used prescription stimulants for improving their capacities for learning in the past year; on some campuses even up to 25% (Greely et al. 2008). Also the results of a German study of the group of the psychiatrist Klaus Lieb look alarming at first glance: According to this study, 20% of students in Germany used cognitive-enhancing drugs in the past year (Dietz et al. 2013). But it must be noted that caffeine tablets were included in the authors' definition of cognitive-enhancing drugs, although they are nonprescription drugs in Germany, and although it is discussed controversially whether caffeine belongs to the category of cognitive-enhancing drugs at all. Unfortunately, this paper does not allow for retrieving the prevalence of using prescription psychotropic drugs for cognitive enhancement. According to a survey published by the German health insurance DAK, 5% of the employed persons in Germany use pharmaceutical drugs for enhancing their cognitive performance (DAK 2009).

These preliminary data allow for the conclusion that in countries such as the USA and Germany, a rather small minority of people use prescription psychotropic drugs for cognitive enhancement. Although their fraction is small, they receive some attention both in the academic discourse and in the mass media, whereby the prevalence rates are often exaggerated (Lucke et al. 2011). Ironically, the often sensationalist media reports and liberal arguments from the bioethical debate may contribute to increasing prevalence rates of psychotropic drug use, at least when they exaggerate the prevalence and the benefits of the drugs, and downplay their risks and adverse effects. Because the debate influences drug consumption patterns, we feel that the issue of cognitive enhancement is not only an abstract problem, but a real bioethical problem, which needs more input from neuroscience and clinical experience.

While many authors strongly oppose neuroenhancement, others are convinced that they have disproved every single objection against it. Since the latter believe that the advantages of neuroenhancement outweigh its disadvantages, they demand its legalization. The main issue of the proponents of neuroenhancement is a liberal argument, namely, that everyone has the right to decide autonomously what to do with his or her body and mind. These liberal arguments pro-neuroenhancement can be found, among others, in the article of Greely and colleagues mentioned earlier (Greely et al. 2008) and in a position paper ("memorandum") of a German research group (Galert et al. 2009).

The advocates of neuroenhancement do not claim that today's medications, which are regarded as "cognitive enhancers," are optimal. Rather, they admit that they are not as efficient as desired. Although many enhancement advocates recommend methylphenidate and/or modafinil, these drugs are not the cognitive enhancers they expect for the future and on which their positive judgments are based. Rather, their argumentations refer to fictional neuroenhancers with reliable benefits and without adverse effects. Their central argument often is a variation of the following: if there were substances that verifiably improve certain mental abilities without adverse effects or with an acceptable risk–benefit profile, then no convincing arguments against the use of these substances or procedures can be made. Obviously, this argument is hypothetical and based on an if–then structure. The conditional portion of this argument ("if") is not meant in a purely speculative way. Rather, its proponents assume that such pharmaceuticals are principally possible and could be available in the foreseeable future (for a detailed discussion see Heinz et al. 2012).

While several participants in the discussion mainly criticize the conclusion of the pro-neuroenhancement argument ("then"), we also question its conditional part ("if"). We ask how probable it is that substances will be available in the foreseeable future that enhance cognition but do not have any adverse effects, particularly any addiction potential. The second part of the argument ("then"), namely, the ethical evaluation of neuroenhancement, comes back into play when we consider whether research on neuroenhancement should be allowed and founded by society (Heinz et al. 2012).

In the following, we will argue that the proponents of neuroenhancement base their considerations on several false assumptions, namely:

1. Underestimation of the risk of addiction to cognitive enhancers
2. Underestimation of further medical risks of cognitive enhancers
3. Overestimation of the benefits of cognitive enhancers.

As we will show, these false assumptions (which are interrelated) contribute to the empirical implausibility of the conditional part of the pro-neuroenhancement argument.

5.2 Underestimation of the risk of addiction to cognitive enhancers

Proponents of neuroenhancement regularly downplay the risk of addiction to neuroenhancers. Thereby, different authors use different strategies or combinations of them: (1) the belittlement of addiction; (2) an out-of-date differentiation between physical dependence and psychological dependence; and (3) the underestimation of the addictive potential of neuroenhancers (particularly by comparisons with the addictive potential of harmless (food) items or gadgets instead of the addictive potential of illicit drugs). A consequence of the underestimation of the drugs' addictive potential is the overestimation of the capability for autonomy of habitual consumers of such drugs. These strategies are used to question one of the main arguments against cognitive enhancement, namely, that it implies a significant risk of addiction.

5.2.1 Belittlement of addiction

The first and most radical strategy downplays addictions in principle: the "transhumanist" and ultraliberal bioethicists Bennett Foddy and Julian Savulescu reject both the understanding of addiction as a brain disease and as a moral condition. Rather, they assert that addiction is "nothing but the drive to experience pleasure," "merely a form of pleasure-seeking," "a strong preference, similar to appetitive preferences" (Foddy and Savulescu 2007). According to them, characterization of addictions as diseases is "an example of the medicalization of socially unacceptable behavior" "like drapetomania" (i.e., the tendency of slaves to run away): the "fiction of affliction allows us to treat them, often against their will." Rather than seeing addictions as "compulsions," Foddy and Savulescu plead for "a fuller conception of autonomy" that "would allow for the possibility that a rational person might simply value the pleasure of heroin use above the value of their health or their jobs."

If one would follow Foddy and Savulescu, the possible risk of addiction to cognitive enhancers is invalid, just because addictions are not diseases but just a freely chosen way of living.

However, the whole argument is based on a false dichotomy between "compulsive" drug taking and subjective (free) choices and ignores more subtle ways by which drugs of abuse interact with decision-making and bias addicts toward unfavorable choices (Heinz et al. 2011).

5.2.2 Out-of-date differentiation between physical dependence and psychological dependence

Since most people will not follow this radical proposition, the second strategy is based on a differentiation between severe and relatively harmless forms of addictions. For that, its proponents differentiate between "physical dependence" and "psychological dependence" (defined as an irrational desire for an object and the extreme discomfort caused by the absence of it). For example, the authors of the "memorandum" state that the risk of "physical dependence" would be a convincing argument against the liberalization of the sale of neuroenhancers, whereas the risk of only "psychological dependence" would not be a sufficient argument, since the latter is a widespread phenomenon, like the "dependence" on cell phones (Galert et al. 2009).

In the same vein, Shaw (2012) has argued against our argumentation (Heinz et al. 2012) that cognitive enhancers imply a risk of addiction: he speculated that "psychological dependence" of neuroenhancers might exist, which does not meet the definition of addiction. Such arguments fail to understand current definitions of addiction, stating that "proponents of enhancement appear to be correct in asserting that one could be psychologically dependent on a drug without meeting this definition" (i.e., current definitions of addiction, as proposed by the American Association for Addiction Medicine). He even introduces the speculative idea of addictions that are not harmful (whereby he comes close to Foddy and Savulescu's claim that addictions are not diseases). However, the idea of unharmful addictions is a self-contradiction, since the definition of addiction contains elements of impairment, dysfunction, and disability as harmful consequences of the disease. Harmless forms of dependence are not called addictions, but passions (Heinz et al. 2014).

The distinction between physical and psychological dependence is based on a scientifically outdated mind–body dualism. Recent definitions of addiction do not distinguish between "psychological" and "physical" dependence, because all mental symptoms (including craving for a drug and loss of control) have correlates in the brain. Modern addiction research traces the particular symptoms back to their different neurobiological correlates (Heinz et al. 2009). Development of tolerance and withdrawal symptoms normally occur due to neuroadaptive processes in inhibitory neurotransmitter systems, where a sudden discontinuation of substance consumption can lead to excitation and observable "physical" withdrawal symptoms such as seizures or vegetative symptoms. Symptoms, classified by some authors as a "psychological" dependence such as the desire for a substance and the loss of control when exposed to it, are associated with dysfunctions in other neural (e.g., motivational) systems and their dopaminergic modulation (Everitt and Robbins 2005; Hyman 2005; Kalivas and Volkow 2005). This holds also true for behavioral addictions, such as pathological gambling, which are also characterized by a strong desire for the activity and a diminished control over its practice. Imaging studies among patients with pathological gambling revealed impaired functional activation in the ventral striatum during the anticipation and arrival of reward, an observation that was also reported in patients with alcohol addiction (Reuter et al. 2005).

Indeed, the dopaminergic neurotransmitter system plays a key role in modern addiction theories. In fact, all known substances with addiction potential release dopamine in the reward system and thus boost further consumption; a key step in addiction development (Di Chiara and Bassareo 2007; Everitt et al. 2008). This is mediated by dopaminergic effects on learning processes, which ultimately make drugs and drug-related stimuli particularly desirable, while so-called naturally rewarding experiences such as food intake and social interaction lose their appeal (Di Chiara and Bassareo 2007; Park et al. 2010; Volkow et al. 2004). Such alterations in reward networks directly contributing to motivation, learning, and decision-making help to explain why chronic drug consumption, while far from promoting an absolutely irresistible "compulsion," nonetheless biases addicts toward future drug intake despite better (conscious) decisions to remain abstinent.

5.2.3 Underestimation of the addictive potential of cognitive enhancers

The third strategy of the proponents of cognitive enhancement consists in downplaying or underestimating the addictive potential of cognitive enhancers. For evaluating the addiction potential of the psychotropic drugs in question, their effects on the dopamine system are of particular interest.

Several proponents of cognitive enhancement argue that also "natural" activities such as eating and drinking, particularly drinking coffee, can influence neurotransmitter systems, particularly the dopamine system (Galert et al. 2009; Greely et al. 2008). That is of course true: since the dopamine system is part of reward-related learning, it rewards all activities that are good for the individual or for reproduction including food intake. But there are substantial differences between dopamine release caused by "natural" activities and the dopamine release caused by drugs of abuse as well as psychotropic substances used for the purpose of cognitive enhancement: First, the effect size is fundamentally different: sensory environmental stimuli associated with sex or communication never have a direct effect on the brain, but are always mediated through sense organs, which transform the stimulus constellations into neuronal codes. However, dopamine concentrations, which are triggered by food, increase only by approximately 50 to 100%, while drugs (e.g., amphetamine, cocaine, but also the psychostimulants methylphenidate and modafinil) induce dopamine releases ranging from 175 to 1000% (Acquas et al. 2002; Di Chiara and Imperato 1988; Heinz et al. 2009; Martel and Fantino 1996). As a consequence of the "unphysiological" extreme rises of dopamine concentrations, the consumers develop craving for the drugs of abuse that are much more effective in causing alterations in this key motivational system, whereas they lose interest for nondrug-associated stimuli. Second, a significant difference consists in the habituation effect; for example, the repeated blockade of dopamine transporters by psychotropic substances with addiction potential consistently leads to an increase in the release of dopamine due to direct pharmacological effects, whereas dopamine release triggered by novel food or environmental stimuli rapidly habituates (Di Chiara and Bassareo 2007; Hyman 2005).

Also, the repeated use of such pharmaceuticals or drugs induces counter-regulatory neuroadaptive processes (e.g., a reduction of dopamine D_2 receptors in the ventral striatum). Thus, responses to nondrug-associated reinforcers are reduced and the goal-directed behavior driven by natural reinforcers is impaired (Hyman 2005, 2007; Kalivas and Volkow 2005). Consequently, the repeated consumption of drugs of abuse that strongly stimulate dopamine release increases the desire for further drug consumption, again shifting behavior from goal-directed activities toward habitual drug intake (Everitt and Robbins 2005; Park et al. 2010). Therefore, chronic consumption of drugs of abuse can severely impair behavioral flexibility and voluntary behavioral control, and impair motivation and the ability to learn from nondrug-related feedback (Hyman 2007; Park et al. 2010).

Some proponents of neuroenhancement emphasize differences between psychostimulants such as amphetamines and modafinil. For example, Galert et al. (2009) reject amphetamines as neuroenhancers because of their known addiction potential and serious side effects, whereas they regard modafinil as a neuroenhancer without such adverse effects.

However, such a categorical separation of modafinil and amphetamines is empirically not justified. Although modafinil (unlike methylphenidate) cannot be injected intravenously (because it is unstable when heated), it can be addictive: modafinil has stronger effects on dopaminergic neurotransmission than previously thought (Volkow et al. 2009), and its oral administration causes substantial dopamine release in the range observed with other psychostimulants (Volkow et al. 2009). Furthermore, modafinil induces dopamine release in the ventral striatum, a core area of the reward system, which is highly relevant for potentially addictive effects of any pharmaceutical agent; for modafinil, ventral striatal dopamine release is much stronger than in any other area of the striatum, which generally predicts addictive properties of a drug. Indeed, the addiction risk of modafinil is supported by a report of the European Medicine Agency (2011), which listed 485 cases of modafinil abuse, tolerance development, and addiction.

Therefore, there is an evidence-based concern that more widespread and uncontrolled use of modafinil for cognitive enhancement could place many subjects at the risk of developing addictive disorders. Reasonably, recent pharmacological regulations in Germany explicitly warn against using modafinil in subjects at risk of developing addiction (European Medicine Agency 2011).

In the strategy of downplaying the addiction risks of cognitive enhancers, coffee plays a prominent role. Some definitions of cognitive enhancement are indeed so broad that they include drinking coffee (and even reading books), whereas others do not include drinking coffee. For our argumentation, it is not necessary to decide which definition is preferable. Proponents of cognitive enhancement (e.g., Greely et al. 2008) claim that coffee consumption has a similar effect to that of the pharmaceuticals being discussed for cognitive enhancement. However, this is only true with regard to some positive effects of coffee: after 85 hours of sleep deprivation, 600 mg caffeine has the same enhancing effect as 20 mg D-amphetamine or 400 mg modafinil (Lieb 2010). Nevertheless, the fact that three different substances have the same enhancing effect does not mean that they also have the same adverse effects. But this exact error in reasoning is made by some of the proponents of the use of modafinil for cognitive enhancement. For example, Shaw (2012) argues as follows: coffee is a harmless neuroenhancer. Since coffee is harmless, harmless neuroenhancers do exist. And then modafinil might be harmless as well. Shaw's argument is based on a faulty syllogism. Indeed, caffeine can enhance and maintain alertness and performance (Brice and Smith 2001; the study is supported by the Institute for Scientific Information on Coffee, whose members are seven of the major European coffee companies). Thus, caffeine and modafinil have a similar effect on the phenomenological level. But behind this phenomenological alikeness lies a neurobiological difference: Unlike modafinil, caffeine does not stimulate dopaminergic neurotransmission in the ventral striatum, including the nucleus accumbens. Since all known drugs of abuse activate dopamine in this core area of the brain reward system, caffeine's lack of effect in this brain region is consistent with its lack of addiction liability according to DSM-IV criteria (Acquas et al. 2010; De Luca et al. 2007). Therefore, there is a neurobiologically definable difference between the consumption of coffee, on the one hand, and of drugs with an addiction risk such as modafinil, on the other, which explains the difference in their addictive potential. It is faulty to conclude from the harmlessness of caffeine to the harmlessness of drugs with a quite different neurobiological mechanism, particularly modafinil (Heinz et al. 2014). Since dopamine-related learning and decision-making are directly implicated in cognitive capacities such as fluid IQ (Schlagenhauf et al. 2013), all neuroenhancers may affect this transmitter system and hence have the risk of addiction.

Shaw (2012) claims that further "neuroenhancers might be invented that do not involve dopamine." This is a typical philosophical armchair argument which ignores neuroscientific research; it is comparable to the claim "A perpetuum mobile might be invented in future that does not need any energy supply." It is easy to imagine that something might be invented, and to demand counterarguments from those who reject the fantasy; however, more concrete hypotheses are needed about the properties and mechanisms of the putatively riskless neuroenhancer that does not involve dopamine, to strengthen such speculative arguments.

We hypothesize that the addiction potential of cognitive enhancers is not a contingent fact of today's stimulant drugs but one of their fundamental properties: if they have the potential of enhancing cognitive capacities, they have to modulate systems which are relevant for learning and memory, the same systems that are implicated in drug addiction (Hyman 2005). The following facts support this hypothesis (Heinz et al. 2012):

1. Basic learning mechanisms such as Pavlovian and operant conditioning are strongly driven by reward and punishment, which are neuronally encoded by monoaminergic systems including dopamine (Hyman 2005; Montague et al. 2004; Schultz et al. 1997).
2. Fluid IQ, which is the cognitive capacity for flexible behavior adjustment and problem solving, is directly associated with dopamine function in the ventral striatum, a core area of the brain reward system (Schlagenhauf et al. 2013).
3. Drugs of abuse induce addiction precisely because of their strong interaction with the monoaminergic systems involved in learning, particularly with dopamine neurotransmission. Thereby, they induce learning mechanisms that promote further drug abuse, first through craving for the drug (elicited by Pavlovian conditioned drug cues), second through operant reinforcement of drug seeking (Di Chiara and Bassareo 2007; Kalivas and Volkow 2005; Park et al. 2010).

Addictive behavior develops when drugs exert strong effects on dopamine and related neurotransmitter systems and thus induce counter-regulative adaptive processes, particularly the downregulation of dopamine receptors in the ventral striatium, a core part of the brain reward system. Thus, natural reinforcers lose their impact and produce craving for the nonphysiologically strong effects of drugs of abuse (Di Chiara and Bassareo 2007; Everitt and Robbins 2005; Heinz et al. 2009; Hyman 2005; Kalivas and Volkow 2005). Indeed, persistence of such alterations in dopamine synthesis and D_2 receptors during abstinence is directly correlated with drug craving and the risk of relapse (Beck et al. 2012; Heinz et al. 1996, 2005).

In principle, cognitive-enhancing drugs could affect dopamine neurotransmission only moderately and comparably to the effects of natural reinforcers. But if this is true they will probably not have an advantage over traditional practice and study. If they exceed physiological effects and induce counteradaptive neuroadaptation, then there is a substantial risk of inducing addictive behavior, as suggested by neurobiological research (Di Chiara and Bassareo 2007; Everitt and Robbins 2005; Heinz et al. 2009; Hyman 2005; Kalivas and Volkow 2005) and clinical evidence (Heinz et al. 1996, 2005; Pélissier-Alicot et al. 2006). Developers of riskless cognitive enhancers face a dilemma: to do better than natural ways to reinforce learning and increase cognitive speed, one will have to interfere with systems that increase the risk of addiction.

5.3 Underestimation of further medical risks of neuroenhancers

Beside the risk of addiction, the pharmaceuticals currently proposed as cognitive enhancers, particularly methylphenidate and modafinil, have a number of risks, adverse effects, and contraindications:

5.3.1 Methylphenidate

Methylphenidate can have severe psychiatric or neurological adverse effects. According to the drug manufacturer Novartis Pharma Schweiz AG (2013), in methylphenidate consumers, sleeplessness and nervousness are very frequent (> 1/10); abnormal behavior, aggression, agitation, anxiety, depression, irritability, headaches, sleepiness, vertigo, and dyskinesia are frequent (< 1/10, > 1/100); hyperactivity, visual and acoustic hallucinations, psychosis, tics, convulsions, cerebrovascular events, suicidal ideation, and (attempted) suicides are very rare (< 1/10,000). Frequent cardiovascular adverse effects are tachycardia, palpitations, arrhythmia, and changes in blood pressure or frequency; rare (< 1/1,000, > 1/10,000) are angina pectoris; very rare is cardiac arrest (Novartis Pharma Schweiz AG 2013). According to a recent FDA communication, methylphenidate is associated with peripheral vasculopathy, including Raynaud's phenomenon (FDA 2013). Further frequent side effects include stomach ache, nausea, vomiting, and reduced appetite. In long-term use in children, a rare side effect is growth disturbances (of both length and weight).

Because of the risk of sudden cardiac arrests, a careful anamnesis and examination is necessary in order to detect possible structural heart anomalies or cardiovascular diseases (Novartis Pharma Schweiz AG 2013).

Methylphenidate has a high abuse potential in oral, intranasal, and intravenous ways of application; it is often used together with alcohol or cannabis (Kraemer et al. 2010). The combination with alcohol is particularly problematic: according to Novartis Pharma Schweiz AG (2013), alcohol can increase the adverse effects of methylphenidate on the central nervous system.

A large-scale survey (545 patients enrolled in an ADHD treatment center, 89.2% with ADHD) has revealed a widespread abuse of methylphenidate and other prescription stimulants: 14.3% of respondents abused prescription stimulants. The medications abused most often are the prescription amphetamine Adderall (54.2%) and methylphenidate (15.0%). The most common manner of stimulant abuse was crushing pills and inhaling/snorting (75.0%), followed by crushing and injecting (6.3%), microwaving/melting to snort (6.3%), and other methods (12.5%). Choice of illicit drug was based on rapidity of high onset, ease of acquisition, ease of use, and cost. The reinforcing effects of stimulants and the "high" experience are particularly strong when drugs are administered intravenously (Bright 2008).

Because of the potential abuse of ADHD medications as recreational drugs, investigations of their addictive potential must not be restricted to its use as a prescribed medication; rather its addictive potential should be investigated in combination with other drugs, particularly alcohol and cannabis.

Besides the addiction risk, we consider the risk of stimulant-induced psychosis as relevant. Methylphenidate can induce psychoses, both in adult ADHS patients, who are treated with the drug, and in persons who abuse it. Unfortunately, to date the problem of methylphenidate-induced psychosis in ADHD patients is mostly neglected in science, and almost no data are available (Kraemer et al. 2010). Reliable predictors for the development of methylphenidate-induced psychosis are to date not known (Kraemer et al. 2010). We expect that the intravenous administration of methylphenidate bears a particular risk for the development of psychoses, since acute increases in dopamine can cause psychosis (Carlsson and Carlsson 2006; Heinz 2002; Howes and Kapur 2009).

5.3.2 Modafinil

According to the safety information of the drug Provigil (brand name of modafinil in the UK, USA, etc.), common side effects are headache, nausea, feeling nervous, stuffy nose, diarrhea, back pain, feeling anxious, trouble sleeping, dizziness, and upset stomach. Possible side effects are psychiatric symptoms (including mania, delusions, hallucinations, suicidal ideation, and aggression), some resulting in hospitalization (also of patients without psychiatric history). Further possible side effects are symptoms of a heart problem (including chest pain, abnormal heart beats, and trouble breathing). The drug may cause a serious rash or a serious allergic reaction that may affect the liver or blood cells; these side effects may need to be treated in a hospital and may be life-threatening (http://www.provigil.com; http://www.provigil.com/media/PDFs/prescribing_info.pdf).

Even a single dose of modafinil may increase anxiety and aggressive mood. According to a double-blind randomized study with healthy young persons, the "somatic anxiety" was significantly increased in the group who had taken 100 mg modafinil as compared to the placebo group and the 200 mg modafinil group. After stress, the 100 mg modafinil group also showed increased ratings of "psychological anxiety" and "aggressive mood" (Randall et al. 2003).

Case reports describe severe side effects: pathological gambling (with a clear-cut temporal relationship with modafinil treatment) (Tarrant et al. 2010), modafinil-induced psychosis (Mariani and Hart 2005), and even death after several weeks of consumption of modafinil together with several amphetamines that were illicitly prescribed by a practitioner (Pélissier-Alicot et al. 2006).

5.4 Overestimation of the benefits

For an evaluation of cognitive enhancers, we have to consider, on the one hand, their risks and side effects, and, on the other hand, their benefits. In the following, we will summarize the scientific findings of the benefits of methylphenidate and modafinil; we will not speculate about possible benefits of drugs that might be invented in future.

5.4.1 Methylphenidate

A meta-analysis of studies with non-sleep-deprived individuals, who received a single dose of methylphenidate, revealed a significant positive effect only in memory, whereas the effects on attention, mood, and executive functions were not significant (Repantis et al. 2010). The main results of five studies with sleep-deprived individuals studies are that a single dose of methylphenidate after a night of sleep deprivation did not have cognitive-enhancing effects, but a negative effect on self-monitoring. In studies with either 36 h of sleep deprivation or with partial sleep deprivation (4 h of sleep), repeated administration of methylphenidate had subjective stimulating effects and a moderate improvement of attention. During a period of 64 h of sleep deprivation, repeated administration did not effectively reduce sleepiness (Repantis et al. 2010).

Repantis and his coauthors (who are proponents of neuroenhancers) conclude that the studies provide no consistent evidence for neuroenhancing effects of methylphenidate, although they revealed a positive effect on memory in healthy individuals. Particularly, the popular opinion that methylphenidate enhances attention was not verified (Repantis et al. 2010).

A recent review (Linssen et al. 2014) that has compared 59 studies came to different results. According to this review, methylphenidate improves cognitive performance of healthy persons in the domains of working memory and speed of processing, and to a lesser extent in the domains verbal learning and memory, attention and vigilance, and reasoning and problem solving, but not visual learning and memory. Important to note is, first, that methylphenidate does not improve cognitive performance globally, but may enhance the performance in certain domains. Second, the effects are dose dependent and the dose–response relationship differs between cognitive domains. For example, in some cognitive domains low doses are more effective than high doses. In some domains, higher doses even reduce performance. The reason for that may be that the relationship between the methylphenidate dose and cognitive performance follows an inverted-U-shaped curve. If methylphenidate increases the dopamine concentration above the optimal level, performance may decrease. Linssen and coauthors summarize that methylphenidate effects are relatively small, and they consider that they are associated with side effects and potential abuse (Linssen et al. 2014).

5.4.2 Modafinil

A double-blind study that randomly allocated 60 healthy, non-sleep-deprived students to placebo or 100 or 200 mg modafinil revealed that effects of modafinil were limited to a few special domains and were not clearly dose related. Modafinil was without effect on reaction time and attention, but the 200 mg group was faster at simple color naming of dots and in the rapid visual information processing (RVIP) test of sustained attention. With regard to memory, modafinil was without effect on verbal short-term memory, but 100 mg improved digit span forward, and both doses improved pattern recognition. However, there were no significant effects on long-term memory, executive function, or visuospatial and constructional ability of category fluency. The authors of the study argue that these effects are insufficient for considering modafinil as a cognitive enhancer in non-sleep-deprived individuals (Randall et al. 2005a).

Another study with a similar design investigated non-sleep-deprived middle-aged persons (Randall et al. 2004). In this age group, modafinil did not have significant effects on most of the cognitive tests. As in the students, only a few special benefits of modafinil were observed: The 200 mg group was significantly faster in a simple color naming of dots and significantly better in a test of constructional ability (clock drawing test) than the placebo group. But the 200 mg group also made significantly more total errors in the intra/extradimensional set shift task than both the other groups. Thus, this study found limited evidence of cognitive-enhancing properties of modafinil in healthy middle-aged volunteers.

In a further study with 89 subjects, several interactions between modafinil and IQ emerged. Modafinil significantly improved the performance in three tests (target sensitivity in the RVIP test, speed of responding in a color naming of dots, and clock drawing) only in the group of "lower" IQ (106 ± 0.6), but not in the "higher" IQ group (115.5 ± 0.5) (Randall et al. 2005b).

In summary, these studies suggest that modafinil can increase the performance in some special cognitive tests mainly in individuals whose performance is suboptimal.

A study from another research group (Müller et al. 2004) found that modafinil had subtle stimulating effects when performing relatively difficult and monotonous working tasks, whereas attentional control tasks were not affected; in addition, modafinil has effects especially in lower performing subjects.

A meta-analysis of Repantis et al. (2010) has come to the following results. After a single dose of modafinil in non-sleep-deprived individuals, a moderate, positive effect on attention was found, whereas the outcomes mood, memory, and motivation remained without any significant systematic changes. In sleep-deprived individuals, a single dose had a very strong and persistent impact on execution function. Furthermore, memory and wakefulness were significantly improved, whereas there were no significant effects in the outcomes attention and mood. Repeated modafinil administration in sleep-deprived individuals increased only wakefulness significantly, but not executive functions and attention (Repantis et al. 2010). The authors summarize that a single dose of modafinil seems to have a cognitive-enhancing effect in cases of moderate sleep deprivation. However, repeated doses during a longer period of sleep deprivation do not boost cognitive performance, but only maintain wakefulness. They doubt (particularly with regard to a possible effect of overconfidence on one's cognitive performance) whether repeated doses of modafinil make sense for preserving performance over a longer time of sleep deprivation.

5.5 Conclusion

For the currently most discussed "cognitive enhancers," namely modafinil and methylphenidate, it is difficult to detect any relevant cognitive-enhancing effects. Although the performance in some cognitive domains can be improved by the intake of the drugs, the improvement seems to come at the price of decreasing performance in other domains. Therefore, for these drugs, it is not justified to speak of cognitive enhancement, but rather of a temporary shifting of performance between different cognitive domains. For certain tasks, such shifts may offer limited advantages. However, the practical problem will remain to find the perfect administration scheme for, e.g., improving the working memory. Furthermore, today's cognitive enhancers are more effective in increasing the performance in individuals whose performance is suboptimal, particularly because of sleep deprivation. But whether this justifies speaking of enhancement is questionable.

The possible small, specific, and temporary benefits must be weighed against the known psychiatric and cardiovascular risks, as well as the risk of addiction. As we argued earlier, addiction is fundamentally more linked to pharmacological cognitive enhancement than its proponents admit. We believe that, in general, the risk–benefit-ratio is too unfavorable to recommend the use of the currently most discussed cognitive enhancers, namely, modafinil and methylphenidate, for the purpose of cognitive enhancement.

Finally, we will discuss two practical questions:

1. Can the use of cognitive enhancers, particularly methylphenidate and modafinil, be recommended for specific professional groups or for specific situations?
2. Should research on cognitive enhancement receive public funding?

5.5.1 Cognitive enhancement for specific professional groups or for specific situations

The US Air Force has approved the use of modafinil for long-lasting missions (US Air Force 2003, quoted in Repantis et al. 2010). Another group for whom cognitive enhancement has been advocated is surgeons (e.g., by Greely et al. 2008). If drugs could enable surgeons to operate longer periods without pauses for sleeping, this might be beneficial for patients whose operations are extremely time consuming, e.g., the separation of Siamese twins. Although such types of surgery are extremely rare, and although several surgeons are always engaged in these operations, it may be that a specialist cannot be replaced by another surgeon with the same degree of skill and experience. We feel that in such cases, the specialist should weigh up the (small) risks of one dose of modafinil for himself against the possible benefits for the patient, but also the risk of overestimating his/her abilities when operating under a psychostimulant with the potential benefit of wakefulness. Also if after a disaster, e.g., after multiple collisions or after a natural catastrophe, not enough first-aiders and physicians may be on-site, we feel that medication which can maintain their wakefulness and concentration can make sense. However, the acceptability of use in these unusual circumstances does not provide support for the habitual use of stimulants by surgeons. The organization of work must respect both the patient's needs and the surgeon's need for recovery. The German Society of Surgeons has warned surgeons against using modafinil or methylphenidate. They argue that these substances might have a negative impact on the power of judgment, the resolution, and the necessary distance from the surgery. They explicitly criticize the "memorandum" which demands the liberalization of use of prescription psychostimulants (Deutsche Gesellschaft für Chirurgie 2009; Galert et al. 2009).

Here an underlying concern regarding the use of neuroenhancement becomes visible: will neuroenhancers be used to "optimize" work performance of specialists at the price of their health? How will such professionals be able to resist the pressure to optimize work performance, particularly when others start taking drugs to do so?

5.5.2 Public funding for research on cognitive enhancement

Several proponents of cognitive enhancement advocate for more research on the effectiveness and safety of drugs for neuroenhancement (Galert et al. 2009; Greely et al. 2008; Shaw 2012). This request is supported by the fact that knowledge about the positive and negative effects of drugs such as modafinil and methylphenidate is insufficient, at least with regard to its nontherapeutic use by persons who consume them for the purpose of enhancement. This holds particularly for the addiction potential of today's and future neuroenhancers, which indeed is discussed controversially.

Two different research strategies could provide answers: first, double-blind, randomized long-term studies with healthy volunteers, and, second, epidemiological studies.

5.5.2.1 Clinical studies

Clinical studies, which investigate efficacy and adverse effects, particularly de novo addictions, can be performed with young volunteers, who would be randomly assigned to different groups (e.g., different drugs, different doses of a drug, placebo). The research protocol will demand that the subjects do not take other drugs during the study. For both scientific and ethical reasons, persons with mental disorders, known risk factors for mental disorders, a psychiatric history, or problematic drug consumption will most certainly be excluded. Since clinical studies need the approval of an ethics commission, the studies must fulfill at least the following criteria: a reasonable goal must be investigated, study participants must give informed consent, and an acceptable risk–benefit profile of the investigation as well as a high scientific quality must be guaranteed (Emanuel et al. 2000). To fulfill the criterion of an acceptable risk–benefit profile is tricky: to investigate the addictive effects of cognitive enhancers, larger groups of previously healthy individuals with varying vulnerabilities for addiction would have to be exposed to a potentially addictive drug. The risk of developing addiction may be substantially lower when subjects with an increased risk of developing addiction are excluded. However, to date no single variable has been identified that reliably predicts addiction to neuroenhancers. To identify such predictive factors, a substantial number of subjects would have to be exposed to potentially addictive drugs in prospective studies. Since these healthy volunteers cannot expect any relevant benefits from study participation (except payment), the risk–benefit ratio for such exposures is rather unfavorable. Even if possible benefits for society are considered, the risk–benefit ratio does not become much better because of the social costs of potential side effects. In contrast to clinical studies that investigate the effectiveness of medication in treating serious illnesses, the investigation of neuroenhancers at best addresses a "luxury problem," namely, the improvement of abilities that are already above average. For these reasons, it is questionable whether an ethics commission would allow for a study that could provide an answer to the question whether the use of cognitive enhancers increases the risk for addiction.

But even if such a study would be permitted, another problem could occur: if the study did not find a significantly increased addiction, then we would be uncertain about whether results in healthy volunteers under optimal conditions and with medical surveillance can be transferred to the use in the general population that includes people with different diseases, risk factors, and mental disorders, and many of whom consume different kinds of medications and recreational drugs. The risk of addiction can be much higher than under study conditions. So if a controlled study suggests that there is no significant risk for drug addiction, this could open the door for an uncontrolled social experiment.

In sum, the ethical objections to clinical research on cognitive enhancers do not justify an absolute ban, but argue against public funding of this research, first because of the poor risk–benefit ratio for the research subjects, and, second, because public resources should not be used for solving luxury problems.

5.5.2.2 Epidemiological studies

Epidemiological studies could collect information about the benefits and risks of drugs used for cognitive enhancement. Their advantage is that they could provide a realistic picture of current patterns of use, since they can include people who are more representative of the whole society. The analysis of demographic data together with data on the consumption of prescription psychotropic drugs, recreational drugs, medications, etc. can reveal risk factors for adverse effects of cognitive enhancers, including addiction. Their main disadvantage is the heterogeneity of the subjects, which has the consequence that large numbers of subjects are required. Furthermore, epidemiological studies can rarely produce conclusive evidence of causal relationships.

In sum, epidemiological studies do not pose additional risks to the subjects, and insofar, they are ethically unproblematic (provided that confidentiality and data security issues, etc. are addressed). Therefore, we support epidemiological studies on the use of prescription psychotropic drugs for the purpose of cognitive enhancement.

References

Acquas, E., G. Tanda, and G. Di Chiara. 2002. Differential effects of caffeine on dopamine and acetylcholine transmission in brain areas of drug-naive and caffeine-pretreated rats. *Neuropsychopharmacology* **27**:182–93.

Acquas, E., S. Vinci, F. Ibba, S. Spiga, M. A. De Lucan, and G. Di Chiara. 2010. Role of dopamine D1 receptors in caffeine-mediated ERK phosphorylation in the rat brain. *Synapse* **64**(5):341–9.

Beck, A., T. Wüstenberg, A. Genauck, J. Wrase, F. Schlagenhauf, M. N. Smolka, K. Mann, and A. Heinz. 2012. Effect of brain structure, brain function, and brain connectivity on relapse in alcohol-dependent patients. *Archives of General Psychiatry* **69**(8):842–52.

Brice, C., and A. Smith. 2001. The effects of caffeine on simulated driving, subjective alertness and sustained attention. *Human Psychopharmacology* **16**:523–31.

Bright, G. M. 2008. Abuse of medications employed for the treatment of ADHD: results from a large-scale community survey. *Medscape Journal of Medicine* **10**(5):111.

Carlsson, A., and M. L. Carlsson. 2006. A dopaminergic deficit hypothesis of schizophrenia: the path to discovery. *Dialogues in Clinical Neuroscience* **8**(1):137–42.

DAK. 2009. Gesundheitsreport 2009. Analyse der Arbeitsunfähigkeitsdaten. Schwerpunktthema Doping am Arbeitsplatz. Available at http://www.dak.de/content/filesopen/Gesundheitsreport_2009.pdf (accessed 2 March 2011).

De Luca, M. A., V. Bassareo, A. Bauer, and G. Di Chiara. 2007. Caffeine and accumbens shell dopamine. *Journal of Neurochemistry* **103**:157–63.

Deutsche Gesellschaft für Chirurgie. 2009. Pressekonferenz der Deutschen Gesellschaft für Chirurgie, 9.12.2009. Available at https://www.thieme.de/specials/presseservice/dgch/pdf/pressemappe_pk_09_final.pdf. Assessed: 21 March 2014.

Di Chiara, G., and V. Bassareo. 2007. Reward system and addiction: what dopamine does and doesn't do. *Current Opinion in Pharmacology* **7**(1):69–76.

Di Chiara, G., and A. Imperato. 1988. Drugs abused by humans preferentially increase synaptic dopamine concentrations in the mesolimbic system of freely moving rats. *Proceedings of the National Academy of Sciences of the United States of America* **85**:5274–8.

Dietz, P., H. Striegel, A. G. Franke, K. Lieb, P. Simon, and R. Ulrich. 2013. Randomized response estimates for the 12-month prevalence of cognitive enhancing drug use in university students. *Pharmacotherapy* **33**(1):44–50.

Emanuel, E. J., D. Wendler, and C. Grady. 2000. What makes clinical research ethical? *Journal of the American Medical Association* **283**: 2701–11.

European Medicine Agency. 2011. Assessment report for modafinil containing medical products. EMA/4038/2011. London.

Everitt, B. J., D. Belin, D. Economidou, Y. Pelloux, J. W. Dalley, and T. W. Robbins. 2008. Neural mechanisms underlying the vulnerability to develop compulsive drug-seeking habits and addiction. *Philosophical Transactions of the Royal Society of London, Series B, Biological Sciences* **363**(1507):3125–35.

Everitt, B. J., and T. W. Robbins. 2005. Neural systems of reinforcement for drug addiction: from actions to habits to compulsion. *Nature Neuroscience* **8**:1481–9.

Foddy, B., and J. Savulescu. 2007. Addiction is not an affliction: Addictive desires are merely pleasure-oriented desires. *American Journal of Bioethics* 7(1):29–32.

Galert, T., C. Bublitz, I. Heuser, R. Merkel, B. Schöne-Seifert, and D. Talbot. (2009). Das optimierte Gehirn. *Gehirn & Geist* **11**:40–8.

Greely, H., B. Sahakian, J. Harris, R. C. Kessler, M. Gazzaniga, P. Campbell, and M. J. Farah. 2008. Towards responsible use of cognitive enhancing drugs by the healthy. *Nature* **456**(7223):702–5.

Heinz, A. 2002. Dopaminergic dysfunction in alcoholism and schizophrenia psychopathological and behavioral correlates. *European Psychiatry* **17**(1):9–16.

Heinz, A., A. Beck, S. M. Grüsser, A. A. Grace, and J. Wrase. 2009. Identifying the neural circuitry of alcohol craving and relapse vulnerability. *Addiction Biology* **14**(1):108–18.

Heinz, A., A. Beck, A. Meyer-Lindenberg, et al. 2011. Cognitive and neurobiological mechanisms of alcohol-related aggression. *Nature Review Neuroscience* **12**(7):400–13.

Heinz, A., P. Dufeu, S. Kuhn, M. Dettling, K. Gräf, I. Kürten, H. Rommelspacher, and L. G. Schmidt. 1996. Psychopathological and behavioral correlates of dopaminergic sensitivity in alcohol-dependent patients. *Archives of General Psychiatry* **53**(12):1123–28.

Heinz, A., R. Kipke, H. Heimann, and U. Wiesing. 2012. Cognitive neuroenhancement: false assumptions in the ethical debate. *Journal of Medical Ethics* **38**:372–5.

Heinz, A., R. Kipke, S. Müller, and U. Wiesing. 2014. True and false concerns about neuroenhancement. A response to "Neuroenhancers, addiction and research ethics" by D. M. Shaw. *Journal of Medical Ethics* **40**(4):286–7.

Heinz, A., T. Siessmeier, J. Wrase, H. G. Buchholz, G. Gründer, L. G. Kumakura, P. Cumming, M. Schreckenberger, M. N. Smolka, F. Rösch, K. Mann, and P. Bartenstein. 2005. Correlation of alcohol craving with striatal dopamine synthesis capacity and D2/3 receptor availability: a combined [18F]DOPA and [18F]DMFP PET study in detoxified alcoholic patients. *American Journal of Psychiatry* **162**:1515–20.

Howes, O. D., and S. Kapur. 2009. The dopamine hypothesis of schizophrenia: version III—the final common pathway. *Schizophrenia Bulletin* **35**(3):549–62.

Hyman, S. E. 2005. Addiction: a disease of learning and memory. *American Journal of Psychiatry* **162**:1414–22.

Hyman, S. E. 2007. The neurobiology of addiction: implications for voluntary control of behavior. *American Journal of Bioethics* 7(1):8–11.

Kalivas, P. W., and N. D. Volkow. 2005. The neural basis of addiction: a pathology of motivation and choice. *American Journal of Psychiatry* **162**:1403–13.

Kraemer, M., J. Uekermann, J. Wiltfang, and B. Kis. 2010. Methylphenidate-induced psychosis in adult attention-deficit/hyperactivity disorder: report of 3 new cases and review of the literature. *Clinical Neuropharmacology* **33**(4):204–6.

Lieb, K. 2010. *Hirndoping: Warum wir nicht alles schlucken sollten*. Mannheim: Artemis and Winkler.

Linssen, A. M. W., A. Sambeth, E. F. P. M. Vuurman, and W. J. Riedel. 2014. Cognitive effects of methylphenidate in healthy volunteers: a review of single dose studies. *International Journal of Neuropsychopharmacology* **17**(6):961–77.

Lucke, J., S. Bell, B. Partridge, and W. Hall. 2011. Bursting the neuroenhancement bubble in bioethics. *American Journal of Bioethics Neuroethics* **2**:38–43.

Mariani, J. J., and C. L. Hart. 2005. Psychosis associated with modafinil and shift work. *American Journal of Psychiatry* **162**(10):1983.

Martel, P., and M. Fantino. 1996. Influence of the amount of food ingested on mesolimbic dopaminergic system activity: a microdialysis study. *Pharmacology, Biochemistry, and Behavior* **55**:297–302.

Montague, P. R., S. E. Hyman, and J. D. Cohen. 2004. Computational roles for dopamine in behavioural control. *Nature* **431**:760–7.

Müller, U., N. Steffenhagen, R. Regenthal, and P. Bublak. 2004. Effects of modafinil on working memory processes in humans. *Psychopharmacology (Berlin)* **177**(1–2):161–9.

Novartis Pharma Schweiz AG. 2013. Ritalin®/-SR/-LA. Fachinformation des Arzneimittel-Kompendium der Schweiz, 28.11.2013. Available at http://www.adhs-schweiz.ch/Bilder/FrmMainMonographie.pdf. Assessed 20 March 2014.

Park, S. Q., T. Kahnt, A. Beck, M. X. Cohen, R. J. Dolan, J. Wrase, and A. Heinz. 2010. Prefrontal cortex fails to learn from reward prediction errors in alcohol dependence. *Journal of Neuroscience* **30**:7749–53.

Pélissier-Alicot, A. L., M. D. Piercecchi-Marti, C. Bartoli, E. Kuhlmann, P.-E. Coiffait, A. Sanvoisin, D. Giocanti, and G. Léonetti. 2006. Abusive prescription of psychostimulants: a study of two cases. *Journal of Forensic Sciences* **51**(2):407–10.

Randall, D. C., N. L. Fleck, J. M. Shneerson, and S. E. File. 2004. The cognitive-enhancing properties of modafinil are limited in non-sleep-deprived middle-aged volunteers. *Pharmacology, Biochemistry and Behavior* **77**:547–55.

Randall, D. C., J. M. Shneerson, and S. E. File. 2005b. Cognitive effects of modafinil in student volunteers may depend on IQ. *Pharmacology, Biochemistry and Behavior* **82**:133–9.

Randall, D. C., J. M. Shneerson, K. K. Plaha, and S. E. File. 2003. Modafinil affects mood, but not cognitive function, in healthy young volunteers. *Human Psychopharmacology, Clinical and Experimental* **18**:163–73.

Randall, D. C., A. Viswanath, P. Bharania, S. M. Elsabagh, D. E. Hartley, J. M. Shneerson, and S. E. File. 2005a. Does modafinil enhance cognitive performance in young volunteers who are not sleep-deprived? *Journal of Clinical Psychopharmacology* **25**(2):175–9.

Repantis, D., P. Schlattmann, O. Laisney, and I. Heuser. 2010. Modafinil and methylphenidate for neuroenhancement in healthy individuals: a systematic review. *Pharmacological Research* **62**:187–206.

Reuter, J., T. Raedler, M. Rose, I. Hand, J. Gläscher, and C. Büchel. 2005. Pathological gambling is linked to reduced activation of the mesolimbic reward system. *Nature Neuroscience* **8**:147–8.

Schlagenhauf, F., M. A. Rapp, Q. J. M. Huys, A. Beck, T. Wüstenberg, L. Deserno, H.-G. Buchholz, J. Kalbitzer, R. Buchert, M. Bauer, T. Kienast, P. Cumming, M. Plotkin, Y. Kumakura, A. A. Grace, R. J. Dolan, and A. Heinz. 2013. Ventral striatal prediction error signaling is associated with dopamine synthesis capacity and fluid intelligence. *Human Brain Mapping* **34**(6):1490–9.

Schultz, W., P. Dayan, and P. R. Montague. 1997. A neural substrate of prediction and reward. *Science* **275**:1593–9.

Shaw, D. M. 2012. Neuroenhancers, addiction and research ethics. *Journal of Medical Ethics* **38**:605–8.

Tarrant, N., A. E. Cavanna, and H. Rickards. 2010. Pathological gambling associated with modafinil. *Journal of Neuropsychiatry and Clinical Neurosciences* **22**(1):**123**:E27–E28.

US Air Force. 2003. Memorandum: Modafinil and management of aircrew fatigue, 2nd December 2003 (quoted in Repantis et al. 2010).

US Food and Drug Administration (FDA). 2013. Ritalin (methylphenidate hydrochloride) tablets, Ritalin SR (extended-release tablets), and Ritalin LA (extended-release capsules). Available at http://www.fda.gov/Safety/MedWatch/SafetyInformation/ucm359945.htm. Last updated: 7 November 2013. Accessed: 20 March 2014.

Volkow, N. D., J. S. Fowler, J. Logan, D. Alexoff, W. Zhu, F. Telang, et al. 2009. Effects of modafinil on dopamine and dopamine transporters in the male human brain: clinical implications. *Journal of the American Medical Association* **301**(11):1148–54.

Volkow, N. D., J. S. Fowler, G. J. Wang, and J. M. Swanson. 2004. Dopamine in drug abuse and addiction: results from imaging studies and treatment implications. *Molecular Psychiatry* 9(6):557–69.

Internet

http://www.provigil.com (accessed 18 March 2014).

http://www.provigil.com/media/PDFs/prescribing_info.pdf (accessed 18 March 2014).

Chapter 6

The effects of Modafinil on creativity: results from a randomized controlled trial

Ahmed Dahir Mohamed[1]

6.1 What is creativity?

Creativity is defined as an activity that involves thinking that is aimed at producing ideas or products that are relatively novel and that are, in some respect, compelling (Kaufman and Sternburg 2006). It involves a novel product, idea, or solution to a problem that is of value to the individual and/or to the larger social group (Hennessey and Amabile 2010).

Creativity researchers typically view creativity as involving two forms of thinking. One form is called convergent thinking, which involves the ability to problem solve in a highly domain-specific manner. This form of creative performance expects participants to have a set answer, or to arrive at a particular goal, and assesses associations formed between two disparate concepts. Hennessey and Amabile (2010) argue that convergent thinking is focused on narrowing possibilities to a workable solution. The RAT, which gives participants three words and expects them to provide one correct answer, can therefore be considered to test a form of convergent thought (Mednick 1962).

On the other hand, divergent thinking assesses the ability to generate multiple abstract ideas or solutions to a problem (Guilford 1957). It has consistently remained as a conceptually, internally, and externally valid element of the creative process (Gibson et al. 2009). Divergent thinking has been argued to consist of spontaneous, free-flowing thinking with the goal of generating many different ideas (Hennessey and Amabile 2010, 574).

6.2 The effects of psychostimulants on creativity

The effects of psychostimulants on creativity have been investigated in patients. For example, methylphenidate (Ritalin) has been shown to improve divergent thinking of creativity in patients diagnosed with attention deficit hyperactive disorder (ADHD). Solanto and Wender (1989) used the Wallach Kogan Tasks (Wallach and Kogan 1965) and tested 19 ADHD children aged between 6 and 10 years, firstly at baseline, followed by placebo, and then followed by three doses (3, 6, and 10 mg) of methylphenidate. The study reported enhanced divergent thinking under methylphenidate in ADHD participants. In contrast, participants showed a significant decline in divergent thinking across non-drug conditions. The study reported that healthy control participants on placebo were similarly impaired on the divergent thinking tasks. Although the authors did not investigate whether drug and/or disorder-related performance was a function of baseline of creativity personality trait, they reported that, after methylphenidate administration, there was a large increase in divergent thinking performance in some participants. Solanto and Wender (1989) argued that these increased performances were representative of atypical subgroups of ADHD children who become enhanced after methylphenidate.

However, a previous study using the same methodology with ADHD children reported a dose-dependent improvement by methylphenidate on divergent thinking and mental flexibility tasks (Douglas et al. 1995). Hence, the effects of methylphenidate, a conventional psychostimulant known to increase dopamine (DA) in the brain (Wang et al. 1999), on divergent thinking may depend on individual differences in baseline of cognitive function and also on optimal dosing. Indeed, a recent study investigating differences in creativity between patients diagnosed with ADHD and healthy participants found that the ADHD patients were significantly better on the divergent thinking task, while healthy controls were significantly better on the convergent thinking task of creativity (White and Shah 2006). However, in that study, the ADHD participants were non-medicated.

There is a dearth of research investigating the effects of psychostimulant drugs on creativity in healthy individuals. To our knowledge, only two studies have been conducted thus far. The results of a recent double-blind, placebo-controlled study showed that Adderall, an amphetamine-based psychostimulant licensed for the treatment of ADHD, enhanced performance on convergent thinking tasks of creativity in healthy participants with low baseline of creative performance, and impaired performance in those with higher baseline of creative performance (Farah et al. 2009). In that study, Adderall had no effect on divergent thinking tasks of creativity. These results point to the view that psychostimulants might modulate creativity performance according to one's baseline, either improving or impairing one's divergent and convergent thinking of creativity.

A recent review by Rosenthal and Westreich (2010) indicated the lack of well-controlled studies on the effects of psychostimulant drugs on creativity. This raises the important question of whether popular drugs like modafinil, a putative cognitive enhancer reportedly used by healthy individuals for cognitive enhancement (Franke et al. 2013; Mohamed and Sahakian 2012), boosts brain systems associated with creativity performance in the same way it has been reported to affect executive functions and working memory in healthy individuals (Turner et al. 2003).

Modafinil is a drug licensed for the treatment of narcolepsy (Ballon and Feifel 2006). It has been reported to be used by healthy people for cognitive enhancement purposes (Franke et al. 2013; Mohamed 2014; Mohamed and Sahakian 2012). In 2004, it was estimated that 90% of modafinil was being used off-label by healthy, non-sleep-deprived individuals (Baranski et al. 2004). The global revenue of modafinil is more than US$700 million per year (Norman and Berger 2008). These reported off-label and revenue figures are contested but they are nonetheless consistent with the claim that healthy physicians on call, students, and academics are increasingly using modafinil to enhance cognitive abilities (Franke et al. 2013; Greely 2006).

However, it is currently unclear as to whether modafinil affects creativity performance in healthy individuals the same way as Adderall. In one study, modafinil improved alertness, vigilance, and verbal tasks of creative thinking, but only after sleep deprivation (Walsh et al. 2004). No changes were reported in the non-verbal creative thinking tasks. Hence, no previous study has directly investigated the effects of modafinil on divergent and convergent thinking tasks of creativity in healthy, non-sleep-deprived volunteers. The aim of this study was to shed light on this empirical question.

6.3 Theoretical rationale

The theoretical rational for investigating modafinil comes from the evidence that creative outputs are modulated by mesolimbic dopaminergic increase (Flaherty 2005). It has recently been argued that there is a link between increased DA in the above areas and creativity (Batir 2009), and there is evidence suggesting a unique role for a specific DA receptor gene (DRD2 Taq1A) for creativity (Reuter et al. 2006). These researchers showed that specific alleles of the gene were associated with higher verbal creativity (A1+ as compared with A1− allele of *DRD2*), while carriers of the A allele of *TPH1* showed significantly higher scores in figural and numeric creativity.

The nucleus accumbens (NAcc) DA release has been shown to increase novelty seeking and creative drive in humans (Flaherty 2005). For example, there is evidence that creative persons have a higher DA compared with Parkinson's disease patients (Batir 2009), a disorder associated with a significant loss of dopaminergic neurons (Beal 2001). These patients are significantly impaired in the performance of creative tasks when subjected to subthalamic deep brain stimulation, but performance is restored to normal levels when treated with DA replacement therapy (Batir 2009).

Currently, the mechanisms by which modafinil exerts its wakefulness effects and how it acts as a cognitive-enhancing agent in healthy individuals are unclear. Several studies indicate that modafinil increases DA neurotransmitters (Mignot et al. 1994; Minzenberg and Carter 2008; Volkow et al. 2009) in several regions of the brain, including the prefrontal cortex (PFC) and the striatum. Molecular studies show that modafinil binds to and requires intact DA transporters (DAT; Qu et al. 2008). Furthermore, modafinil has been shown to increase DA in rats, mice, monkeys, and healthy human brains (Minzenberg and Carter 2008). The results of studies of healthy humans have shown that modafinil increases DA and activates the mesolimbic areas of the brain (Volkow et al. 2009), which are involved in novelty seeking and creative thinking (Flaherty 2005; Reuter et al. 2006).

For instance, a recent double-blind placebo-controlled crossover design study with healthy males investigated the effects of modafinil on DA neurotransmitter and DA transporter (DAT; Volkow et al. 2009). The study employed a positron emission tomography technique with [^{11}C]raclopride, a D2/D3 radioligand which is sensitive to changes in endogenous DA, and [^{11}C]cocaine, a DAT radioligand which is sensitive to changes on extracellular DA and on DAT. After the administration of 200 and 400 mg of modafinil, the drug both blocked DAT and increased extracellular DA in the mesolimbic areas of the brain, more specifically in the caudate, putamen, and NAcc in healthy men. This was possible because modafinil decreased both the mean [^{11}C]raclopride and [^{11}C]cocaine binding potential in the caudate, putamen, and NAcc, reflecting increases in extracellular DA and increased occupancy of DAT. This indirectly led to higher DA availability in the above brain areas, which, among others, contain a cluster of distinct areas associated with creative output (Flaherty 2005). Given that DA increases in these areas have been shown to increase creativity and novelty seeking in humans (Bardo et al. 1996; Heilman et al. 2003), it is possible that modafinil might influence creative thinking through the modulation of DA in the caudate, putamen, and NAcc.

One important empirical question that has yet to be addressed in research on cognitive-enhancing effects of modafinil is whether modafinil influences creativity performance in healthy young adults. To date, no research has investigated the effects of modafinil on divergent and convergent thinking tasks of creativity in healthy participants.

The aim of this study was to investigate the effects of modafinil on creative performance in healthy, non-sleep-deprived volunteers using validated, reliable measures of divergent and convergent thinking of creativity (Kim 2011). Given the evidence reviewed so far in this section, it was expected that modafinil would enhance both convergent and divergent thinking of creativity. Furthermore, given the effects reported in the Farah et al. (2009) study, it was expected that modafinil would increase creativity as a function of baseline creativity trait.

6.4 Methods

6.4.1 Ethical approval

The study was approved by the Cambridge Local Research and Ethics Committee and by the Medicines and Health Care Products Regulatory Agency, the national drug licensing agency, London, United Kingdom.

6.4.2 Drug formulation

The drug and the placebo tablets, which were identical, were synthesized in the Royal Free Hospital, London, United Kingdom.

6.4.3 Sample size analysis

A sample of 30 per group was estimated to provide sufficient statistical power based on previous within-participant medication studies (Turner et al. 2003). For a large effect size of modafinil study ($d = 0.52$), the statistical power software (G-Power) recommended 30 participants to provide a power of 0.95 to detect a large effect size with 0.05 error (i.e., type I false-positive error of 5%). A stronger interaction between creativity performance and group (creativity scores 9 modafinil vs. placebo) with a large effect size ($d = 0.52$), for false positives ($a = 0.05$) with 30 participants in each group, will therefore yield a power of 0.95.

6.4.4 Participants

Sixty-four healthy male ($n = 31$) and female ($n = 33$) volunteers (mean age ± SD = 25.34 ± 3.95, range 19–36 years) were identified via the City of Cambridge, the Department of Psychiatry participant panel at the University of Cambridge, and local advertisements. Participants between the ages of 18 and 40 years were chosen to avoid any effects of cognitive decline. The larger age range was chosen to be representative of a wide range of ages and socioeconomic statuses.

An experienced psychologist screened all volunteers for the presence of any of the pre-specified exclusion criteria. Participants were excluded if they had any significant psychiatric history, had visual or motor impairment, or concurrently used any psychotropic medications or any medication contraindicated for use with modafinil. In addition, participants with a history of hypertension, cardiac disorders, epilepsy, and drug or alcohol abuse were excluded. All participants were advised not to consume alcohol or caffeine for 12 h before the testing sessions. All participants were questioned about compliance with alcohol and caffeine restrictions before inclusion into the study. A light breakfast or snack and juice were allowed before, but not during the experimental session. After participants were assessed by the experienced psychologist, a written informed consent was obtained.

Evidence suggests that modafinil improves function in a dose-dependent manner (Chen et al. 2007), and the best therapeutic (Minzenberg and Carter 2008) and experimental dose is 200 mg (Turner et al. 2003; Wong et al. 1998). Hence, when participants completed baseline physiological measures (blood pressure and pulse), they were given a single oral dose of 200 mg of modafinil or placebo with a small glass of water.

They were then asked to rest in a quiet room. Peak plasma concentrations of modafinil have been obtained within 2–3 h after oral administration with an elimination half-life of 12–15 h (Wong et al. 1998). Therefore, 2 h post-drug administration, participants completed a baseline measure of creativity personality trait and tasks that measured divergent and convergent thinking of creativity (see sections 6.4.6 and 6.4.7 for detailed descriptions of the tasks). After completing each creativity task, participants also completed a range of cognitive and motivational tasks for another experiment (the summary of the results are reported in the Discussion). After the completion of the study, participants were debriefed by the psychologist. They received monetary compensation of £25, plus local transport expenses. They were discharged by a clinical research nurse.

6.4.5 Experimental design

The study employed a randomized, double-blind placebo-controlled method with a parallel group design, deliberately chosen to avoid problems with practice effects common in crossover design studies with participants performing creativity (Rose and Lin 1984) and working memory tasks (Lowe and Rabbitt 1998). Participants in the study were randomly allocated to receive either a single oral dose of 200 mg of modafinil or a lactose placebo.

In order to balance drug conditions for gender, males and females were separately randomized for medications. Unblinding of the medication was undertaken after the first data analysis.

All volunteers were asked to spend the waiting time with low arousing activities (reading, watching TV, or napping) in a dayroom, and were monitored by research nurses. The testing of the creativity tasks was performed exactly 2 h after drug administration in a silent consultation room at the Wellcome Trust Clinical Research Facility at Addenbrooke's Centre for Clinical Investigation, Cambridge, United Kingdom.

6.4.6 Creativity measures

6.4.6.1 Creative Personality Scale (CPS)

The CPS is an adjective checklist questionnaire, which measures creativity personality trait (Gough 1979). The CPS was used to measure participants' creativity baseline trait. The CPS contains 18 creative adjective checklists and 12 conservative adjective checklists. A positive point is awarded when a participant endorses a creative checklist (e.g., individualistic), while a negative point is given when the participant chooses a conservative checklist (e.g., conventional). The CPS is a reliable and valid measure for the identification of creative personality (Carson et al. 2005; Kaduson and Schaefer 1991), and has been used consistently in recent research on creativity (Batey and Furnham 2008; Dewett and Gruys 2007). Furthermore, the CPS was validated with ratings from over 1000 creativity participants (Gough 1979).

6.4.7 Convergent thinking tasks

6.4.7.1 Group Embedded Figures Task (GEFT)

The GEFT (Figure 6.1) is a nonverbal convergent thinking task that requires participants to regroup the elements of a geometric design in ways that reveal simple shapes embedded in a complex figure (Witkin et al. 2002). During the experiment, participants were provided with seven simple shapes on an A4 paper, followed by nine complex test items.

Participants were instructed to identify and trace a simple geometric shape embedded within a complex figure. They had 3 min to complete all of the nine in any order. A point was given for each correctly traced item, with a maximum score of nine points.

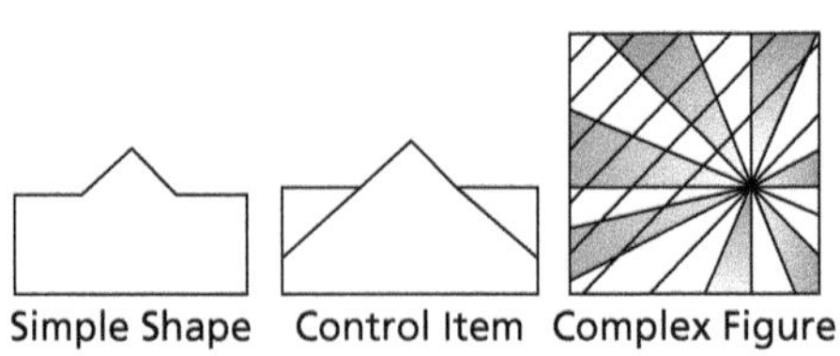

Figure 6.1 The Group Embedded Figure Task (GEFT). This figure shows a simple shape, a control item, and a complex figure. Participants had to find the simple shape in complex figure and trace it.

6.4.7.2 Remote Associates Test (RAT)

Mednick (1962; Table 6.1) theorized that creativity requires the capacity to generate remote associations that can connect hitherto disparate ideas and devised the RAT to assess this capacity. Since then, the RAT has been shown to measure creativity trait (Eysenck 1995, 80; Katz and Pestell 1989) and creative potential (Martindale and Dailey 1996). Experimentally, it has been used in variety of clinical studies on manic depression (Fodor and Greenier 1995) and schizotypy (Weinstein and Graves 2001). Psychometrically, it has been tested with Japanese (Baba 1982), Jamaican (Hamilton 1982), and Hebrew (Nevo 1978) populations, and has been found to have excellent cross-cultural validity.

The RAT is the most reliable creativity test that provides a measure of convergent thinking of creativity and insightful problem solving in the verbal domain (Mednick 1962). During the experiment, participants were presented with 15 triads, each containing three words, and were asked to identify and supply a word that has an associative linkage with three separate words. For example, if the given words were wheel, electric, and high, participants had 30 sec to supply the associated word (i.e., chair). Further examples are presented in Table 6.1.

6.4.8 Divergent thinking tasks

6.4.8.1 Drawing Task from the Abbreviated Torrance Test for Adults (ATTA)

The ATTA (Figure 6.2) is a standardized, abbreviated form of the Torrance Test of Creative Thinking (Goff 2002). This task was used to assess divergent thinking of creativity. It includes two picture-drawing items and a single verbal task. Only the picture items were used in this experiment. Participants were given a piece of paper that had seven incomplete figures (Figure 6.2).

Participants were instructed to use the incomplete figures to make an unusual picture, by adding lines to the incomplete figures, and by sketching some interesting objects and pictures. They were instructed to think of a picture or an object that no one else would think of, and they had to make the picture tell as complete and interesting story as they could, by adding lines to and building the first figure. They had to make up an interesting title for each of their drawings and write it at the bottom of each block, next to the number of the figure. They were given 7 min to complete the whole task. Two independent judges, who were blind to the study objective and participant conditions, scored the task according to the criteria set by Goff (2002).

Table 6.1 Examples from the Remote Associates Test (RAT)

Triad	Answer
Political-Surprise-Line	Party
Preserve-Ranger-Tropical	Forest

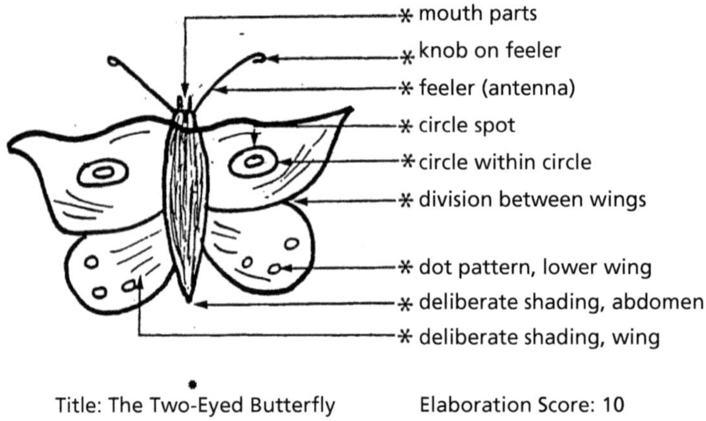

Figure 6.2 A completed example of the Abbreviated Torrance Test for Adults (ATTA). On the task, participants were scored for flexibility and elaboration.

6.4.8.2 Alternative Uses Task (AUT)

The AUT measures divergent thinking of creativity (Guilford 1957), and has been used in recent cognitive and human psychopharmacological studies that have investigated the effects of stimulants on creativity (Farah et al. 2009). Participants were given the name of an object and were asked to come up with as many alternative uses as they could think of for the named object within 1 min and 20 sec. Participants were given an example during the instruction. For example, if the word "tissue" was given as the object, an example of an appropriate alternative use would be a "blanket for a doll." They were told that the alternative use must make sense, so a response such as "eat as food" would not count. The objects used in this experimental work were shoe, button, key, barrel, car tire, and pencil. The responses of the participants were recorded and scored by two individuals who were double-blind to the study objective and participant condition. The responses were scored for originality, fluency, flexibility, and detail, according to the criteria of Guilford (1957). Scoring was comprised of the following four components:

Originality: Each response is compared to the total amount of responses from all of the participants. Responses that were given by only 5% of the group count as unusual (one point) and responses given by only 1% of them count as unique (two points).

Fluency: The total of all responses.

Flexibility: The number of different categories used.

Elaboration: The amount of detail (e.g., "a doorstop" counts zero, whereas "a doorstop to prevent a door slamming shut in a strong wind" counts two (one point for explanation of door slamming and another for further detail about the wind).

6.4.8.3 Wallach and Kogan pattern meaning (PM) and line drawing (LD) tasks

The Wallach and Kogan (1965) test of divergent thinking assesses the generation of associations through pattern meaning (Figure 6.3) and line drawings. These tasks were chosen

Figure 6.3 An example of the Pattern Meaning (PM) section in the Wallach and Kogan task.

because they reliably measure originality and fluency and they have been used to assess divergent thinking of creativity in high-functioning adults (Claridge and McDonald 2009) and in ADHD patients (Solanto and Wender 1989). These tasks have demonstrated high reliability and high internal consistency (Cropley and Mansley 1969).

Furthermore, the tasks are poorly correlated with intelligence tests (Cropley and Mansley 1969) and can be scored objectively and reliably without the need for an external evaluator (Claridge and McDonald 2009; Cropley and Mansley 1969; Plucker and Renzulli 1999). Participants were instructed to look at six patterns and six line drawings. With each one, they were asked to list all the things they could think each of the complete patterns and drawings might be. For instance, participants were shown Figure 6.3 and were awarded one point if they came up with a speedometer or a hedgehog as an answer. They were allowed to turn or move the drawings in any way to help their efforts. Dependent variables were the total number of generated responses and unique responses that no other participant had given.

6.5 Results

6.5.1 Convergent thinking tasks of creativity

6.5.1.1 GEFT

To test the hypothesis of whether there was a significant difference between the modafinil and the placebo group on the performance of the GEFT, a one-way analysis of variance (ANOVA) was conducted and found a marginal effect of drug on the GEFT ($p = 0.09$). Participants on modafinil performed marginally better than participants on placebo.

To test the hypothesis of whether creativity personality baseline and drug condition (modafinil or placebo) were predictors of the mean GEFT performance, a multiple linear regression analysis was conducted and found the overall regression nonsignificant ($p = 0.28$).

6.5.1.2 RAT

To test the hypothesis of whether there was a significant difference between the modafinil and placebo group on the RAT, a one-way ANOVA was conducted and was found to be nonsignificant ($p = 0.72$).

To test the hypothesis of whether creativity personality trait and drug condition (modafinil or placebo) predicted the performance of the mean RAT, a multiple linear regression analysis was conducted and found the overall regression nonsignificant ($p = 0.85$).

6.5.1.3 High versus low creativity participants

To investigate whether modafinil had similar effects on the RAT as Adderall (Farah et al. 2009), a median split was carried out on participants' responses to the CPS questionnaire, and those above the median (> 8) were classified as high in creativity baseline trait, and those below the median (< 8) were classified as low in creativity baseline trait.

To test the hypothesis of whether baseline of creativity (high vs. low) and drug (modafinil vs. placebo) had an effect on creativity performance, a two-way ANOVA using drug (modafinil vs. placebo) × creativity baseline (high vs. low) as independent variables and the RAT scores as a dependent variable was conducted, and found a significant drug and creativity baseline interaction ($p = 0.023$) = 5.318 (Figure 6.4). Modafinil participants

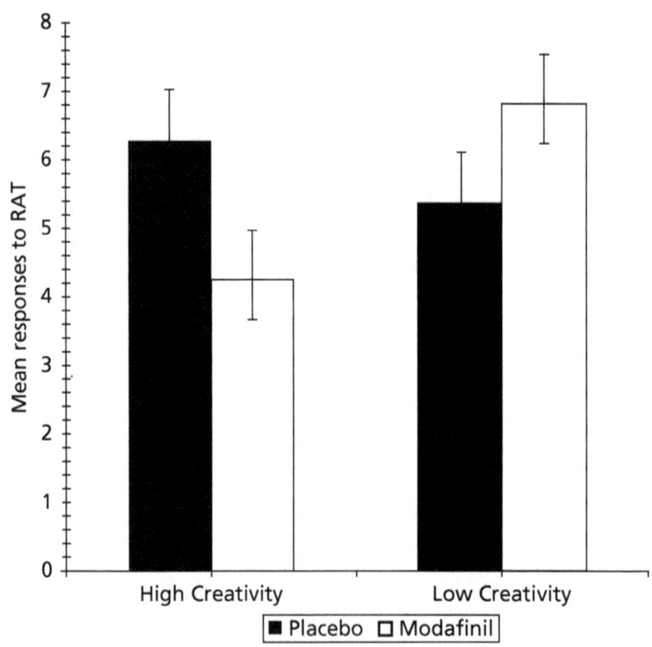

Figure 6.4 The effect of drug on high and low creativity participants on the RAT. In the modafinil group, participants low in creativity trait were significantly improved by modafinil, compared with those high in creativity baseline trait ($p = 0.04$). Error bars represent the SEM.

low in creativity trait had significantly higher RAT scores than those high in creativity trait (Figure 6.4).

A further one-way ANOVA confirmed a significant difference between the high and low creativity participants on modafinil ($p = 0.04$); simple effects confirmed that the modafinil participants scoring low on trait baseline creativity had improved responses on the RAT, while the performance of modafinil participants high in trait creativity was impaired on the task.

6.5.2 Divergent thinking of creativity

6.5.2.1 ATTA flexibility scores

To test the hypothesis of whether there was a significant difference between the modafinil and the placebo group on the performance of the flexibility scores in the divergent thinking task, a one-way ANOVA was conducted on the ATTA flexibility scores. It found a main effect of drug for total flexibility ($p < 0.01$), and a main effect of drug for mean flexibility ($p < 0.01$) see Figure 6.5). Participants on modafinil had significantly lower mean flexibility scores than participants on placebo.

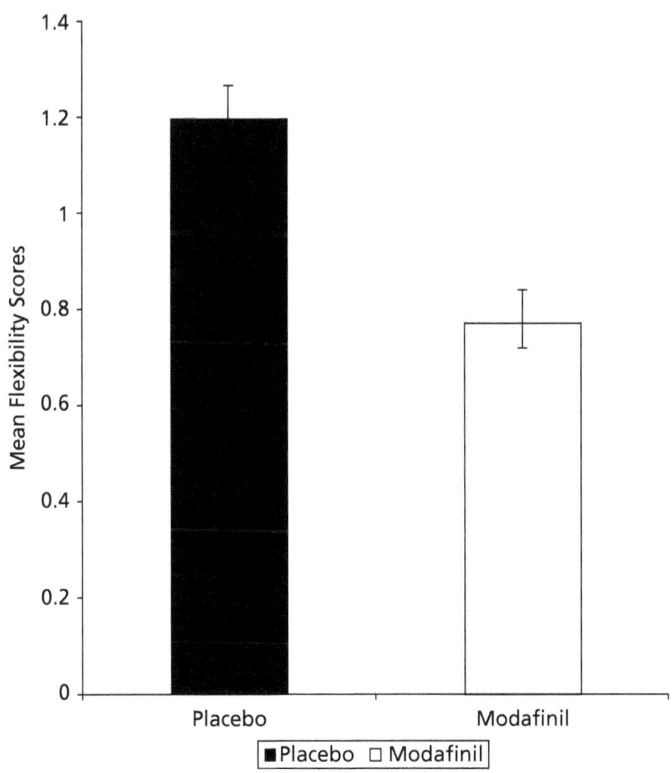

Figure 6.5 The effect of drug on the mean flexibility scores on the ATTA. Participants on placebo obtained significantly higher mean flexibility scores on the task than those on modafinil ($p < 0.01$). Error bars represent the SEM.

6.5.2.2 ATTA elaboration scores

To test the hypothesis of whether there was a significant difference between the modafinil and placebo group on the performance of the divergent thinking task a one-way ANOVA was conducted. It found a trend toward significance of drug on mean elaboration ($p = 0.053$), and a marginal effect of drug on total elaboration ($p = 0.056$). Participants on modafinil had marginally significantly lower elaboration scores than participants on placebo (see Figure 6.6).

To test the hypothesis of whether creativity personality trait and drug condition (modafinil or placebo) predicted mean elaboration scores and mean flexibility scores from the ATTA, separate multiple linear regression analyses were conducted. The regressions were marginally significant for elaboration ($p = 0.068$), and significant for flexibility ($p < 0.01$), respectively.

To test the hypothesis of whether there was a significant difference between the modafinil and the placebo group on the overall mean of the ATTA divergent thinking scores, a one-way ANOVA was conducted. It found a significant effect of drug on the ATTA mean scores ($p < 0.01$). Participants on modafinil had significantly lower mean divergent thinking scores than participants on placebo (see Figure 6.7).

To test the hypothesis of whether creativity personality trait and drug condition (modafinil or placebo) predicted the overall mean ATTA performance, a multiple linear

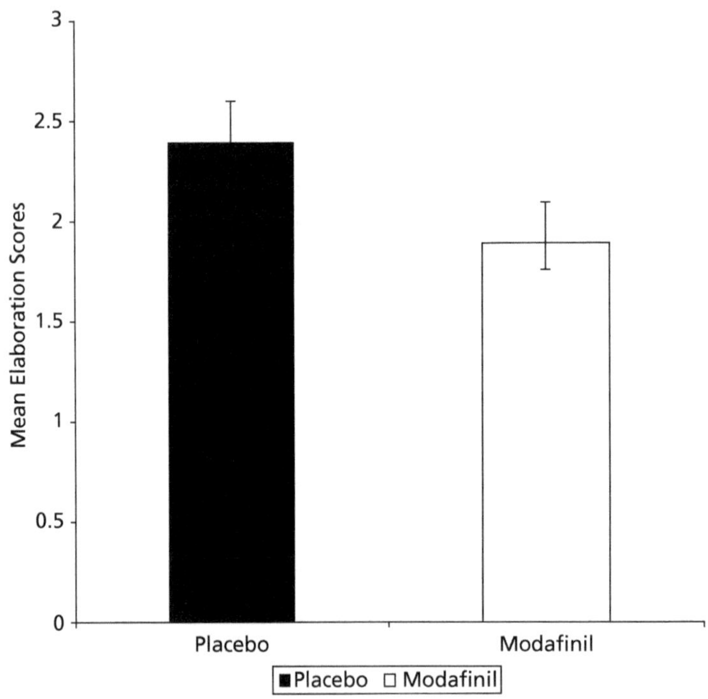

Figure 6.6 The effect of the drug on the mean elaboration scores on the ATTA. On average, participants on placebo were marginally significantly better on elaboration scores than those on modafinil ($p = 0.05$). Error bars represent the SEM.

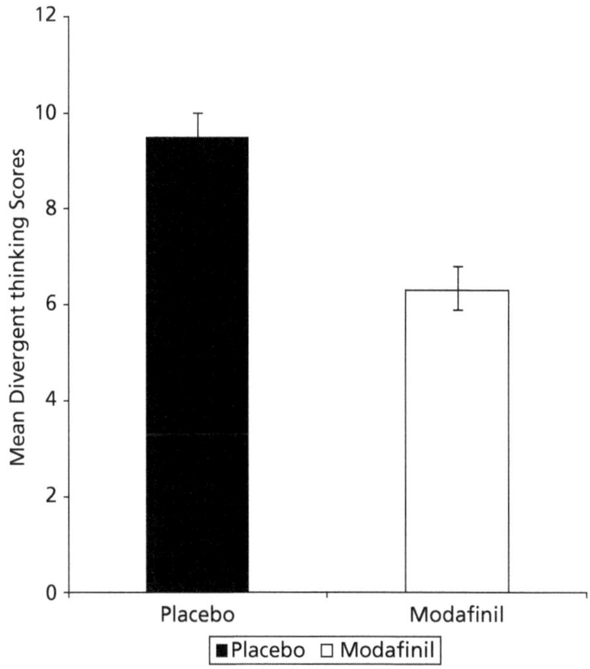

Figure 6.7 The effect of drug on the mean divergent scores on the ATTA. On average, participants on placebo were marginally significantly better on divergent thinking scores than those on modafinil ($p < 0.01$). Error bars represent the SEM.

regression analysis was conducted and found to be significant ($p < 0.01$). Creativity personality baseline trait did not predict the mean ATTA performance and was not significant ($p = 0.2$) while drug condition was a significant negatively predictor of the mean ATTA performance ($p < 0.01$).

This analysis shows that the creativity baseline trait itself was not a significant predictor of the mean divergent thinking performance but that drug, namely modafinil administration, was a significant negative predictor of the divergent thinking performance as measured by the ATTA.

6.5.2.3 High and low creativity group

To test the hypothesis of whether creativity baseline trait (high vs. low) and drug condition (modafinil or placebo) had a significant effect of main divergent thinking scores, a planned comparison ANOVA was conducted and found to be significant ($p < 0.01$). There was a main effect of drug condition on ATTA mean creativity performance ($p < 0.01$), no effect of creativity baseline on ATTA mean creativity performance ($p = 0.13$), and no significant interaction between creativity baseline and drug condition ($p = 0.45$).

To further investigate the main effects of drug, a post hoc analysis revealed that participants high in creativity who were on modafinil had a significant lower mean ATTA scores than participants on placebo ($p = 0.03$). Similarly, participants low in creativity baseline

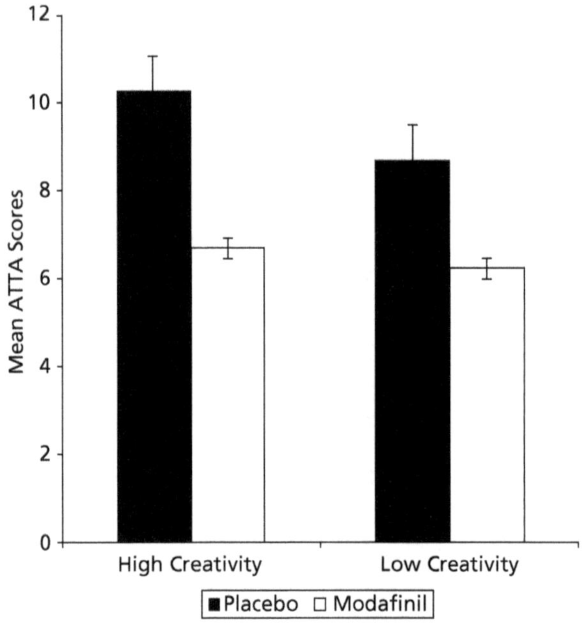

Figure 6.8 The effect of drug on the mean ATTA scores in the high and low creativity group. In both the high and the low creativity groups, participants on placebo obtained significantly higher mean ATTA scores than those on modafinil ($p < 0.01$). Error bars represent the SEM.

trait who were on modafinil had a significant lower mean ATTA score than participants on placebo ($p = 0.006$) (see Figure 6.8). This suggests that modafinil had a uniform negative effect on the performance of the ATTA scores regardless of the participants' creativity baseline trait.

6.5.2.4 AUT

To test whether there was a significant difference between the modafinil and placebo on the AUT, a one-way ANOVA was conducted and found to be not significant ($p = 0.99$). Regression analysis using creativity baseline trait and drug as predictors of creativity performance was also found to be not significant ($p = 0.76$).

6.5.2.5 PM and LD

To test whether there was a significant difference between the modafinil and placebo on the PM and the LD, a one-way ANOVA was conducted for both tasks. There was no significant effect for both analyses ($p = 0.6$, $p = 0.10$), respectively.

Regression analysis using creativity baseline trait and drug as predictors of creativity performance found no significant effect for both tasks ($p = 0.9$, $p = 0.5$), respectively.

6.5.3 Creativity and drug interactions

Does creativity personality trait interact with the drug to predict participants' performance on creative thinking tasks?

When interaction terms for drug and creativity baseline trait were combined and investigated with separate regression analyses for each creativity task, these analyses were not significant and did not predict the performance for the mean GEFT ($p = 0.79$), mean RAT($p = 0.5$), total flexibility ($p = 0.45$), mean flexibility ($p = 0.45$), total elaboration ($p = 0.22$), mean elaboration ($p = 0.198$), mean line drawing ($p = 0.62$), mean pattern meaning ($p = 0.8$), and mean AUT performance ($p = 0.5$). Finally, the regression analysis for predicting mean ATTA divergent thinking performance using drug and creativity personality trait as predictors was significant only for the drug ($p < 0.01$) but not for drug and creativity interaction ($p < 0.23$). Hence, drug and CPS interaction did not predict the mean ATTA performance but drug (i.e., modafinil) was the main negative predictor of ATTA divergent thinking performance ($p < 0.001$).

6.6 **Discussion**

The aim of this study was to investigate the effects of modafinil on reliable measures of divergent and convergent thinking tasks of creativity in healthy volunteers. For the convergent thinking tasks, the results show that modafinil had a marginally positive effect on the GEFT. However, this was not significant. There was no main effect of drug on the RAT. Hence, overall, modafinil had no main effect on the convergent thinking tasks of creativity.

When the creativity personality trait scores and drug treatment were combined to see whether they predicted the convergent thinking performance, the results from the regression analysis showed no significant effect, highlighting that the interaction between creativity personality trait and drug treatment (either modafinil or placebo) did not predict performance on the convergent thinking tasks.

We investigated whether modafinil had similar effects on the RAT as was reported by Farah et al. (2009) for Adderall. When participants were divided into high and low creativity group, using a median split analysis, improvements were seen in the RAT among those receiving modafinil, but only for participants who had low creativity personality trait at baseline. This means that modafinil reduced the performance of the convergent thinking task in participants who initially scored high on the creativity personality trait.

Similarly, Farah et al. (2009) reported improvements of performance on Adderall in participants who were, at baseline, below the median of performance on the RAT when they were on placebo. The results from the current study are consistent with earlier evidence suggesting that psychostimulants impair performance in participants who are high in cognitive function, whereas they might improve performance in those with low cognitive function (e.g., patients diagnosed with ADHD). For example, the DA agonist bromocriptine improved performance on executive function tasks in individuals with low working memory capacity, but impaired performance in those with high working memory capacity (Kimberg et al. 1997). The results on the RAT are consistent with the empirical evidence showing that those with lower baseline of working memory performance are moderately improved by psychostimulants (Mehta et al. 2004; Randall et al. 2005). The results corroborate arguments put forward by other contributors to this book (see

Hertwig and Hills (Chapter 11), Hall and Strang (Chapter 19), and Schleim and Quednow (Chapter 10) in this volume).

When the effects of modafinil on the performance of the divergent thinking task as measured by the ATTA were investigated, modafinil significantly reduced both the mean flexibility performance and the total mean divergent thinking performance. When we assessed whether the drug and creativity baseline trait together predicted the performance of the task, the results showed that the creativity personality trait was not a significant predictor but modafinil was a significant negative predictor of divergent thinking performance. The results suggest that modafinil significantly reduced the performance of the divergent thinking task, and this was independent of the participants' creativity personality trait. These findings are consistent with recently reviewed evidence that in non-ADHD adults, psychostimulant may actually impair performance on tasks that require adaptation and flexibility (Advokat 2010).

Interestingly, the results from the other cognitive experiments, which are not reported here, showed that modafinil failed to improve the CANTAB Paired Associate Learning task, which assesses short-term memory, and the Digit Span test (from the Wechsler IQ test), which assesses auditory memory. Modafinil also impaired word retrieval, as measured by the Hayling Sentence Completion Task, which assesses response initiation, response inhibition, and language retrieval (Mohamed and Lewis 2014). Furthermore, participants on modafinil reported significantly increased subjective pleasure as measured by a visual analogue scale for mood. These subjective increases in pleasure were independent of the actual performance on the creativity and cognitive tasks. On a task of motivational reinforcement learning, modafinil increased latency of response and impaired the learning of motivational contingencies associated with positive reward (Mohamed 2015). Taken together, these results suggest that in healthy individuals, modafinil might have some subtle negative or nuanced effects on different cognitive processes, particularly by reducing performance in the domains of higher order cognitive processes such as convergent and divergent thinking of creativity. Hence, these results argue against the notion that modafinil is a general cognitive-enhancing drug.

It is likely that modafinil did not have robust effects on all the creativity tasks because it modulates not only DA but it also increases multiple other neurotransmitters such as norepinephrine, serotonin, and glutamate in several regions of the brain, including the PFC, hippocampus, hypothalamus, and the striatum (see Minzenberg and Carter 2008).

An alternative explanation of the lack of the robust effects of modafinil on all the creativity tasks could be that the study sample was too small to detect a small to moderate effect size. Hence, the findings from this study are not conclusive and should be interpreted with caution. It is important that further trials are conducted to investigate the effects of modafinil on divergent and convergent thinking tasks of creativity.

6.6.1 **Limitations**

Although the findings from this study are consistent with Farah et al. (2009)'s study, and with previous psychopharmacology research, the results from the RAT should be interpreted with caution. This is because median split analyses split groups arbitrarily and assumed that the high and low creativity groups were inherently different. However, the participants in the placebo and in the modafinil groups who participated in this study were demographically similar. Moreover, the regression analysis, which inquired whether the combination of creativity trait and drug effect predicted the performance of the RAT, did not show significance. Hence, in order to replicate Farah et al.'s (2009) findings, it is important to investigate further the effects of modafinil on the RAT with a larger number of participants.

Moreover, apart from the ATTA, most of the divergent thinking tasks were not significant. Furthermore, contrary to the second hypothesis, apart from the RAT, modafinil did not increase creativity performance as a function of creativity baseline trait. Therefore, although the results from this study show that modafinil significantly reduced divergent thinking of creativity in healthy individuals, they are not conclusive enough to determine the effects of modafinil on creativity in healthy individuals. Hence, future studies need to collect more data on larger samples to gain a clear understanding of the effects of modafinil on creativity. Furthermore, as dose-dependent effects of modafinil on the creativity tasks were not investigated, future studies need to investigate the effects of 100–400 mg of modafinil on creativity tasks.

6.7 Conclusions

The current study investigated the effects of modafinil on highly reliable divergent and convergent thinking tasks on creativity. Modafinil had no main effect on convergent thinking tasks as measured by the GEFT and the RAT. However, a median split analysis showed that modafinil participants low in creativity trait had significantly higher RAT scores than those high in creativity trait.

Modafinil significantly reduced flexibility and marginally reduced elaboration as measured by the ATTA. Modafinil significantly reduced the overall performance of the ATTA, which assessed divergent thinking processes of creativity. The main creativity personality trait did not interact with the drug effects to predict creativity performance. Rather, modafinil administration was a significant negative predictor on its own in the divergent thinking performance.

The results from this study suggest that, rather than acting as a general cognitive enhancer, modafinil has more subtle negative or nuanced effects on different cognitive processes, particularly in the domains of convergent and divergent thinking of creativity. However, the results need to be interpreted with caution until replicated in a larger sample of participants.

6.8 Acknowledgments

Dr Mohamed would like to thank the Wellcome Trust for the neuroscience doctoral studentship, which funded this work. Dr Mohamed would also like to thank Dr James Rowe and Dr Timothy Rittman for medical standby during the trial; Professor Simon Baron

Cohen and Professor Anthony J. Holland for the excellent support and supervision they provided during the doctoral training at the University of Cambridge; and Dr Anthony E. Phillips for wonderful support in both Cambridge and Oxford. Finally, thanks to the participants who participated in the trial and the clinical nurses who monitored and discharged the participants.

Note

1. This work is drawn from Mohamed, A. D. 2014. The effects of modafinil on convergent and divergent thinking of creativity: a randomized controlled trial. *Journal of Creative Behavior*, doi: 10.1002/jocb.73.

References

Advokat, C. 2010. What are the cognitive effects of stimulant medications? Emphasis on adults with attention-deficit/hyperactivity disorder (ADHD). *Neuroscience and Biobehavioral Reviews* **34**:1256–66.

Baba, Y. 1982. An analysis of creativity by means of the Remote Associates Test for Adult Revised in Japanese (JARAT FORM A). *Japanese Journal of Psychology* **52**:330–6.

Ballon, J., and D. Feifel. 2006. A systematic review of modafinil: Potential clinical uses and mechanisms of action. *Journal of Clinical Psychiatry* **67**:554–66.

Baranski, J. V., R. Pigeau, P. Dinich, and I. Jacobs. 2004. Effects of modafinil on cognitive and meta-cognitive per-formance. Clinical Trial Comparative Study. *Human Psychopharmacology* **19**:323–32.

Bardo, M. T., R. L. Donohew, and N. G. Harrington. 1996. Psychobiology of novelty seeking and drug seeking behavior. *Behavioural Brain Research* **77**:23–43.

Batey, M., A. and Furnham. 2008. The relationship between measures of creativity and schizotypy. *Personality and Individual Differences* **45**:816–21.

Batir, A. 2009. Creativity induced by dopamine agonists in Parkinson's disease: Abstract. MDs 13th International Congress of Parkinson's Disease and Movement Disorders: Abstract Mo-163. Presented June 8, 2009.

Beal, M. F. 2001. Experimental models of Parkinson's disease. *Nature Reviews Neuroscience* **2**:325–34.

Carson, S. H., J. B. Peterson, and D. M. Higgins. 2005. Reliability, validity, and factor structure of the creative achievement questionnaire. *Creativity Research Journal* **17**:37–50.

Chen, C. R., W. M. Qu, M. H. Qiu, X. H. Xu, M. H. Yao, Y. Urade, and Z. L. Huang. 2007. Modafinil exerts a dose-dependent antiepileptic effect mediated by adrenergic [alpha]1 and histaminergic H1 receptors in mice. *Neuropharmacology* **53**:534–41.

Claridge, G., and A. McDonald. 2009. An investigation into the relationship between convergent and divergent thinking, schizotypy, and autistic traits. *Personality and Individual Differences* **49**:794–9.

Cropley, A. J., and G. W. Mansley. 1969. Reliability and factorial validity of the Wallach-Kogan creativity tests. *British Journal of Psychology* **60**:395–8.

Dewett, T., and M. L. Gruys. 2007. Advancing the case for creativity through graduate business education. *Thinking Skills and Creativity* **2**:85–95.

Douglas, V. I., R. G. Barr, J. Desilets, and E. Sherman. 1995. Do high doses of stimulants impair flexible thinking in attention-deficit hyperactivity disorder? *Journal of the American Academy of Child and Adolescent Psychiatry* **34**:877–85.

Eysenck, H. J. 1995. *Genius: The natural history of creativity*. Cambridge, UK: Cambridge University Press.

Farah, M., H. Caroline, G. Sankoorikal, M. E. Smith, and A. Chatterjee. 2009. When we can enhance cognition with Adderall, do we sacrifice creativity? *Psychopharmacology (Berl)* **203**:541–7.

Flaherty, A. W. 2005. Frontotemporal and dopaminergic control of idea generation and creative drive. *Journal of Comparative Neurology* **493**:147–53; doi:10.1002/cne.20768.

Fodor, E. M., and K. D. Greenier. 1995. The power motive, self-affect, and creativity. *Journal of Research in Personality* **29**:242–52.

Franke, A., C. Bagusat, P. Dietz, I. Hoffmann, P. Simon, R. Ulrich, and K. Lieb. 2013. Use of illicit and prescription drugs for cognitive or mood enhancement among surgeons. *BMC Medicine* **11**:102.

Gibson, C., B. S. Folley, and S. Park. 2009. Enhanced divergent thinking and creativity in musicians: A behavioral and near-infrared spectroscopy study. *Brain and Cognition* **69**:162–9.

Goff, T. 2002. *Abbreviated Torrance test for adults manual.* Bensenville, IL: Scholastic Testing Service.

Gough, H. G. 1979. A creative personality scale for the Adjective Check List. *Journal of Personality and Social Psychology* **37**:1398–1405.

Greely, H. T. 2006. The social effects of advances in neuroscience: legal problems, legal perspectives. In *Neuroethics: Definining the issues in theory, practice, and policy*, ed. J. Illes, 246–63. New York: Oxford University Press.

Guilford, J. P. 1957. Creative abilities in the arts. *Psychological Review* **64**:110–18.

Hamilton, M. A. 1982. "Jamaicanizing" the Mednick Remote Associates Test of creativity. *Perceptual and Motor Skill* **55**:321–2.

Heilman, K. M., S. E. Nadeau, and D. O. Beversdorf. 2003. Creative innovation: possible brain mechanisms. *Neurocase* **9**:369–79.

Hennessey, B. A., and T. M. Amabile. 2010. Creativity. *Annual Review of Psychology* **61**:569–98.

Kaduson, H. G., and C. E. Schaefer. 1991. Concurrent validity of the Creative Personality Scale of the Adjective Check List. *Psychological Reports* **69**:601–2.

Katz, A. N., and D. Pestell. 1989. Attentional processes and the finding of remote associates. *Personality and Individual Differences* **10**:1017–25.

Kaufman, J. C., and R. J. Sternburg. 2006. *The international handbook of creativity.* Cambridge, UK: Cambridge University Press.

Kim, K. H. 2011. The APA 2009 Division 10 debate: Are the Torrance tests of creative thinking still relevant in the 21st century? *Psychology of Aesthetics, Creativity, and the Arts* **5**:302–8.

Kimberg, D. Y., M. D'Esposito, and M. J. Farah. 1997. Effects of bromocriptine on human subjects depends on working memory capacity. *NeuroReport* **8**:3581–5.

Lowe, C., and P. Rabbitt. 1998. Test/re-test reliability of the CANTAB and ISPOCD neuropsychological batteries: theoretical and practical issues. *Neuropsychologia* **36**:915–23.

Martindale, C., and A. Dailey. 1996. Creativity, primary process cognition and personality. *Personality and Individual Differences* **20**:409–14.

Mednick, S. 1962. The associative basis of the creative process. *Psychological Review* **69**:220–32.

Mehta, M. A., I. M. Goodyer, and B. J. Sahakian. 2004. Methylphenidate improves working memory and set-shifting in AD/HD: relationships to baseline memory capacity. *Journal of Child Psychology and Psychiatry* **45**:293–305.

Mignot, E., S. Nishino, C. Guilleminault, and W. C. Dement. 1994. Modafinil binds to the dopamine uptake carrier site with low affinity. *Sleep* **17**:436–7.

Minzenberg, M. J., and C. S. Carter. 2008. Modafinil: A review of neurochemical action and effects on cognition. *Neuropsychopharmacology* **33**:1477–1502.

Mohamed, A. D. 2014. Reducing creativity with psychostimulants may debilitate mental health and well-being. *Journal of Creativity in Mental Health* **9**:146–63.

Mohamed, A. D. 2015. The effects of modafinil on motivation and on saliency of pleasure in healthy individuals: quantitative evidence from the cognitive neurosciences. *AJOB Neuroscience* **6**(1):15–17.

Mohamed, A. D., and C. R. Lewis. 2014. Modafinil increases the latency of response in the Hayling sentence completion test in healthy volunteers: a randomised controlled trial. *PLoS ONE* **9**:e110639.

Mohamed, A. D., and B. J. Sahakian. 2012. The ethics of elective psychopharmacology. *International Journal of Neuropsychopharmacology* **15**:559–71.

Nevo, B. L. I. 1978. Remote Associates Test: assessment of creativity in Hebrew. *Megamot* **24**:87–98.

Norman, C., and M. Berger. 2008. Neuroenhancment: status quo and perspectives. *European Archives of Psychiatry and Clinical Neuroscience* **258**:110–14.

Plucker, J. A., and J. S. Renzulli. 1999. Psychometric approaches to the study of human creativity. In *Handbook of creativity*, ed. R. J. Sternberg, 35–64. New York: Cambridge University Press.

Qu, W.-M., Z.-L. Huang, X.-H. Xu, N. Matsumoto, and Y. Urade. 2008. Dopaminergic D1 and D2 receptors are essential for the arousal effect of modafinil. *Journal of Neuroscience* **28**:8462–9.

Randall, D. C., J. M. Shneerson, and S. E. File. 2005. Cognitive effects of modafinil in student volunteers may depend on IQ. *Pharmacology Biochemistry and Behaviour* **82**:133–9.

Reuter, M., S. Roth, K. Holve, and J. Hennig. 2006. Identification of first candidate genes for creativity: a pilot study. *Brain Research* **1069**:190–7.

Rose, L. H., and H.-T. Lin. 1984. A meta-analysis of long-term creativity training programs. *Journal of Creative Behavior* **18**:11–22.

Rosenthal, R., and L. M. Westreich. 2010. *Cosmetic psychopharmacology: drugs that enhance wellbeing, performance and creativity.* Cambridge, UK: Cambridge University Press.

Solanto, M. V., and E. H. Wender. 1989. Does methylphenidate constrict cognitive functioning? *Journal of the American Academy of Child and Adolescent Psychiatry* **28**:897–902.

Turner, D., T. Robbins, L. Clark, A. Aron, J. Dowson, and B. Sahakian. 2003. Cognitive enhancing effects of modafinil in healthy volunteers. *Psychopharmacology (Berl)* **165**:260–9.

Volkow, N., J. S. Fowler, J. Logan, D. Alexoff, W. Zhu, F. Telang, G.-J. Wang, M. Jayne, J. M. Hooker, C. Wong, B. Hubbard, P. Carter, D. Warner, P. King, C. Shea, Y. Xu, L. Muench, and K. Apelskog-Torres. 2009. Effects of modafinil on dopamine and dopamine transporters in the male human brain: clinical implications. *Journal of the American Medical Association* **301**:1148–54.

Wallach, M. A., and N. Kogan. 1965. *Modes of thinking in young children.* New York: Holt, Rinehart and Winston.

Walsh, J. K., A. C. Randazzo, K. L. Stone, and P. K. Schweitzer. 2004. Modafinil improves alertness, vigilance, and executive function during simulated night shifts. *Sleep* **27**:431–9.

Wang, G.-J., N. D. Volkow, J. S. Fowler, J. Logan, N. R. Pappas, C. T. Wong, et al. 1999. Reproducibility of repeated measures of endogenous dopamine competition with [11C]raclopride in the human brain in response to methylphenidate. *Journal of Nuclear Medicine* **40**:1285–91.

Weinstein, S., and R. E. Graves. 2001. Creativity, schizotypy, and laterality. *Cognitive Neuropsychiatry* **6**:131–46.

White, H. A., and P. Shah. 2006. Uninhibited imaginations: creativity in adults with attention-deficit/hyperactivity disorder. *Personality and Individual Differences* **40**:1121–31.

Witkin, H., P. Oltman, E. Raskin, and S. Karp. 2002. *Group embedded figures test manual.* Redwood City, CA: Mind Garden.

Wong, Y., S. Gorman, G. MCCormick, and P. Grebow. 1998. Single-dose pharmacokinetics of modafinil and methylphenidate given alone in and in combination to healthy male volunteers. *Journal of Clinical Pharmacology* **38**:276–82.

Chapter 7

Does modafinil improve cognitive functioning in healthy individuals?

Ahmed Dahir Mohamed

7.1 Introduction

Drugs that are used to treat cognitive deficits in persons with clinical neuropsychiatric disorders such as Alzheimer's disease (AD), attention deficit hyperactivity disorder (ADHD), Parkinson's disease, and schizophrenia alter neurotransmitter modulation of cognition and improve attention, learning, memory, and executive functions (EF) in these patients. EFs are mental processes that enable humans to plan, solve problems, and think flexibly about how to conduct complex everyday cognitive processes. Working memory (WM) is the ability to temporarily hold relevant information active over a short interval and to manipulate it online (Baddeley 2010).

These drugs improve cognition by balancing neurotransmitter systems (e.g., dopamine (DA) and noradrenaline (NA)) in ways that restore some aspects of cognition in these patients. However, in experimental settings, evidence from healthy volunteers shows that these drugs improve robust neuropsychological cognitive tasks in both steep and inverted U-shaped fashion, with optimal enhancement dependent on taking the right dose and having the right baseline of cognitive function (de Jongh et al. 2008).

7.1.1 Current trends

Current trends show strong positive attitudes toward the use of these drugs by the general population (Husain and Mehta 2011). There is evidence that psychostimulant drugs are being used by some healthy individuals for cognitive enhancement (Farah et al. 2004; Farah 2010) and for improvements in academic performance (Franke et al. 2011; Smith and Farah 2011).

7.1.2 Cognitive enhancement

Cognitive enhancement has been defined as the amplification or extension of core capacities of the mind through improvement or augmentation of internal and external information processing systems (Sandberg and Savulescu 2011). Hence, because healthy individuals are using these drugs to amplify their cognitive capacities (Mohamed and Sahakian 2011), they are now conceptualized as a form of cognitive enhancement and have been termed pharmacological cognitive enhancers (PCEs) (Greely et al. 2008). For instance, in a recent online survey conducted by *Nature* on 1600 academics, one in five respondents used drugs to boost cognition and 44% reported taking the drug modafinil, mainly to improve concentration. However, the participants were not drawn from a random sample and the results might only be representative of those who regularly interact with the journal. Given that we do not know the opinion of those who did not respond to the survey, it is also highly likely that the survey overestimates the prevalence of the use of modafinil among academics. Furthermore, it is unclear how much of the indicated use of modafinil for concentration was for cognitive enhancement in the strict sense or whether the responders were using it to offset cognitive impairments from sleep deprivation caused by long haul flights, jet lag, or demanding academic pressures, or even mild cognitive impairment caused by ageing, stress, or undiagnosed psychological disorders.

7.1.3 The aims of this chapter

This chapter focuses on modafinil, a medication licensed for narcolepsy which has multiple mechanisms of action (Minzenberg and Carter 2008), and it is reportedly being used off-label by healthy adults with the aim of improving their concentration and enhancing attention. The chapter reviews evidence from double-blind, placebo-controlled studies and provides up-to-date information about modafinil's modes of action, cognitive effects, and safety information.

7.2 Modafinil

Modafinil is a wake-promoting agent that is licensed for use in treating narcolepsy (Benerjee et al. 2004). It has recently been approved for the treatment of "shift work sleep disorder." Modafinil is said to be a popular drug among healthy individuals aiming to enhance cognition (Lynch et al. 2011). In 2004, it was estimated that 90% of modafinil was being used off-label by healthy, non-sleep-deprived individuals (Baranski et al. 2004). The global revenue of modafinil is more than US$700 million per year (Norman and Berger 2008). This off-label and revenue figures are consistent with the claim that healthy physicians on call, students, and academics are increasingly using modafinil to enhance cognitive performance (Franke et al. 2011, 2013; Greely 2006a,b; Greely et al. 2008). However, it is important to mention that the above-mentioned estimates and the revenues, which are extremely high given that they do not include those who use modafinil as shift workers, have not been credited to a specific source. Furthermore, some of the claims made by opinion articles are not supported by the references cited therein, and the scientific literature (Partridge et al. 2011) shows mixed evidence on whether people report using modafinil as a medication or even as a recreational drug.

7.2.1 Molecular mechanisms of modafinil

Currently, the mechanisms by which modafinil exerts its wakefulness effects and how it acts as a cognitive-enhancing agent in healthy individuals are unclear. Several studies indicate that modafinil increases DA (Volkow et al. 2009), NA (Hilaire et al. 2001), serotonin [5-HT] (Ferraro et al. 2005), and glutamate (Ferraro et al. 1997) in several regions of the brain, including the prefrontal cortex (PFC), hippocampus, hypothalamus, and striatum. However, there is stronger evidence that it mainly modulates NA and DA neurotransmitters (Minzenberg and Carter 2008; Volkow et al. 2009). For instance, modafinil inhibits NA transporters (NET) (Madras et al. 2006) and elevates extracellular NA levels in various brain regions including the PFC (Hilaire et al. 2001). It also binds to and requires intact DA transporters (Qu et al. 2008; Turner et al. 2004a) and α-adrenergic receptors (Yang et al. 2007). These respectively lead to DA increase in the caudate, putamen, and nucleus accumbens (NAcc) (Volkow et al. 2009) and NA increase in the PFC and in the locus coeruleus (LC) (Hilaire et al. 2001). These are areas in the brain associated with reward and pleasure and increased attention and vigilance, respectively.

7.2.2 Modafinil increase noradrenaline to improve wakefulness

Noradrenaline is a major synaptic catecholamine neurotransmitter that is produced mainly in the LC. Noradrenergic release from the LC has been implicated in modulating diverse functions including vigilance (i.e., attention), autonomic arousal, reward function, and motivation (Breedlove et al. 2010). Conversely, NA release in the PFC has been associated with improvements in memory, EF (Breedlove et al. 2010), including optimal attention to relevant stimuli (Ramos and Arnsten 2007), and better response inhibition (Arendt et al. 2007). The results of studies using in vivo positron emission tomography (PET) and in vitro trace methods with rhesus monkeys have shown that modafinil occupies NET, mainly in the thalamus (Madras et al. 2006). Consistent with this, the results of functional magnetic resonance imaging (fMRI) studies show that modafinil modulates NA neurotransmitter and increases PFC efficiency in healthy subjects. For example, during an event-related fMRI study, Minzenberg et al. (2008) administered 200 mg of modafinil to healthy subjects and reported that modafinil shifted LC–NA to low tonic–high phasic pattern to optimize task performance. Modafinil administration was associated with a decrease in task-irrelevant LC activity and increased task-relevant activity in PFC, with enhanced LC–PFC functional connectivity, indicating that modafinil produces adaptive shifts in cortical brain areas that optimize information processing and attention. However, the interpretation of these results has been contested (Astafiev et al. 2010).

Another recent neuroimaging study showed that a chronic low dose of modafinil significantly decreased blood-oxygen-level dependent (BOLD) fMRI amygdala reactivity to fearful stimuli by comparison with placebo. Rasetti et al. (2010) explored the effect of 100 mg of modafinil per day for one week on emotional and cognitive information processing circuits in a sample of healthy non-sleep-deprived volunteers. They observed a reduced BOLD signal in the PFC and anterior cingulate cortex during EF tasks, a WM task, and a variable attentional control task. In that study, although modafinil seems to have reduced the ability to process fearful emotional stimuli, it enhanced the efficiency of PFC cortical cognitive information processing and increased attention and ease of processing of cognitively demanding load. Hence, if increasing NA improves attention (Breedlove et al. 2010), then modafinil's action on neurotransmitter NA may explain its attention-enhancing effects.

7.2.3 **Modafinil amplifies dopamine to increase its rewarding and motivating effects**

Modafinil also increases DA, which is another major synaptic catecholamine neurotransmitter produced in basal forebrain and diencephalon (Breedlove et al. 2010). The hypothesis that modafinil increases DA is consistent with evidence that modafinil requires intact DA–adrenergic interactions rather than simply increasing NA to promote wakefulness (Wisor and Eriksson 2005). Neurotransmitter DA is active in the basal ganglia and the olfactory system and is limited in parts of the cerebral cortex. However, DA has many functions in the brain and because of these many projections, its role is complex and its actions have been implicated in the gating of inputs and modulation of states of neural activity in postsynaptic follower cells (Cooper et al. 2003). For instance, the mesocortical and mesolimbic pathways have abundant dopaminergic innervations in many parts of the brain including the ventrotagemental area and the substantia nigra. These target the caudate, putamen, and limbic cortex including the medial PFC, the cingulate, the entorhinal cortex, and the limbic structures (Cooper et al. 2003), all areas that play a significant role in motivation, reinforcement of reward, and addictive states (Koob and Volkow 2010; Volkow and Baler 2014; Volkow et al. 1999, 2004; Wise 2004). Hence, drugs that increase DA in these areas modulate motivational and reward inputs and act as positive reinforcers, but they also act on different parts of the brain at the same time, making their actions complex. This supports the notion that DA is involved in many complex actions, one of which involves mediating both immediate pleasurable aspects of natural rewards and arousal that predicts impending rewards (Kupfermann et al. 200).

7.2.4 **Molecular studies**

Molecular studies show that modafinil increases DA levels in the brain areas listed in section 7.2.1. For example, mice lacking DAT or D_1 and D_2 receptors ($D_{R1/2}$) do not respond to the wake promoting effects of modafinil. Qu et al. (2008) used DA D_{R2}-deficient mice along with D_{R1}- and D_{R2}-specific antagonists to clarify the role of DA receptors in the arousal effects of modafinil. In wild-type mice, modafinil induced wakefulness in a dose-dependent manner. However, pretreatment with either D_{R1} antagonist [SCH23390] or D_{R2} antagonist raclopride at 2 mg/kg blocked the arousal effects of modafinil. Similarly, D_{R2} knockout (KO) mice exhibited less arousal from modafinil, and pretreatment with a D_{R1} antagonist abolished the arousal effects of modafinil. These findings indicate that dopaminergic D_{R1} and D_{R2} are essential for the wakefulness induced by modafinil. More specifically, in animals modafinil indirectly increased extracellular levels of DA in the NAcc (Murillo-Rodríguez et al. 2007; Nguyen et al. 2011). Furthermore, PET studies with healthy volunteers have shown that modafinil competes with DAT and indirectly increases DA in the caudate, putamen, and NAcc (Volkow et al. 2009). DA releases in these areas have been associated with positive reinforcement, incentive salience of reward (Berridge and Robinson 1998), and addiction (George and Koob 2010; Koob and Volkow 2010; Volkow and Baler 2014). For the effects of modafinil on dopamine and motivation see (Mohamed 2012, 2015).

7.3 Cognitive modes of action

This section reviews experimental evidence showing that modafinil moderately improves WM processes and EFs in healthy individuals low in cognitive function and in patients with neuropsychiatric disorders (Mohamed and Sahakian 2011). This indicates that the drug might be more beneficial for clinical rather than for cognitive-enhancing purposes.

7.3.1 Clinical applications

Clinically, modafinil has been suggested as an effective treatment of ADHD (Turner et al. 2003b, 2004a), a condition characterized by impulsivity, inattention, and hyperactivity. For instance, clinical studies show that modafinil significantly improves short-term memory span, visual memory, spatial planning, and stop-signal motor inhibition in ADHD adults. Turner et al. (2004a) found that 200 mg of modafinil improved performance on tests of short-term memory span, visual memory, spatial planning, and response inhibition in these patients. On several measures, modafinil increased sustained attention and accuracy while slowing response latency. Similarly, modafinil has been reported to improve cognition in patients with schizophrenia (Turner et al. 2004b), depression (Vaishnavi et al. 2006), Parkinson's disease (Happe et al. 2001; Tyne et al. 2010), and addiction (McGough et al. 2009). However, there is a mixed evidence of effiacy in those suffering from multiple sclerosis (Kraft and Bowen 2005; Stankoff et al. 2005; Zifko et al. 2002). The results of a recent study that used a large neuropsychological battery of tests showed that modafinil was only able to improve the recognition of sad faces in patients who experienced the first episode of psychosis (Scoriels et al. 2011). Nonetheless, some of these studies seem to suggest that modafinil might be beneficial for the treatment of fatigue-related symptoms that occur in a range of neuropsychiatric disorders.

7.3.2 Ethical issues in cognitive enhancement

As outlined in sections 7.1 and 7.2 some healthy individuals with no psychiatric disorders are using modafinil in an attempt to boost their cognitive performance (Vastag 2004). This has raised a range of ethical questions including whether modafinil is safe for healthy individuals and whether it may negatively affect other forms of cognitive and brain functioning (Mohamed 2012, 2014c; Mohamed et al. 2011). A further issue relates to whether it is ethical to use healthy subjects in cognitive enhancement research. As we discussed this issue elsewhere (Mohamed and Sahakian 2011), the current chapter focuses on the effects of modafinil on cognition in healthy subjects.

7.3.3 Modafinil moderately improves attention in healthy individuals

There is some evidence of cognitive-enhancing effects of modafinil in healthy humans (Minzenberg and Carter 2008). More specifically, modafinil has been shown to enhance discrete aspects of cognition such as mnemonic function-digit span (DS), delayed-matching-to-sample, pattern recognition memory (PRM), planning, and the stop-signal reaction task (SSRT), a test of response inhibition (Müller et al. 2004). Moreover, Turner et al. (2003a) showed that a single oral dose of modafinil (100 or 200 mg) significantly improved performance on tests of DS, visual PRM, spatial planning, and SSRT in healthy volunteers. The same doses also improved the response time in tests on decision-making, delayed-matching-to-sample, and spatial planning. However, this evidence also suggests that modafinil increases performance accuracy at the cost of slowing response latency on several tasks (Mohamed and Lewis 2014; Turner et al. 2003a). This has prompted suggestions that modafinil may act to increase vigilance (Baranski et al. 2004). However, Müller et al. (2004) reported that modafinil significantly reduced error rates in a long-delay visuospatial task and under manipulation conditions, without producing a speed–accuracy trade-off. More recently, Winder-Rhodes et al. (2009) argued that the influence of modafinil may be subtle and restricted to challenging tasks, and its effects are limited in healthy, non-sleep-deprived and high performing individuals. At the same time, modafinil improved accuracy in an attention-shifting task, without reaction time (RT) trade-off, especially in tasks on which participants' resources were most challenged (Marchant et al. 2009). The results of a study by Baranski et al. (2004) showed that modafinil improved fatigue levels, subjective vigour, RT, and vigilance in 18 healthy non-sleep-deprived subjects. In that study, modafinil improved serial RT, logical reasoning, and 1-back WM task, a task requiring the detection of repeated numbers. Furthermore, modafinil reduced RT responses in the Stroop color-naming task and improved accuracy in the clock-drawing task (Randall et al. 2005c) and the digit backward and forward span, in which subjects immediately repeat digits either backward or forward (Randall et al. 2005a,b,c). Recently, modafinil improved performance on tests of EF and WM in sleep-deprived physicians (Sugden et al. 2012), functions that are important in conducting surgical operations under stress and after limited sleep. If these findings are accurate, they demonstrate that modafinil has the potential to have positive effects on certain discrete cognitive functions and/or promote wakefulness in healthy individuals.

However, modafinil administration does not always lead to improvements in cognition in healthy individuals. In fact, to date improvements have been observed only in experimental settings and they have been reported to be very small in size (Randall et al. 2004, 2005b).

7.3.4 Lack of effects of modafinil on cognition

Research from the same studies of healthy volunteers also failed to find cognitive-enhancing effects of modafinil on several cognitive tasks. For example, Müller et al. (2004) failed to find effects of 200 mg of modafinil on simple digit maintenance, letter cancelation, and trail making, while Turner et al. (2003a) reported that 100 or 200 mg of modafinil did not have any significant effects on spatial memory span, spatial working memory (SWM), rapid visual processing, or attentional-set-shifting task. Additionally, no effects were observed on the paired associates learning (PAL) task, which is a test of memory, and the "one-touch" Tower of London spatial planning task. Furthermore, Turner et al. (2003a) reported a lack of effect of modafinil on the Cambridge Gambling Task. Consistent with these findings, the results of a series of studies by Randall et al. (2003, 2004, 2005a,b,c) failed to find any effects of 100 or 200 mg of modafinil on SWM, logic memory, the Paced Auditory Serial Addition Test, symbol copy, digit cancelation, verbal fluency, intradimensional/extradimensional (ID/ED) attentional-set-shifting task, trail A/B task, verbal fluency, clock drawing, visual delayed-matching-to-sample, spatial planning, DS, sustained attention, logic memory, and Stroop and verbal fluency tasks, including the Hayling Sentence Completion Test (Mohamed and Lewis 2014). As also confirmed by systematic reviews such as Repantis et al. (2010), this evidence indicates that modafinil does not always boost cognitive functions in healthy individuals, especially if these functions involve sophisticated problem-solving abilities that require more than improved attention.

7.3.5 The U-inverted function of neurotransmitter modulation of cognition

Can insights from neuropsychopharmacology explain the inconsistent evidence reviewed in the previous sections? Evidence from neuropsychopharmacology suggests that drug improvements relate to optimal balance of neurotransmitter modulation of cognition and function (Iverson et al. 2009), indicating that it is often not beneficial to over- or understimulate brain neurotransmitter systems. For example, the effects of some well-known psychostimulants, including modafinil, follow the Yerkes–Dodson law (Yerkes and Dodson 1908). This law explains the relationship between arousal and performance by hypothesizing that there is an optimum level of arousal above and below which performance deteriorates. This principle might apply to several neurotransmitter systems involved in cognitive function which often follows an inverted-U-shaped curve in which deviations from the optimal level in either direction produce suboptimal performance. For instance, low levels of NA release in the brain engage α_2-adrenergic receptors and improve EF, whereas higher levels of NA release engage α_1-adrenergic receptors and cause significant stress in humans and animals (Finlay et al. 1995) and impair PFC functioning (Ramos and Arnsten 2007). A similar U-inverted relationship is reported between DA and WM function (Vijayraghavan et al. 2007) in that both marked increases and decreases of DA in the PFC have been associated with suboptimal performance (Iversen 2001). Consistent with this hypothesis, methylphenidate, a DA agonist, improves cognitive performance in individuals with low WM function (Husain and Mehta 2011), while guanfacine, an α_2-adrenergic receptor agonist, has beneficial effects on WM and attentional functions in patients with ADHD, but does not improve WM or EF in healthy male volunteers (Müller et al. 2004). Thus, the effects of pharmacological substances on cognition depend on baseline cognitive functioning, optimal dosing, and one's genetic predisposition to metabolize a given drug (Evans and Relling 1999). The U-inverted function of drug and neurotransmitter actions has recently been hypothesized as reflecting our evolutionary developmental limitations (see Chapter 11, and Hills and Hertwig 2011). If true, this might explain why pharmacological substances fail to improve cognitive function in a linear fashion, particularly in healthy individuals.

7.3.6 Ceiling effects

The inconsistent results in modafinil studies might also reflect different experimental procedures and differential effects on neurotransmitter modulation (Mohamed and Sahakian 2011). For instance, in the Turner et al. (2003a) study, the lack of any effect of modafinil might have reflected ceiling effects since the sample comprised high-functioning healthy students from Cambridge University who were not sleep deprived and who were tested with easier versions of these tasks. For instance, the SWM task that was used in that study would have been easy for high-performing subjects. This might partially explain why some studies have failed to detect any improvement on cognitive tasks in healthy volunteers given modafinil.

7.3.7 Baseline of IQ and WM

Not all null results reported in studies of normal subjects were due to ceiling effects. For instance, baseline IQ is another factor that influences the cognitive-enhancing effects of modafinil in healthy subjects. To illustrate this, Randall et al. (2003, 2004) reported a lack of effect of modafinil on cognitive function in high-performing healthy subjects, but did not take into account subjects' baseline IQ. A later analysis of their results revealed that the cognitive benefits from modafinil were most prominent in subjects with low IQ and were limited in those with high IQ (Randall et al. 2005b). Consequently, modafinil might improve cognition in individuals low in IQ (Randall et al. 2005b) and WM (Husain and Mehta 2011) but might not benefit healthy individuals high in cognitive function.

7.3.8 Dose-dependent effects

Evidence suggests that modafinil improves cognitive function in a dose-dependent manner in mice (Chen et al. 2007) with the best average therapeutic dose equivalent to 200 mg in humans (BNF 2010). Consistent with this argument, in Turner et al.'s (2003a) study, modafinil effects were seen in a dose-dependent manner where greater improvement were seen after 200 mg were administered. For instance, on the SSRT, the greatest capacity to inhibit responding was observed in the 200 mg group, who also performed optimally in the task, with the least number of errors, compared with those groups given placebo and a low dose (100 mg) of modafinil (Turner et al. 2003a). This is consistent with the results of Winder-Rhodes et al.'s (2009) study, which failed to replicate the cognitive-enhancing effects of modafinil with a higher dose (300 mg) on a battery of cognitive tasks. Furthermore, Baranski et al. (2004) also used high doses of modafinil

(4 mg/kg or 300 mg) with healthy volunteers and found no cognitive-enhancing effects of modafinil on addition, line discrimination, and subjective judgment in performance on cognitive tasks. Likewise, Wesensten (2006) reported a lack of any effects of 400 mg of modafinil over and greater than 600 mg of caffeine on the Stroop task, verbal fluency, simple RT, and Wisconsin's Card Sorting task. This suggests that 200 mg of modafinil may be the best therapeutic dose, with higher and lower doses not improving cognitive performance.

7.4 Modafinil and creativity

Cognition is a multifaceted construct that encompasses numerous mental functions including both "cold" forms of cognition, such as attention, planning, problem solving, and response inhibition, and "hot" forms of cognition, such as creativity (Hennessey and Amabile 2010), risky decision-making, and motivation (Rees et al. 1999). Pharmacological drugs may differentially modulate hot and cold cognitive processes in healthy subjects. For example, an association has been found between increased DA in the mesolimbic areas and creativity performance (Batir 2009). There is also emerging evidence for the role of DA D_{R2} receptor gene (DRD2 Taq1A) in creativity (Reuter et al. 2006). DA release in the NAcc has been show to increase novelty seeking and creative drive in humans (Flaherty 2005). Farah et al. (2009) showed that Adderall, a mixed amphetamine salt licensed for the treatment of ADHD that increases DA in the brain, enhanced performance on the Remote Associates Task (RAT), a test of convergent thinking of creativity. It enhanced performance in healthy individuals with low creativity performance but impaired performance in healthy individuals with high creativity performance.

How does modafinil affect these cognitive processes? So far there are no studies investigating the effects of modafinil on creativity in healthy non-sleep deprived subjects. Walsh et al. (2004) reported improvements with modafinil on alertness, vigilance, and the verbal rather than figural versions of the Torrance Test of Creative Thinking Task during simulated nightshift study. The study reported that, after sleep deprivation, subjects on placebo were significantly worse in verbal flexibility but not in verbal fluency. However, the aim of Walsh et al.'s (2004) study was to investigate the effectiveness of modafinil after sleep deprivation resulting from shift work, not to study the effects of modafinil on creativity. Furthermore, no effects were reported on the nonverbal, divergent thinking tasks which are an important component of creativity process.

We also recently conducted a double-blind, placebo-controlled randomized trial that investigated the effects of modafinil on divergent and convergent thinking tasks of creativity in heathy non-sleep-deprived neurotypical participants (Mohamed 2014; see also Chapter 6 in this volume). The results of this trial show that modafinil significantly impaired the performance of the convergent thinking tasks of creativity, as measured by the RAT, in participants who scored high in the creativity personality trait while also significantly increasing the performance of the convergent thinking tasks in participants who scored low in the creativity personality trait. Modafinil administration reduced the performance of the divergent thinking tasks of creativity in all participants who took 200 mg modafinil, indicating that modafinil might reduce creativity thinking in healthy individuals. Hence, consistent with the Farah et al. (2009) study, our results show that, like Adderall, modafinil improved creativity as measured by the RAT in subjects having a low baseline creativity trait and impaired performance in subjects having a high baseline creativity trait (Mohamed 2014a; Chapter 6 in this volume).

As healthy individuals fall into a wide spectrum of normality, these results suggest that creativity may be improved by modafinil in some individuals and impaired in others. The theoretical rational for why psychostimulant drugs like modafinil might impair creativity in healthy individuals has been discussed elsewhere by the author (Mohamed 2014b,d).

7.4.1 Effects of modafinil on mood and subjective well-being

Peak plasma concentrations of modafinil have been obtained 2–3 h after oral administration, with an elimination half-life of 15 h (Wong et al. 1997). Modafinil can also have effects on mood and bodily symptoms. Indeed, Randall et al. (2003) administered 100 or 200 mg of modafinil and after 3 h tested healthy subjects' responses to scales investigating mood and bodily functions, before and after an extensive battery of cognitive tests. They reported that modafinil significantly elevated somatic anxiety, producing shaking, palpitations, dizziness, restlessness, muscular tension, physical tiredness, irritability, and greater increases of aggressive mood after undertaking cognitive tests. However, there is also evidence that modafinil increases subjective reports of vigour in patients with narcolepsy (Becker et al. 2004) and in healthy subjects (Baranski et al. 2004). There is currently no evidence on how in the long term modafinil impacts on subjective well-being in healthy individuals. Conclusive evidence on the effects of modafinil on mood is lacking and warrants further research.

7.5 Meta-analytic reviews

Two recent reviews of modafinil show that in single doses modafinil had moderately beneficial effects on attention and wakefulness (Minzenberg and Carter 2008; Repantis et al. 2010). They indicate that modafinil leads to moderate improvements in EF in non-sleep-deprived subjects and significant improvement in sleep-deprived subjects, while not producing improvements in mood, memory, and motivation. In repeated doses, modafinil had no effect on attention but elevated anxiety (Repantis et al. 2010).

A recent review of popular cognitive enhancers concluded that, in studies of non-ADHD adults, psychostimulant medications do not promote acquisition of new information and might impair performance on tasks that require adaptation, flexibility, and planning (Advokat 2010). The review concluded that the evidence did not support the claim that stimulants are cognitive enhancers (Advokat 2010). Hence, as real-world evidence on the enhancing potential of modafinil in healthy individuals is currently lacking, there is a need to develop targeted and sophisticated neuropsychological tasks that reliably measure real-world achievements, such as improvements in well-being, examination results, or driving performance. Until then it is prudent to conclude that the effects of modafinil are at best small in healthy individuals and that any effects on everyday cognitive performance are rather limited.

7.5.1 The toxicity and side effects of modafinil

Experimental studies have reported that modafinil has low liability for abuse (Deroche-Gamonet et al. 2002), and a lower risk of generating anxiety (Simon et al. 1994) and adverse cardiovascular effects than dexamphetamine. However, according to Cephalon, which used to manufacture modafinil, the drug may cause serious side effects that include a rash and allergic reactions (Kaper 2010; Provigil 2011), although the frequency of these effects is not documented. Other potential side effects from using modafinil include dry mouth, appetite restriction, gastrointestinal disturbances—including nausea, diarrhoea, constipation, and dyspepsia—abdominal pain, tachycardia, vasodilatation, chest pain, palpitation, headache including migraine, anxiety, sleep disturbances, dizziness, drowsiness, depression, confusion, paraesthesia, asthenia, visual disturbances, Stevens-Johnson syndrome, and toxic epidermal necrolysis (BNF 2010; Provigil 2011; Ryan et al. 2011). Any of these may need to be treated in a hospital and may be life threatening. However, the percentage of patients who discontinued modafinil because of side effects was only 0.8% (13 per 1585). This is particularly low given that these patients were taking modafinil on a daily basis (Provigil 2011). Nevertheless, there may be unknown long-term adverse effects, particularly when modafinil is used by adolescents whose brains are still in development. To date, no toxicity data exist for healthy individuals using modafinil as a cognitive enhancer. In the USA, Spiller et al. (2009) recently performed a retrospective multi-poison center chart review of patients from 11 US states who had a single substance ingestion of modafinil with follow-up to a known outcome for the years 2000–7. The findings indicated that the most frequently reported clinical effects were tachycardia, insomnia, agitation, dizziness, and anxiety. Seventeen percent of patients who took modafinil required a medical admission. In this case series study, clinical toxicity effects of modafinil overdoses were generally mild, predominantly tachycardia and central nervous system (CNS) toxicity. However, in a minority of patients, clinically significant effects occurred that warranted specific therapy. Since there is evidence that side effects of modafinil are higher in other patient groups relative to narcolepsy patients (Lavault et al. 2011), it is unclear how these toxicity results could be generalized to healthy, normally functioning individuals using modafinil for a variety of purposes. Moreover, safety concerns may have particular importance when modafinil is used by healthy individuals, given that beneficial effects in such individuals could be relatively small (Schöne-Seifert and Talbot 2010). There have been no systematic longitudinal scientific studies of the long-term effects of modafinil use in healthy people (Mohamed and Sahakian 2011). There are still further safety concerns about the risks of purchasing toxic or impure substances advertised as modafinil over the internet. It is still early days in the widespread use of modafinil to judge the risks of long-term use. In general, the early reports of all drug effects are likely to be positive. A drug needs to have been used by large numbers of people, often enough and for substantial periods of time, before less common, and any potentially more serious adverse effects of regular use, can be detected (Bell et al. 2012; Lucke et al. 2011). Existing systems of passive pharmacovigilance are poor at detecting these adverse effects, especially if a drug is being used off-label (Lucke et al. 2011). Hence, the above-reviewed evidence suggests that modafinil might have adverse effects on the healthy individual although the long-term consequence and to what extent that these adverse effects are life threatening are currently unknown.

7.6 **Conclusions**

This chapter has provided up-to-date information about modafinil's modes of action and its cognitive and affective effects in animals and in healthy individuals. Modafinil has small to moderate attention-enhancing effects on EF and WM in healthy individuals low in WM but does not seem to enhance cognition in healthy individuals who are functioning optimally or are high in IQ. The effects of modafinil on cognition are small, or at best moderate, and there is a lack of ecological validity for the transfer of cognitive-enhancing effects in laboratories to the real world. There is evidence that modafinil does not improve several well-known neuropsychological tasks in healthy individuals. Furthermore, some experimental studies suggest that modafinil increases latency of response in healthy individuals and in patients with ADHD. The effects of modafinil seem to be dose dependent and to obey the Yerkes–Dodson Law in that both high and low dose might either have no effect or impair cognition in healthy individuals. Modafinil might have a negative effect on mood. In repeated doses, modafinil may not improve attention. The current evidence suggests both negative and positive effects of modafinil on cognition, with very small cognitive benefits in healthy individuals. In conclusion, this work does not support the view that modafinil is a cognitive-enhancing drug. Instead, it suggests that modafinil might be a conventional psychostimulant. Further rigorous research is required on how modafinil affects diverse cognitive and affective functions including social cognition, subjective well-being, and ethical and moral reasoning in healthy individuals.

7.7 **Acknowledgments**

Dr. Ahmed Dahir Mohamed is a registered psychologist and is a post-doctoral fellow and adjunct lecture in Developmental psychology (at the School of Psychology in the University of Nottingham Malaysia Campus. He is currently a visiting fellow in neuroethics at the Bioethics Centre in the University of Otago, New Zealand. Dr. Mohamed would like to thank the Wellcome Trust for funding this research (Grant 086041/Z/08/Z). Many thanks to Professor Simon Baron Cohen and Professor Anthony J. Holland for the comments they provided on earlier drafts of this chapter and for the excellent supervision and support they provided during the author's doctoral training at the University of Cambridge and at the University of Oxford.

References

Advokat, C. 2010. What are the cognitive effects of stimulant medications? Emphasis on adults with attention-deficit/hyperactivity disorder (ADHD). *Neuroscience and Biobehavioral Reviews* **34**:1256–66.

Arendt, M., R. Rosenberg, I. Fjordback, J. Brandholdt, L. Foldagek, L. Sher, and P. Munk-Jørgensen. 2007. Testing the self-medication hypothesis of depression and aggression in cannabis-dependent subjects. *Psychological Medicine,* **37**:935–45.

Astafiev, S. V., A. Z. Snyder, G. L. Shulman, and M. Corbetta. 2010. Comment on "Modafinil shifts human locus coeruleus to low-tonic, high-phasic activity during functional MRI" and "Homeostatic sleep pressure and responses to sustained attention in the suprachiasmatic area". *Science* **328**:309.

Baddeley, A. 2010. Working memory. *Current Biology* **20**:R136–40.

Baranski, J. V., R. Pigeau, P. Dinich, and I. Jacobs. 2004. Effects of modafinil on cognitive and meta-cognitive performance. *Human Psychopharmacology: Clinical and Experimental* **19**:323–32.

Batir, A. 2009. Creativity induced by Dopamine Agonists in Parkinson's Disease: Abstract. *MDs 13th International Congress of Parkinson's Disease and Movement Disorders: Abstract Mo-163. Presented June 8, 2009.*

Becker, P. M., J. R. Schwartz, N. T. Feldman, and R. J. Hughes. 2004. Effect of modafinil on fatigue, mood, and health-related quality of life in patients with narcolepsy. *Psychopharmacology* **171**:133–9.

Bell, S., J. C. Lucke, and W. Hall. 2012. Lessons for enhancement from the history of cocaine and amphetamine use. *AJOB: Neuroscience* **3**:24–9.

Benerjee, D., M. V. Vitiellob, and R. R. Grunsteinc. 2004. Psychopharmacology for excessive daytime sleepiness. *Sleep Medicine Reviews* **8**:339–54.

Berridge, K., and T. Robinson. 1998. What is the role of dopamine in reward: hedonic impact, reward learning, or incentive salience? *Brain Research: Brain Research Reviews* **28**:309–69.

British National Formulary (BNF). 2010. British National Formulary-60. Available at http://bnf.org/bnf/bnf/current/ Accessed November 2010.

Breedlove, S. M., N. V. Watson, and M. R. Rosenzweig. 2010. *Biological psychology: an introduction to behavioral, cognitive, and clinical neuroscience,* sixth edition. Sinauer Associates.

Chen, C. R., W. M. Qu, M. H. Qiu, X. H. Xu, M. H. Yao, Y. Urade, and Z. L. Huang. 2007. Modafinil exerts a dose-dependent antiepileptic effect mediated by adrenergic [alpha]1 and histaminergic H1 receptors in mice. *Neuropharmacology* **53**:534–41.

Cooper, J. R., F. E. Bloom, and H. R. Roth. 2003. Dopamine. In *The bioechemical basis of neuropharmacology,* ed. J. R. Cooper, E. F. Bloom, and H. R. Roth. Oxford: Oxford University Press.

Deroche-Gamonet, V., M. Darnaudéry, L. Bruins-Slot, F. Piat, M. Le Moal, and P. V. Piazza. 2002. Study of the addictive potential of modafinil in naive and cocaine-experienced rats. *Psychopharmacology* **161**:387–95.

Evans, W. E., and M. V. Relling. 1999. Pharmacogenomics: translating functional genomics into rational therapeutics. *Science* **286**:487–91.

Farah, M. 2010. *Neuroethics: an introduction with readings.* Hong Kong: MIT Press.

Farah, M., H. Caroline, G. Sankoorikal, M. E. Smith, and A. Chatterjee. 2009. When we can enhance cognition with Adderall, do we sacrifice creativity? *Psychopharmacology* **203**:541–7.

Farah, M. J., J. Illes, R. Cook-Deegan, H. Gardner, E. Kandel, P. King, E. Parens, B. Sahakian, and P. R. Wolpe. 2004. Neurocognitive enhancement: what can we do and what should we do? *Nature Reviews: Neuroscience* **5**:421–5.

Ferraro, L., T. Antonelli, W. T. O'Connor, S. Tanganelli, F. Rambert, and K. Fuxe. 1997. The antinarcoleptic drug modafinil increases glutamate release in thalamic areas and hippocampus. *Neuroreport,* **8**, 2883–2887.

Ferraro, L., K. Fuxe, L. Agnati, S. Tanganelli, M. Tomasini, and T. Antonelli. 2005. Modafinil enhances the increase of extracellular serotonin levels induced by the antidepressant drugs fluoxetine and imipramine: a dual probe microdialysis study in awake rat. *Synapse* **55**:230–41.

Finlay, J. M., M. J. Zigmond, and E. D. Abercrombie. 1995. Increased dopamine and norepinephrine release in medial prefrontal cortex induced by acute and chronic stress: Effects of diazepam. *Neuroscience* **64**:619–28.

Flaherty, A. W. 2005. Frontotemporal and dopaminergic control of idea generation and creative drive. *Journal of Comparative Neurology* **493**:147–53.

Franke, A., C. Bagusat, P. Dietz, I. Hoffmann, P. Simon, R. Ulrich, and K. Lieb. 2013. Use of illicit and prescription drugs for cognitive or mood enhancement among surgeons. *BMC Medicine* **11**:102.

Franke, A. G., C. Bonertz, M. Christmann, M. Huss, A. Fellgiebel, E. Hildt, and K. Lieb. 2011. Non-medical use of prescription stimulants and illicit use of stimulants for cognitive enhancement in pupils and students in Germany. *Pharmacopsychiatry* **44**:60–6.

George, O., and G. F. Koob. 2010. Individual differences in prefrontal cortex function and the transition from drug use to drug dependence. *Neuroscience and Biobehavioral Reviews* **35**:232–47.

Greely, H. 2006a. The social effects of advances in neuroscience: Legal problems, legal prespectives. In *Neuroethics: defining the issues in Theory, Practice, and Policy*, ed. J. Illis. New York: Oxford University Press.

Greely, H., B. Sahakian, J. Harris, R. C. Kessler, M. Gazzaniga, P. Campbell, and M. J. Farah. 2008. Towards responsible use of cognitive-enhancing drugs by the healthy. *Nature* **456**:702–5.

Greely, H. T. 2006b. Neuroethics: defining the issues in theory, practice and policy. European Molecular Biology Organization.

Happe, S., W. Pirker, C. Sauter, G. Klosch, and F. Zeitlhofer. 2001. Successful treatment of excessive daytime sleepiness in parkinson's disease with modafinil. *Journal of Neurology* **248**:632–4.

Hennessey, B. A., and T. M. Amabile. 2010. Creativity. *Annual Review of Psychology* **61**:569–98.

Hilaire, Z. D. S., M. Orosco, C. Rouch, G. Blanc, and S. Nicolaidis. 2001. Variations in extracellular monoamines in the prefrontal cortex and medial hypothalamus after modafinil administration: a microdialysis study in rats. *Neuroreport* **12**:3533–7.

Husain, M., and M. A. Mehta. 2011. Cognitive enhancement by drugs in health and disease. *Trends in Cognitive Sciences* **15**:28–36.

Iversen, L. 2001. *Drugs: a very short story.* Oxford: Oxford University Press.

Iverson, L., S. Iverson, F. Bloom, and R. Roth. 2009. *Introduction to neuropsychopharmacology,* New York: Oxford University Press.

de Jongh, R., I. Bolt, M. Schermer, and B. Olivier. 2008. Botox for the brain: enhancement of cognition, mood and pro-social behavior and blunting of unwanted memories. *Neuroscience and Biobehavioral Reviews* **32**:760–76.

Kaper, R. 2010. Important safety information for Provigil and Nuvigil. Available at http://nuvigilrems.com/pdf/Dear%20HCP%20Letter.pdf. Accessed August 2011.

Koob, G. F., and N. D. Volkow. 2010. Neurocircuitry of Addiction. *Neuropsychopharmacology* **35**:217–38.

Kraft, G. H., and J. Bowen. 2005. Modafinil for fatigue in MS: a randomized placebo-controlled double-blind study. *Neurology* **65**:1995–7; author reply 1995–7.

Kupfermann, I., E. Kandel, and S. Iversen. 2000. Motivational and addictive states. In *Principles of neural science*, international edition, ed. E. R. Kandel, J. H. Schwartz, and T. M. Jessell, 998–1013. New York: McGraw-Hill.

Lavault, S., Y. Dauvilliers, X. Drouot, S. Leu-Semenescu, J.-L. Golmard, M. Lecendreux, P. Franco, and I. Arnulf. 2011. Benefit and risk of modafinil in idiopathic hypersomnia vs. narcolepsy with cataplexy. *Sleep Medicine* **12**:550–6.

Lucke, J. C., S. Bell, B. Partridge, and W. D. Hall. 2011. Deflating the Neuroenhancement Bubble. *AJOB Neuroscience* **2**:38–43.

Lynch, G., L. C. Palmer, and C. M. Gall. 2011. The likelihood of cognitive enhancement. *Pharmacology Biochemistry and Behavior* **99**:116–29.

McGough, J., M. J. Mancino, Z. Feldman, M. P. Chopra, W. B. Gentry, C. Cargile, and A. Oliveto. 2009. Open-label pilot study of modafinil for methamphetamine dependence. *Journal of Clinical Psychopharmacology* **29**:488–91.

Madras, B., Z. Xie, A. Jassen, H. Panas, L. Lynch, R. Johnson, E. Livni, T. Spencer, A. Bonab, G. Miller, and A. Fischman. 2006. Modafinil occupies dopamine and norepinephrine transporters in vivo and modulates the transporter and trace amine activity in vitro. *Journal of Pharmacology and Experimental Therapeutics* **319**:561–9.

Marchant, N., F. Kamel, K. Echlin, J. Grice, M. Lewis, and J. Rusted. 2009. Modafinil improves rapid shifts of attention. *Psychopharmacology* **202**:487–95.

Minzenberg, M. J., and C. S. Carter. 2008. Modafinil: a review of neurochemical action and effects on cognition. *Neuropsychopharmacology* **33**:1477–1502.

Minzenberg, M. J., A. J. Watrous, J. H. Yoon, S. Ursu, and C. S. Carter. 2008. Modafinil shifts human locus coeruleus to low-tonic, high-phasic activity during functional MRI. *Science* **322**(5908):1700–2.

Mohamed, A. D. 2012. Modafinil has the potential for addiction. *AJOB Neuroscience* **3**:36–8.

Mohamed, A. D. 2014a. The effects of modafinil on convergent and divergent thinking of creativity: a randomized controlled trial. *Journal of Creative Behavior*, doi: 10.1002/jocb.73 .

Mohamed, A. D. 2014b. Neuroethical issues in pharmacological cognitive enhancement. *Wiley Interdisciplinary Reviews: Cognitive Science*, doi: 10.1002/wcs.1306. Available at http://onlinelibrary.wiley.com/doi/10.1002/wcs.1306/full.

Mohamed, A. D. 2014c. Neuroethical issues in pharmacological cognitive enhancement. *Wiley Interdisciplinary Reviews: Cognitive Science* **5**:533–49.

Mohamed, A. D. 2014d. Reducing creativity with psychostimulants may debilitate mental health and well-being. *Journal of Creativity in Mental Health* **9**:146–63.

Mohamed, A. D. 2015. The effects of modafinil on motivation and salience of pleasure in healthy individuals: quantitative evidence from the cognitive neurosciences. *AJOB Neuroscience* **6**:15–17.

Mohamed, A. D., and C. R. Lewis. 2014. Modafinil increases the latency of response in the Hayling Sentence Completion Test in healthy volunteers: a randomised controlled trial. *PLoS ONE* **9**:e110639.

Mohamed, A. D., U. Müller, C. Lewis, J. Rowe, T. Rittman, T. W. Robbins, and B. J. Sahakian. 2011. The effects of modafinil on cold cognition, creativity and motivation in healthy subjects. Presented at British Association of Psychopharmacology Summer Meeting 2011, 24–27 Jul 2011, MB14 Harrogate- July 2011: Winner of the British Association of Psychopharmacology Summer Bursary Award. Available at http://cdn.f1000.com/posters/docs/106020298.

Mohamed, A. D., and B. J. Sahakian. 2011. The ethics of elective psychopharmacology. *International Journal of Neuropsychopharmacology* FirstView, 1–13.

Müller, U., N. Steffenhagen, R. Regenthal, and P. Bublak. 2004. Effects of modafinil on working memory processes in humans. *Psychopharmacology*, **177**, 161–169.

Murillo-Rodríguez, E., R. Haro, M. Palomero-Rivero, D. Millán-Aldaco, and R. Drucker-Colín. 2007. Modafinil enhances extracellular levels of dopamine in the nucleus accumbens and increases wakefulness in rats. *Behavioural Brain Research* **176**(2):353–7.

Nguyen, T. L., Y. H. Tian, I. J. You, S. Y. Lee, and C. G. Jang. 2011. Modafinil-induced conditioned place preference via dopaminergic system in mice. *Synapse* **65**:733–41.

Norman, C., and M. Berger. 2008. Neuroenhancment: status quo and perspectives. *European Archives of Psychiatry and Clinical Neuroscience* **258**:110–14.

Partridge, B. J., S. K. Bell, J. C. Lucke, S. Yeates, and W. D. Hall. 2011. Smart drugs "as common as coffee": media hype about neuroenhancement. *PLoS ONE* **6**:e28416.

Provigil 2011. Modafinil data sheet. Data for the first authorization. Available at http://www.cephalon.com/Documents/pdfs/PROVIGIL_Fact_Sheet.pdf. Accessed July 2011.

Qu, W.-M., Z.-L. Huang, X.-H. XU, N. Matsumoto, and Y. Urade. 2008. Dopaminergic D1 and D2 receptors are essential for the arousal effect of modafinil. *Journal of Neuroscience* **28**:8462–9.

Ramos, B. P., and A. F. T. Arnsten. 2007. Adrenergic pharmacology and cognition: Focus on the prefrontal cortex. *Pharmacology and Therapeutics* **113**:523–36.

Randall, D. C., F. H. Cafferty, J. M. Shneerson, I. E. Smith, M. Llewelyn, and S. E. File. 2005a. Chronic treatment with modafinil may not be beneficial in patients with chronic fatigue syndrome. *Journal of Psychopharmacology* **19**:647–60.

Randall, D. C., N. L. Fleck, J. M. Shneerson, and S. E. File. 2004. The cognitive-enhancing properties of modafinil are limited in non-sleep-deprived middle-aged volunteers. *Pharmacology Biochemistry and Behavior* 77:547–55.

Randall, D. C., J. M. Shneerson, and S. E. File. 2005b. Cognitive effects of modafinil in student volunteers may depend on IQ. *Pharmacology Biochemistry and Behavior* **82**:133–9.

Randall, D. C., J. M. Shneerson, K. K. Plaha, and S. E. File. 2003. Modafinil affects mood, but not cognitive function, in healthy young volunteers. *Human Psychopharmacology: Clinical and Experimental* **18**:163–73.

Randall, D. C., A. Viswanath, P. Bharania, S. M. Elsabagh, D. E. Hartley, J. M. Shneerson, and S. E. File. 2005c. Does modafinil enhance cognitive performance in young volunteers who are not sleep-deprived? *Journal of Clinical Psychopharmacology* **25**:175–9.

Rasetti, R., V. S. Mattay, B. Stankevich, K. Skjei, G. Blasi, F. Sambataro, I. C. Arrillaga-Romany, T. E. Goldberg, J. H. Callicott, J. A. Apud, and D. R. Weinberger. 2010. Modulatory effects of modafinil on neural circuits regulating emotion and cognition. *Neuropsychopharmacology* **35**:2101–9.

Rees, G., P. M. Gollwitzer, and J. A. Bargh. 1999. The psychology of action: linking cognition and motivation to behavior. *Journal of Psychophysiology* **13**:199–200.

Repantis, D., P. Schlattmann, O. Laisney, and I. Heuser. 2010. Modafinil and methylphenidate for neuroenhancement in healthy individuals: a systematic review. *Pharmacological Research* **62**:187–206.

Reuter, M., S. Roth, K. Holve, and J. Hennig. 2006. Identification of first candidate genes for creativity: A pilot study. *Brain Research* 1069:190–7.

Ryan, J. B., A. Katsiyannis, and E. M. Hughes. 2011. Medication treatment for attention deficit hyperactivity disorder. *Theory into Practice* **50**:52–60.

Sandberg, A., and J. Savulescu. 2011. The social and economic impacts of cognitive enhancement. In *Enhancing human capacities*, ed. J. Savulescu, R. ter Meulen, and G. Kahane. Oxford: Wiley-Blackwell.

Schöne-Seifert, B., and D. Talbot. 2010. (Neuro-)enhancement. In *Ethics in psychiatry*, ed. H. Helmchen and N. Sartorius. Amsterdam: Springer.

Scoriels, L., J. H. Barnett, G. K. Murray, S. Cherukuru, M. Fielding, F. Cheng, B. R. Lennox, B. J. Sahakian, and P. B. Jones. 2011. Effects of modafinil on emotional processing in first episode psychosis. *Biological Psychiatry* **69**:457–64.

Simon, P., C. Panissaud, and J. Costentin. 1994. The stimulant effect of modafinil on wakefulness is not associated with an increase in anxiety in mice A comparison with dexamphetamine. *Psychopharmacology* **114**:597–600.

Smith, M. E., and M. J. Farah. 2011. Are prescription stimulants "smart pills"? The epidemiology and cognitive neuroscience of prescription stimulant use by normal healthy individuals. *Psychological Bulletin* **137**:717–41.

Spiller, H. A., D. Borys, J. R. K. Griffith, W. Klein-Schwartz, A. Aleguas, D. Sollee, D. A. Anderson, and T. S. Sawyer. 2009. Toxicity from modafinil ingestion. *Clinical Toxicology* **47**:153–6.

Stankoff, B., E. Waubant, C. Confavreux, G. Edan, M. Debouverie, L. Rumbach, T. Moreau, J. Pelletier, C. Lubetzki, and M. Clanet; French Modafinil Study Group. 2005. Modafinil for fatigue in MS: a randomized placebo-controlled double-blind study. *Neurology* **64**(7):1139–43.

Sugden, C., C. R. Housden, R. Aggarwal, B. J. Sahakian, and A. Darzi. 2011. Effect of pharmacological enhancement on the cognitive and clinical psychomotor performance of sleep-deprived doctors: a randomized controlled trial. *Annals of Surgery* **255**:222–7.

Turner, D. C., L. Clark, J. Dowson, T. W. Robbins, and B. J. Sahakian. 2004a. Modafinil improves cognition and response inhibition in adult attention-deficit/hyperactivity disorder. *Biological Psychiatry* **55**:1031–40.

Turner, D. C., L. Clark, E. Pomarol-Clotet, P. McKenna, T. W. Robbins, and B. J. Sahakian. 2004b. Modafinil improves cognition and attentional set shifting in patients with chronic schizophrenia. *Neuropsychopharmacology* **29**:1363–73.

Turner, D. C., T. W. Robbins, A. R. Aron, J. Dowson, and B. J. Sahakian. 2003a. Cognitive enhancement effects of modafinil in healthy volunteers. *Psychopharmacology* **165**(3):260–9.

Turner, D. C., T. W. Robbins, J. Dowson, and B. J. Sahakian. 2003b. Modafinil and adult ADHD: effects of a novel cognitive enhancer. *Journal of Psychopharmacology* **17**:A71–a71.

Tyne, H. L., J. Taylor, G. A. Baker, and M. J. Steiger. 2010. Modafinil for Parkinson's disease fatigue. *Journal of Neurology* **257**(3):452–6.

Vaishnavi, S., K. Gadde, S. Alamy, W. Zhang, K. Connor, and J. R. T. Davidson. 2006. Modafinil for atypical depression: effects of open-label and double-blind discontinuation treatment. *Journal of Clinical Psychopharmacology* **26**:373–8.

Vastag, B. 2004. Poised to challenge need for sleep, "wakefulness enhancer" rouses concerns. *JAMA* **291**:167–70.

Vijayraghavan, S., M. Wang, S. G. Birnbaum, et al. 2007. Inverted-U dopamine D1 receptor actions on prefrontal neurons engaged in working memory. *Nature Neuroscience* **10**:376–84.

Volkow, N. D., and R. D. Baler. 2014. Addiction science: uncovering neurobiological complexity. *Neuropharmacology* **76**(Part B):235–49.

Volkow, N. D., J. S. Fowler, J. Logan, D. Alexoff, W. Zhu, F. Telang, G. J. Wang, M. Jayne, J. M. Hooker, C. Wong, B. Hubbard, P. Carter, D. Warner, P. King, C. Shea, Y. Xu, L. Muench, and K. Apelskog-Torres. 2009. Effects of modafinil on dopamine and dopamine transporters in the male human brain: clinical implications. *JAMA* **301**:1148–54.

Volkow, N. D., J. S. Fowler, and G.-J. Wang. 1999. Imaging studies on the role of dopamine in cocaine reinforcement and addiction in humans. *Journal of Psychopharmacology* **13**:337–45.

Volkow, N. D., J. S. Fowler, G. J. Wang, and J. M. Swanson. 2004. Dopamine in drug abuse and addiction: results from imaging studies and treatment implications. *Molecular Psychiatry* **9**:557–69.

Walsh, J. K., A. C. Randazzo, K. L. Stone, and P. K. Schweitzer. 2004. Modafinil improves alertness, vigilance, and executive function during simulated night shifts. *Sleep* **27**:431–9.

Wesensten, N. J. 2006. Effect of modafinil on cognitive performance and alertness during sleep deprivation. *Current Pharmaceutical Design* **12**:2457–71.

Winder-Rhodes, S., S. R. Chamberlaine, M. Idris, T. W. Robbins, B. J. Sahakian, and U. Müller. 2009. Effects of modafinil and prazosin on cognitive and physiological functions in healthy volunteers. *Journal of Psychopharmacology* **24**(11):1649–57.

Wise, R. A. 2004. Dopamine, learning and motivation. *Nature Reviews Neuroscience* **5**:483–94.

Wisor, J. P., and K. S. Eriksson. 2005. Dopaminergic—adrenergic interactions in the wake promoting mechanism of modafinil. *Neuroscience* **132**:1027–34.

Wong, Y., S. Gorman, G. McCormick, and P. Grebow. 1997. Single-dose pharmacokinetics of modafinil and methylphenidate given alone in and in combination to healthy male volunteers. *Sleep Research* **26**:133.

Yang, B., R. C. K. Chan, J. Jing, T. Li, P. Sham, and R. Y. L. Chen. 2007. A meta-analysis of association studies between the 10-repeat allele of a VNTR polymorphism in the 3′-UTR of dopamine transporter gene and attention deficit hyperactivity disorder. *American Journal of Medical Genetics Part B: Neuropsychiatric Genetics* **144B**:541–50.

Yerkes, R. M., and J. D. Dodson. 1908. The relation of strength of stimulus to rapidity of habit-formation. *Journal of Comparative Neurology and Psychology* **18**:459–82.

Zifko, U. A., M. Rupp, S. Schwarz, H. T. Zipko, and E. M. Maida. 2002. Modafinil in treatment of fatigue in multiple sclerosis: results of an open-label study. *Journal of Neurology* **249**(8):983–7.

Chapter 8

Cognitive enhancement using noninvasive brain stimulation: weighing opportunity, feasibility, and risk

Priyanka P. Shah-Basak and Roy H. Hamilton

8.1 Introduction

Humans have always sought to create implements that augment our intrinsic abilities. In the distant past we designed simple tools to amplify the strength of our muscles. The creation of writing greatly leveraged our language-processing abilities in order to dramatically increase the scope and impact of communication. In the industrial age machines allowed us to work many times faster than bare hands would ever allow. More recently the information-processing ability of the human brain was extended almost immeasurably by the introduction of the computer, and our communication abilities took another quantum leap forward with the advent of the internet. Now, as both science and society pay increasing attention to the brain as one of the last great frontiers for exploration, we are starting to develop and understand new tools to enhance our abilities. However, these new implements are fundamentally different from all others that preceded them. Whereas the prior products of our ingenuity augmented our abilities by duplicating or extending the work of our hands and minds, these new techniques have the capacity to boost human abilities directly at the source by improving brain function. These approaches include psychopharmacology, cognitive training, and noninvasive brain stimulation (NBS), the subject of this chapter.

NBS techniques have emerged in the past 30 years as powerful neuroscientific tools that allow focal manipulation of brain function and have been advantageous for elucidating structure function relationships in the brain (Miniussi et al. 2013). The ability to focally alter brain activity also allows for direct intervention with respect to neural function. This has led to the emergence of now accepted therapies (for example, treatment for depression) as well as research investigations into a number of promising clinical applications of NBS for both psychiatric and neurologic conditions. However, the ability to noninvasively and focally manipulate neural function—particularly as it pertains to cognitive performance—is also leading a growing number of investigators and an increasingly curious public to explore the possibility of employing NBS to enhance normal human intellect.

In this chapter, we will discuss two emerging questions: First, can NBS be employed reliably to enhance cognition in healthy individuals? Secondly, what are the risks? In addressing the first question, we will introduce readers to various NBS techniques, with a specific focus on transcranial direct current stimulation (tDCS) because it is the technique most likely to be employed for cognitive enhancement for the foreseeable future. We will review experimental data from cognitive neuroscience supporting the notion that NBS, and specifically tDCS, can be employed to transiently enhance aspects of cognition. As part of this discussion, we will consider what is known and not known about the treatment parameters of tDCS. We will summarize a smaller, but potentially promising set of studies that indicate that repeated stimulation may have lasting salutary effects on certain normal cognitive functions. In addressing the possible risks of using NBS to enhance normal cognition, we will first consider the environments in which the demand for optimal performance may lead to the implementation of NBS in healthy individuals. We will then discuss the potential risks associated with cognitive enhancement using NBS and also address some of the limits of what is currently known about these technologies. Finally, we will think about the future of NBS as a cosmetic cognitive enhancement tool and consider the obligation of experts in the fields of neuroscience, public health, and public policy to find the appropriate balance between ensuring public safety and respecting the autonomy of individuals who wish to manipulate their brains using NBS.

8.2 **Noninvasive brain stimulation techniques**

Recent advances in neurotechnologies such as functional magnetic resonance imaging (fMRI) and NBS have greatly enhanced our ability to understand and manipulate brain activity as it pertains to specific cognitive functions. In doing so, these technologies have laid the groundwork for focused cognitive enhancement in healthy individuals (Coffman et al. 2014). In recent years, the body of evidence indicating that NBS can be used to safely enhance cognition has grown steadily. As opposed to invasive techniques like deep brain stimulation, which requires surgical implantation of a stimulation device, NBS allows for manipulation of brain function using components placed outside of the scalp. NBS can be classified into two types of technologies: transcranial magnetic stimulation (TMS) and transcranial electrical stimulation (tES). These techniques differ in several important ways, both mechanistically in terms of their effects on brain function and methodologically with respect to their implementation. Within the category of tES (Paulus 2011), the most commonly employed method is transcranial direct current stimulation, although transcranial random noise stimulation (tRNS) and transcranial alternating current stimulation (tACS) have also recently emerged as promising methods for cognitive enhancement.

In the sections that follow, readers will be introduced to different NBS techniques and the key parameters pertaining to their implementation (e.g., stimulation location, intensity, duration). Importantly, after a brief discussion of TMS, we will focus largely on tDCS, since this technology has emerged in recent years as the main approach being used experimentally by both investigators and amateur stimulation enthusiasts to enhance normal cognition. To that end, this section will include a discussion of both the logistical and technical advantages and challenges surrounding the use of tDCS for enhancement.

8.2.1 Transcranial magnetic stimulation (TMS)

TMS is based on the principle of electromagnetic induction first discovered by Michael Faraday in 1831 (Barker et al. 1985; Walsh and Pascual-Leone 2003). The device consists of a stimulator unit and a TMS coil. The stimulator, made of capacitors that store large electrical charge, is connected to a casing with a coil of copper wires, which form the TMS coil. In order to deliver TMS, the coil is held tangentially outside the scalp near the brain area to be stimulated. Stored charge in the stimulator is discharged to the coil, thereby producing a brief and time-varying magnetic field at the scalp. This magnetic field penetrates the skull, and depending on stimulation intensity and the shape and orientation of the coil, an electrical current is generated in the cortical neurons underneath the coil. This current is sufficient to depolarize neuronal membranes and trigger action potentials. Stimulation can be targeted precisely to specific brain regions using neuronavigational systems configured with magnetic resonance imaging (MRI) scans of subjects' brains. This approach achieves a spatial resolution of a few millimeters. Other less precise approaches to targeting brain regions are also often employed, including demarcation of targets on the scalp based on the 10- to 20-measurement system normally used to place electroencephalogram (EEG) electrodes or based on the presumed distance between the desired cortical target and a site on the brain where TMS can be used to elicit a predictable functional response (for example, the region of the motor cortex representing the hand) (Pascual-Leone et al. 1992; Rossini et al. 1994; Rothwell et al. 1991). TMS can be delivered either via single pulses, one pulse delivered at a time separated by a few seconds or minutes, or repetitively at a fixed number of pulses per second (repetitive TMS or rTMS). Typically, low-frequency rTMS (< 5 hertz [Hz]) is characterized by decreased cortical excitability, whereas high-frequency rTMS (≥ 5 Hz) is characterized by enhanced excitability (Fitzgerald et al. 2006; Pascual-Leone et al. 1998; Rossi et al. 2000). Both high- and low-frequency rTMS have been associated with persistent after-effects on cortical excitability and subsequent neural function (Hoogendam et al. 2010; Ziemann 2001).

8.2.2 Transcranial direct current stimulation

Transcranial direct current stimulation is a relatively new technique, developed in the late 1990s. Typically, tDCS is administered by delivering small electric currents (1–2 milliamperes (mA)) to the scalp using a battery-operated device connected to two relatively large (often 5×7 cm^2 or 5×5 cm^2) saline-soaked surface electrodes (Nitsche and Paulus 2000). The duration of stimulation is often limited to 10–30 min within a session. Weak electrical current delivered in this way must penetrate the skin, skull, and meninges before it reaches the brain. This current is insufficient to generate action potentials, which is one of the major mechanistic differences between TMS and tDCS. The current generated by tDCS instead incrementally alters neuronal transmembrane potentials, thereby changing the likelihood that neurons will trigger action potentials over time. Thus, tDCS is often conceptualized as a neuromodulatory rather than a neurostimulatory technique (Shah et al. 2013).

The direction of the current flow, also referred to as polarity, is dependent on the placement of positive (anode) and negative (cathode) electrodes on the subject. Although current passes through both electrodes, by convention the electrode that is being used to target the brain regions of interest is referred to as the "active" electrode; the other electrode—termed the "reference" electrode by convention—is placed either on a site on the scalp at a distance from active electrode (a so-called "cephalic" reference electrode) or somewhere on the body or on the head but off the scalp (an "extracephalic" electrode). Examples of commonly applied cephalic reference electrode sites are the supraorbital region (forehead) or the mastoid (behind the ear), usually placed contralateral to the active electrode; examples of extracephalic sites (Im et al. 2012) include the deltoid (shoulder) or buccinator muscle (chin). Current passes from the anode, through the cortex, and out at the cathode (Hamilton et al. 2011). Evidence based on polarity-specific tDCS effects in animal and single-cell studies suggests that the neurons (the soma) in close proximity to the anode exhibit increased depolarization across their membranes, while those close to the cathode exhibit increased hyperpolarization. In humans, studies of the effects of tDCS on the primary motor cortex suggest that anodal stimulation increases cortical excitability in underlying brain regions, while cathodal stimulation decreases cortical excitability in underlying regions (Nitsche and Paulus 2000, 2001). However, more recent evidence argues against this simple relationship between polarity and cortical excitability, particularly with current strengths greater than 1 mA and for durations greater than 20 min (Batsikadze et al. 2013).

Most of what is known thus far about polarity-specific effects of tDCS on cortical excitability in humans is based upon induced neuroplastic changes in primary motor areas (Nitsche and Paulus 2000; Stagg and Nitsche 2011). However, recent evidence suggests that the polarity-specific effects of tDCS on complex cognitive tasks may be more variable than on motor physiology (Jacobson et al. 2012), which may have implications for our understanding of tDCS as it relates to cognitive enhancement. Besides polarity, other factors may significantly influence the neural response to stimulation, including the orientation of brain structures such as gyri and sulci with respect to current flow, the electrophysiological state of brain areas *before* stimulation, and the cognitive tasks employed either *during* or *immediately after* stimulation (Dockery et al. 2009b; Javadi and Cheng 2013; Miniussi et al. 2013).

One important difference between TMS and tDCS is the spatial resolution of the two technologies. Because conventional tDCS employs large electrodes, the precision of this device is less than that associated with magnetic brain stimulation. Moreover, as noted in section 8.2.1, TMS is often implemented in conjunction with neuronavigational systems that enable precision at the level of several millimeters. By contrast, electrode placement for tDCS typically follows the less precise 10-20 EEG International System. Therefore, localization of brain areas is not precise. However, it may be the case that the imprecise nature of tDCS may provide some benefit over a more focal TMS method in enhancing cognitive abilities that are represented in the brain in more distributed networks. Recently, high-definition tDCS devices have also been developed, which may enable much more precise targeting of cortical electrical stimulation (Datta et al. 2009; Edwards et al. 2013; Kuo et al. 2013).

Despite its lower spatial resolution, several factors make tDCS a more attractive NBS approach than TMS for boosting cognition. The first is that tDCS has fewer safety concerns. TMS is associated with a small but real risk of inducing seizures. This is in part because TMS relies on suprathreshold stimulation of neuronal populations rather than incremental manipulation of neural excitability. By contrast, tDCS is safe and well tolerated by both healthy subjects and patient populations, and there are no reports of seizures or persistent adverse effects associated with tDCS. Mild side effects such as itching, tingling, mild burning sensations, and less frequently headaches have been reported, but these effects do not last long past the period of stimulation (Brunoni et al. 2012; Kessler et al. 2012). However, as we will discuss in the sections that follow, there may be safety issues imposed by repeated use of tDCS that are not yet well understood. Aside from safety, the ease of using tDCS compares favorably with TMS. While TMS is cumbersome and expensive and can require bulky auxiliary equipment, tDCS is small and portable. It can therefore be applied concurrently with cognitive/behavioral tasks for associative training during stimulation, since subjects can freely move their heads during tDCS. TMS is considerably more limited in this regard, as subjects are generally required to sit still. Given its many advantages over TMS, we strongly suspect that tDCS (rather than TMS) will be employed as a tool for cognitive enhancement in the foreseeable future. Therefore, in the following sections, we will focus on experimental evidence surrounding the use of tDCS to boost cognitive function in healthy individuals.

8.2.3 Transcranial random noise and alternating current stimulation

tRNS stimulation differs from tDCS in that instead of constant direct current delivery, current levels are randomly generated, with a normal distribution around a specific mean intensity. Other parameters related to the stimulation electrodes, like position and size, are similar to tDCS (Antal et al. 2008; Paulus 2011; Terney et al. 2008). Terney et al. (2008) was the first group to apply tRNS in humans, and successfully enhanced motor cortical excitability for up to an hour after only 10 min of stimulation. The robust changes in cortical excitability observed after tRNS have been be attributed to repeated opening and closing of sodium channels in stimulated neuronal populations. In addition, tRNS may influence cortical oscillations, leading to changes in excitability. These proposed mechanisms are potentially consistent with the observation that reversing electrode polarities in tRNS does not appear to interfere with the augmentation in cortical excitability, suggesting that tRNS-induced changes in cortical excitability are independent of direction of current flow (Paulus 2011; Terney et al. 2008).

Transcranial alternating current stimulation (tACS) is a frequency-specific stimulation method that is also thought to influence oscillatory neuronal activity. This method differs from tRNS in that a sinusoidal current is applied at a fixed frequency rather than a randomly presented range of frequencies. Often, tACS is applied at frequencies that mirror the predominant frequency bands observed in EEG in different regions of the brain (Antal et al. 2008; Pogosyan et al. 2009; Zaehle et al. 2010). The body of research employing tRNS (Cappelletti et al. 2013; Fertonani et al. 2011; Mulquiney et al. 2011) and tACS (Antal and Paulus 2013; Jausovec et al. 2014; Zaghi et al. 2010) to enhance cognition is more limited than that supporting tDCS, but is growing (Kuo and Nitsche 2012).

8.3 **Which abilities can we augment?**

It is important for the purpose of our discussion to clarify what we consider to be cognitive enhancement, and how this differs from clinical interventions designed to improve cognition. At the heart of this distinction is the idea that therapies treat diseases, while enhancement improves normal abilities (Chatterjee 2013). We and others have referred to this notion of boosting the neural functions of healthy individuals as "cosmetic neurology" (Chatterjee 2004, 2013; Hamilton et al. 2011). In the past decade, tDCS has been used increasingly in experimental settings to transiently enhance performance in a variety of cognitive domains. The majority of investigators have been primarily interested in briefly altering brain function as a way to further elucidate structure–function relationships underlying cognition. However, this growing body of experimental evidence also suggests that tDCS-induced neuroplastic changes could potentially be used cosmetically to enhance mental abilities such as executive functions, language, learning and memory, and visuospatial processing.

Many cognitive neuroscience studies involving tDCS share basic similarities in approach. Using a variety of sources of evidence including lesion studies and fMRI, brain areas believed to be integral to the cognitive operation in question are first identified. Stimulation is applied to the regions in question using anodal and/or cathodal stimulation; sham stimulation is often included to control for placebo effects. Manipulation of cognitive function is verified using behavioral tests, typically by comparing performance after active stimulation to performance before stimulation and to sham stimulation. During sham, current is usually turned off within the first 30 sec of stimulation and it remains off throughout the stimulation duration. Since most subjects habituate to the sensations associated with tDCS after a few seconds, subjects typically cannot reliably distinguish between active and sham stimulation, especially at current strengths less than 1.5 mA (Ambrus et al. 2012; Gandiga et al. 2006; O'Connell et al. 2012; Russo et al. 2013).

One presumed advantage of NBS compared with pharmacological approaches is their specificity, in that TMS, tDCS, and related approaches selectively manipulate brain function, stimulating specific neural structures associated with particular behaviors. By contrast, while medications can selectively affect neurotransmitter systems, they generally do not manipulate the activity of circumscribed anatomical regions and can thus have effects on brain areas that are irrelevant to the cognitive operation in question. These agents frequently have systemic effects outside the brain as well. In the sections that follow, we will survey recent cognitive neuroscience studies in which tDCS has been used to manipulate and enhance cognition in a variety of cognitive domains. As we do so, we also will focus on three aspects of tDCS that define the specificity of its effects: (1) polarity (anodal vs. cathodal), (2) cognitive resolution (i.e., the ability of tDCS at a site to manipulate one form of mental activity and not another), and (3) regional specificity (i.e., the ability to affect the brain area being stimulated and not unrelated, distant areas).

8.3.1 Executive functions and affective control

Executive functions broadly describe a number of control processes involved in governing and optimizing complex cognitive operations. Working memory—the ability to hold and manipulate information—is an important and often studied component of executive function (Baddeley 1986). Enhancement in executive functions, especially working memory, is frequently attempted with tDCS, in part because it is believed that strengthening in this cognitive domain may transfer to more global improvement of cognition and function (Klingberg et al. 2002). A sizeable body of evidence indicates that facilitating activity in the left dorsolateral prefrontal cortex (DLPFC)—a region critically involved in working memory—using anodal tDCS (atDCS) improves performance in healthy adults. Fregni et al. (2005) was one of the first groups to report improved working memory performance after a single session of atDCS. Subjects were more accurate on a 3-back letter working memory task. The effects were polarity specific, since accuracy was not affected by cathodal (ctDCS) or by sham tDCS. Anodal tDCS of a control site (motor cortex) also did not impact performance, supporting the regional specificity of the stimulation effect. Several other subsequent tDCS studies have also demonstrated similar benefits in tasks that rely upon working memory (Ohn et al. 2008; Zaehle et al. 2011).

Outside of working memory, the polarity-specific effects of tDCS on other executive functions are less clear. For instance, Dockery et al. (2009b) demonstrated differential aftereffects of atDCS and ctDCS on a cognitive task requiring skill learning and planning abilities in healthy adults. While atDCS improved performance (faster reaction times) during the late learning stages, ctDCS improved performance during the acquisition and early learning stages; these benefits persisted for up to a year after termination of treatment with tDCS. The authors speculated that ctDCS in this case may have acted as a "neuronal noise reducer" during acquisition of a novel task, thereby facilitating learning, while atDCS may have enhanced performance on an already learned task. Therefore, the cognitive task being performed *during* tDCS may impact how different polarities of stimulation influence activity in brain areas being stimulated.

In line with Dockery et al. (2009b), some studies have demonstrated improved performance in executive functions in association with "inhibitory" ctDCS. In one such study (Filmer et al. 2013), the ability to "multitask" was facilitated in healthy individuals with ctDCS administered over the posterior lateral PFC. Multitasking behavior was elicited by engaging subjects in a dual task that required them to respond to both visual and auditory stimuli. Subjects reacted to stimuli in the dual task more rapidly immediately after receiving ctDCS, while atDCS and sham did not provide any benefit. While ctDCS improved performance on a dual task, both atDCS and ctDCS improved performance under a single-task control condition. In other words, *after* administration of tDCS, contrasting polarity-specific effects were observed in response to increasing task demands (single vs. dual task). In another study using ctDCS, Chrysikou et al. (2013) investigated whether inducing hypofrontality could foster flexible and novel thought in healthy individuals. Subjects were presented with pictures of everyday objects and asked to articulate a common or an uncommon use for them. After a single session of ctDCS over left ventrolateral PFC, subjects exhibited improved reaction times and reduced omissions when thinking of uncommon uses for objects; there was no benefit of sham or of ctDCS over the right homolog site. Performance on a control task—forward digit span—that also engages the frontal lobe remained unchanged after ctDCS. These studies and others like them (Dockery et al. 2009a; Filmer et al. 2013) suggest that ctDCS may not always suppress cognitive functions, but may, in the right circumstances, also enable cognitive enhancement.

Some other noteworthy applications of tDCS involving manipulation of executive functions include enhancement of cognitive set-shifting performance using atDCS over left DLPFC (Leite et al. 2011) and the use of ctDCS over the anterior PFC to enhance deceptive behavior (Karim et al. 2010). In addition to unilateral stimulation, dual-hemispheric tDCS, wherein one target receives atDCS while a contralateral homolog site receives ctDCS, has also been shown to manipulate executive functions in healthy individuals. For instance, reduced risk-taking behavior was reported by Fecteau et al. (2007) after bilateral inhibition/excitation of the left and right DLPFC compared with sham.

Stimulation of the PFC is currently being explored as a potential treatment for disorders of mood, affect, and anxiety, and as a technique for alleviating chronic pain (Boggio et al. 2008a; Fregni et al. 2006). However, preliminary evidence suggests that the ability of tDCS to enhance emotional states in healthy individuals is mixed. On one hand, Boggio et al. (2009) found that atDCS of left DLPFC resulted in reduced unpleasant perception of images and reduced emotional discomfort compared with pre-tDCS performance, sham, and two active control sites (left visual and motor cortex); in this and in other studies, use of active stimulation control conditions militates against the notion that improved performance resulted simply from a generalized increase in arousal or attention related to the scalp sensation or other nonspecific phenomena associated with tDCS. On the other hand, more recent evidence suggests that enhancing emotional states in healthy individuals with tDCS may be more challenging. For example, Nitsche et al. (2012) found that atDCS over left DLPFC improved emotion processing but did not improve subjective emotional states. Bilateral modulation of DLPFC in another recent study did not induce any stimulation-specific mood changes in healthy individuals (Plazier et al. 2012). More investigation is required to clarify the degree to which tDCS can be used to alter and enhance emotional states.

8.3.2 **Language**

Transcranial direct current stimulation has been used experimentally in several studies to enhance language processing in healthy individuals. For example, a single session of atDCS applied over left posterior perisylvian region (PPR) has been shown to transiently improve reaction times for picture naming compared with sham and compared with atDCS over a right PPR control site (Sparing et al. 2008). Improved performance and speeded reaction times were also observed on a picture-naming task following atDCS over the left DLPFC compared with sham and ctDCS of the same site (Fertonani et al. 2010). Anodal tDCS over left DLPFC did not alter performance on a control task intended to test for changes in arousal or attention. Enhanced acquisition of novel vocabulary terms was demonstrated after atDCS over left PPR compared with sham and ctDCS of the left PPR (Floel et al. 2008). Other studies have shown that tDCS over inferior frontal areas can also enhance verbal fluency (Cattaneo et al. 2011; Iyer et al. 2005), wherein one study found no effect of 1-mA atDCS (compared with sham and ctDCS) but found a significant improvement when atDCS was administered at 2 mA (Iyer et al. 2005). Taken together, studies like these provide mounting evidence that tDCS, applied with appropriate parameters, can be used to boost the acquisition and retrieval of new words and fluency in cognitively healthy individuals.

8.3.3 Learning and memory

A variety of studies have focused on the impact of tDCS on learning and memory systems. One of the first of these was carried out by Marshall et al. (2004), who showed that atDCS administered to bilateral frontal locations during early slow-wave sleep, a stage of sleep associated with consolidation of declarative memory, improved memory retention. Specifically, atDCS during sleep improved retention of word pairs compared with sham, while atDCS during the wakeful state did not impact retention. Procedural memory was not affected by atDCS during sleep, which reflects the cognitive specificity of stimulation effects.

Recently, one group examined the effects of delivering atDCS and ctDCS to the left posterior perisylvian cortex on associative verbal learning and memory (Javadi and Walsh 2012), wherein tDCS was applied during 2 separate phases of learning—an encoding phase and a recognition phase (took place 60 minutes after encoding). While atDCS during encoding improved verbal memory, ctDCS impaired encoding compared with sham and active control stimulation (over the left motor cortex). Stimulation during recognition phase resulted in a trend toward facilitation of recognition with atDCS, while ctDCS impaired performance compared with sham. These results suggested that atDCS can manipulate encoding of new words and recognition of memorized words. Importantly, ctDCS administered during both encoding and recognition phases impaired performance, which contrasts with the findings of Dockery et al. (2009b), who found polarity-dependent effects on different phases of learning in a skill-learning and planning task (see section 8.3.1). This suggests that polarity-specific effects of tDCS observed in one cognitive domain may not reliably predict effects in other domains.

In a follow-up study, Javadi and Cheng (2013) manipulated the reconsolidation of verbal memory with atDCS over left DLPFC. Accuracy in retrieving memorized words reinforced with atDCS was significantly higher than it was under the ctDCS and sham conditions, or in subjects who did not undergo reconsolidation with tDCS. Taken together with the authors' prior study, superior gains in long-term declarative memory might be achieved by administering atDCS during the encoding or reconsolidation phases of memory formation.

In addition to potentially boosting declarative memory, there is some evidence that tDCS may be useful for enhancing implicit learning and memory. Many studies in this area have focused on the ability of tDCS to enhance performance on implicit motor tasks (Galea and Celnik 2009; Reis et al. 2009; Stagg et al. 2011). One noteworthy exception was Kincses et al. (2003), who demonstrated that atDCS over left PFC improved performance on a probabilistic classification learning, a task which involves the formation of implicit associations between a set of arbitrary geometric shapes and weather outcomes, while ctDCS impaired performance. Neither atDCS nor ctDCS of a control site, the primary visual cortex, altered performance. These findings suggest a potential use of tDCS in enhancing implicit non-motor learning and memory functions in healthy individuals.

8.3.4 Visuospatial processing

Limited evidence suggests that tDCS can also enhance visuospatial processing. In one study (Bolognini et al. 2010), facilitation by atDCS over right posterior parietal cortex (PPC) enhanced multisensory exploration and visuospatial attentional skills. Specifically, atDCS over right but not left PPC induced improvement of visual exploration and covert visual orienting. Another example of improved visuospatial processing was provided by Weiss and Lavidor (2012), who employed an attentional load paradigm. In this study, ctDCS over the right PPC enhanced flanker processing—a measure of attentional capacity—under a high-attentional load condition. According to attentional load theory, under low-attentional load conditions spared resources enable processing of both targets and flanking distractors at the cost of an increase in reaction time. By contrast, under high-load conditions limited attentional resources are primarily allocated to processing of target stimuli rather than flankers; reaction times in this case are minimally affected by the presence of flankers. The study results therefore suggested that subjects were better able to process both targets and flankers after receiving ctDCS, further suggesting that manipulating right PPC with tDCS could enhance visuospatial attentional abilities in healthy individuals. In another study of spatial attention, improved attentional selection was observed in association with ctDCS over right intraparietal sulcus (Moos et al. 2012). Interestingly, this study reported enhancement of top-down attentional control with 2 mA ctDCS but not after atDCS, or ctDCS with current strength of 1 mA.

8.4 Taking cognitive enhancement from the laboratory to the real world

As we have discussed, a growing body of evidence from cognitive neuroscience indicates that tDCS and other NBS approaches can be used in experimental settings to temporarily enhance aspects of cognition in anatomically and functionally selective ways. However, this evidence alone does not prove the case that stimulation is a practical, effective, or useful way to augment human cognition. In order to support those claims, additional questions need to be raised which focus on whether brain stimulation approaches that produce gains in the laboratory are meaningful in the "real world." In the following section we will discuss three issues that are germane to the practical use of tDCS (and other NBS techniques) for cognitive enhancement: Are there environments and scenarios where the types of changes in performance observed in brain stimulation studies would seem to be especially useful? Can the transient effects induced by tDCS be made more long-lasting? Do we understand enough about the technique to elicit reliable and predictable effects?

While it is easy to speculate as to how tDCS approaches could be applied to intellectually demanding environments like school or the workplace, investigators by and large do not execute research studies with these settings in mind. We will therefore consider a few illustrative areas in which recent studies seem to support the use of tDCS for specific practical applications. One such potential application is the enhancement of mental abilities in healthy older adults, in order to mitigate normal age-associated cognitive decline. For instance, after a single session of atDCS, older subjects experienced improvement on a word generation task (Meinzer et al. 2013). Interestingly, in this study enhanced brain activity and resting-state functional connectivity in older subjects was found after tDCS, which resembled patterns seen in younger adults.

Another potential use of tDCS that seems to have especially direct real-world applications involves its ability to modulate craving behaviors in healthy individuals. Fregni et al. (2008) demonstrated that after receiving a single session of simultaneous atDCS of the right DLPFC and ctDCS on left DLPFC, subjects demonstrated a decrease in reported food craving, fixated on food-related stimuli less frequently, and ingested fewer calories compared with those receiving sham tDCS. Related to this work, other investigators have begun to explore whether tDCS can be used to regulate cognition related to other common appetitive behaviors such as smoking and alcohol cravings (Boggio et al. 2008b; da Silva et al. 2013).

A third real-world environment that seems conducive to tDCS is the military. Military training often involves the acquisition of highly specialized skills, the accurate and flexible execution of which can have life or death consequences. For instance, honed visuospatial abilities may be crucial for identifying threats in complex and unknown environments, while acute procedural memory skills may be helpful in the operation of complex machinery. In one recent study, tDCS was applied in the context of a virtual reality environment used in prior studies for military training in which subjects were required to detect threats that were either disguised or concealed (Clark et al. 2012). Accuracy in detecting concealed and disguised objects significantly improved during and immediately after 2 mA atDCS over right frontal areas compared with 0.1 mA atDCS; the effects persisted 90 min after stimulation. Of note, the potential application of tDCS in the military and security services has been a source of some ethical concern (Sehm and Ragert 2013), a topic we will return to later in this chapter.

As we have discussed earlier in this chapter, most studies in which tDCS has been used to boost aspects of cognition have been cognitive neuroscience experiments designed to elucidate structure–function relationships in the brain, not to develop cosmetic applications of brain stimulation for healthy individuals. Therefore, most prior studies have employed very small numbers of stimulation sessions, and have only measured the effects of tDCS on behavior for very short periods following stimulation. Recently, however, there has been a surge in the number of studies exploring long-lasting effects of tDCS. Many of these studies incorporate repeated applications of tDCS over multiple days, pair tDCS with cognitive training, and follow up with subjects weeks or even months after the termination of stimulation. One such recent study (Meinzer et al. 2014) evaluated the effects of 5 days of language training with atDCS as compared with sham on a new word learning paradigm. The investigators found that superior performance and a steeper learning curve were observed after atDCS of left posterior temporoparietal junction compared with the sham condition. Importantly, this learning success was maintained after a week. Similarly, Kadosh and colleagues (2010) provided evidence of heightened numerical learning after 6 days of dual-hemispheric tDCS—anodal stimulation of the right parietal cortex and cathodal stimulation of the left parietal cortex—compared with sham, which persisted for at least 6 months following active stimulation. In one other study, Reis and others (2009) investigated effects of atDCS paired with a challenging motor task to enhance motor skill acquisition and retention. There was greater skill acquisition after receiving atDCS over the primary motor areas (M1) for 5 consecutive days compared with the sham tDCS, and these effects persisted 3 months after the stimulation terminated. These studies suggest that sustained improvement in learning success was maintained over an extended period of time as a direct result of receiving tDCS.

It is noteworthy that not all studies that have explored the long-term effects of tDCS have demonstrated a persistent benefit due to stimulation. For example, Martin and colleagues (2013) recently evaluated whether ten sessions of cognitive training (CT), targeting working memory, paired with atDCS of left DLPFC could enhance performance on an adaptive dual *n*-back task compared with sham tDCS also paired with CT. While accuracy was higher in the active tDCS + CT group compared with the sham tDCS + CT group during stimulation sessions, accuracy levels between these groups did not differ at either 10 days or 4 weeks following stimulation. In addition, the investigators were interested in whether improvement in their task generalized to tasks that call upon the same cognitive domain, but found no evidence for transfer between the adaptive *n*-back task and non-trained tasks such as a digit span task and other tasks assessing attention and working memory.

While a large and growing body of evidence is emerging that demonstrates that NBS technologies like tDCS can induce transient and perhaps long-term enhancements of cognition, there remain many aspects of NBS that we only understand to a limited extent. With respect to tDCS, there are a number of properties and parameters that can be thought of as falling in the category of "known unknowns." One example is the relationship between stimulation intensity and elicited physiologic and behavioral effects.

Recent data suggest that this relationship is not linear, and that at higher stimulation intensities and longer durations, ctDCS can induce changes in motor excitability similar to those seen after atDCS (Batsikadze et al. 2013). Another relative unknown is the specific path(s) of current flow associated with placement of electrodes in a specific montage. While sophisticated computer models can be used to estimate current flow (Datta et al. 2009; Im et al. 2008), models that employ differing approaches may disagree with one another and unfortunately cannot be readily verified independently (Datta et al. 2009; Im et al. 2012; Wagner et al. 2007). Finally, another known unknown is the wide range of psychopharmacologic agents that appear to affect tDCS in somewhat unpredictable ways. Evidence suggests that some commonly used classes of medications can have facilitative, suppressive, or even paradoxical effects on tDCS, but the mechanisms that govern these drug-stimulation interactions are not yet well characterized (Stagg and Nitsche 2011).

In summary, while many cognitive neuroscience studies involving tDCS are not specifically geared toward finding real-world applications, evidence from a variety of different investigations suggests that some of behaviors that can be modified using tDCS are potentially relevant to real-life scenarios and may prove meaningful and desirable to specific populations. Moreover, while data on the long-term effects of tDCS are mixed, at least some preliminary studies suggest that repeated sessions of tDCS has the capacity to augment mental abilities for an extended period of time. However, more studies are required, both in order to further explore the potential long-terms effects of tDCS and also to address important unanswered questions about relevant tDCS parameters and mechanisms.

8.5 Should we worry? Ethical concerns surrounding noninvasive cognitive enhancement

Based on the evidence we have presented, one could imagine that, in the not too distant future, tDCS may be widely used to enhance a variety of skills and abilities. While this notion may initially seem far-fetched, the rising use of pharmacologic agents and other interventions to enhance cognition in healthy individuals prompts us to wonder whether NBS is simply the next step in the already widespread adoption of cosmetic neurology. Currently, medications that had initially been developed to treat cognitive deficits in specific patient populations (e.g., Adderall and Ritalin for attention deficit hyperactivity disorder) are being co-opted by many healthy individuals to boost their performance in schools, workplaces, and other intellectually demanding environments. Given the tremendous pressures that modern society places on individuals to perpetually improve their performance and productivity, it seems entirely possible that an inexpensive, easily implemented, and currently unregulated technology like tDCS may be adopted by a public eager to better their brains. To that end, there is a growing community of amateur "do-it-yourself" tDCS users who take upon themselves to develop tDCS devices and stimulate themselves expressly for the purpose of enhancing their cognitive abilities (Brain blast 2013; Melville 2013; Murphy 2013). In some ways the notion of exceeding the normal limits of human cognition with brain stimulation is very appealing. However, the prospect of widespread cosmetic brain stimulation also raises several ethical concerns that we will address in this section (Hamilton et al. 2011). Broadly, we will group our concerns into four categories: (1) safety (side effects and toxicity, unintended and unknown consequences, and the capacity for addiction); (2) manipulation of human character and selfdom; (3) risks to autonomy; and lastly (4) threats to distributive justice (Chatterjee 2009; Farah 2005; Hamilton et al. 2011).

8.5.1 Safety

Fortunately, no severe adverse events or significant persistent side effects have been noted in any study involving tDCS to date. Several large studies of healthy subjects support these claims and demonstrate that tDCS is both safe and well-tolerated. For example, Poreisz and colleagues (2007) reported on 567 sessions with tDCS to multiple brain regions in 102 healthy subjects and patients. None of the subjects asked for a session to be terminated or required any medical assistance. The most common symptom was a mild tingling sensation, which occurred in 71% of subjects. Fatigue was noted in 35% of subjects during but not after the stimulation, and itching under the electrode was reported by 30%. Burning was reported by 22% and pain by 16%, but, once again, both were regarded as quite mild. In addition to sensory side effects, there are theoretical concerns related to skin irritation, toxicity, and damage in brain tissue with tDCS. Currently accepted current strengths (less than 2 mA) and electrode sizes (25–35 cm^2 for conventional tDCS) have not been associated with tissue damage (McCreery et al. 1990) or with significant heating under the electrodes (Datta et al. 2009; Nitsche et al. 2003; Nitsche and Paulus 2000). Mild skin redness under the electrodes is a commonly reported side effect in healthy individuals (Durand et al. 2002), and some evidence suggests that persons with skin disease or those more susceptible to skin irritation could be prone to skin burning, especially with greater current strengths and daily treatments with tDCS (Palm et al. 2008). Skin burning and irritation can be largely avoided by taking precautions like ensuring that the electrodes are properly soaked in saline and sanitized before reuse. While the safety profile of tDCS is generally reassuring, it is important to recognize that most of the reports concerning the safety and tolerability of tDCS relate to its use in single sessions. Far fewer studies have investigated safety issues related to prolonged use of tDCS (Clark et al. 2012; Palm et al. 2008), making it difficult to know whether repeated use over days or weeks poses an additional risk to subjects.

One intriguing category of potential risk to consider is the unintended cognitive consequences of tDCS. While there is no evidence to date of long-term deleterious cognitive effects of tDCS, there are hypothetical reasons to believe that using tDCS to enhance performance on some aspects of cognition may diminish performance in others (Farah et al. 2009). One argument derives from the notion that the neural systems that govern complex behavior are themselves complex, and that altering the activity of a brain region in a relatively indiscriminate unidirectional way (i.e., excitation or inhibition) may lead to simultaneous enhancement of some behaviors but suppression of others. For example, we have reviewed studies in this chapter in which excitatory atDCS of the left frontal lobe is associated with improved performance on working memory tasks and other tests of executive function (Fregni et al. 2005). However, we have also seen data that suggest that inhibitory ctDCS of the left frontal lobe results in improvement in creativity and flexible thinking (Chrysikou et al. 2013). One might hypothesize that repeated stimulation of the left frontal lobe with a single electrode polarity might be beneficial for mental activity in one domain but deleterious in the other.

While there is currently little empirical evidence to suggest that repeated use of tDCS results in a trade-off between cognitive gains and losses, this may be related in part to the very limited ways in which most investigators who use this technology have assessed its effects on cognition. The vast majority of tDCS studies are not designed to explore its effects on remote brain areas or on cognitive domains beyond those relevant to their specific aims. Like streetlamps that only illuminate small patches of ground but leave other areas in darkness, most tDCS experiments do not allow one to make clear inferences about the effects of stimulation on either the rest of the brain or the rest of behavior. Given the intricate and essential connectivity of brain regions, it seems likely that stimulation at a target site would at least sometimes result in downstream effects on secondary targets, with subsequent behavioral consequences that remain largely unknown.

If tDCS is to be used as a tool for cognitive enhancement in healthy individuals, another safety-related concern that should be considered is its potential for addiction. After all, if tDCS is capable of enhancing cognition, then one could describe it as a psychoactive performance-enhancing intervention, which begs the question of whether individuals who receive tDCS frequently would, over time, find themselves physically or psychologically dependent on it. Before addressing the addictive potential of tDCS, it is useful to review some key features of addiction. The American Society for Addiction Medicine describes the state of addiction as being characterized by "inability to consistently abstain, impairment in behavioral control, craving, diminished recognition of significant problems with one's behaviors and interpersonal relationships, and a dysfunctional emotional response" (American Society of Addiction Medicine 2011). Addictive substances or behaviors usually give rise to an intrinsically positive experience for the user and engender a desire for repetitive use (Hyman and Malenka 2001). Physiologic features often associated with pharmacologic agents that have the capacity for addiction are tolerance (decrease in the effects of an agent with chronic administration) and physical dependence (the emergence of withdrawal symptoms in the absence of an agent). From a neurobiological perspective, these agents typically interact with dopaminergic neurotransmitter systems linked to reward and reinforcement, and also interact with other limbic centers and prefrontal circuits that relate to the ability to exert control over addiction-related behaviors (Heinz et al. 2012; Robbins and Everitt 1996; Wise 1996). Repetitive use of such agents ultimately leads to biochemical and molecular changes in the brain that promote more use and inhibit behavioral control (Hyman and Malenka 2001).

Based on this understanding of addiction, there is no evidence to date to suggest that individuals who undergo tDCS or any other form of NBS have a predilection for becoming addicted to these technologies. Moreover, it is hard to see how tDCS would fit with classical pharmacological notions of addiction. To our knowledge, tDCS is not intrinsically pleasurable, and there is no evidence that it gives rise to cravings for repeated use. Moreover, there are no data to support the idea that repeated sessions of stimulation give rise to tolerance to the effects of tDCS, that individuals become dependent on those effects, or that in the absence of stimulation there are physiologic, neurologic, or behavioral signs of withdrawal. In short, there is no reason currently to suspect that tDCS or other NBS techniques would have the same capacity for addiction as drugs like tobacco, cocaine, or alcohol. On the other hand, it is certainly true that individuals can become addicted to a wide variety of substances and behaviors, some of which seem at least initially counterintuitive. To that end, perhaps individuals who become addicted to cosmetic surgery and other forms of enhancement may represent a cautionary tale that the desire for self-improvement has the capacity to lead to patterns of dysfunctional and potentially self-injurious behavior. Finally, it should be noted that recent preliminary studies suggest that tDCS could potentially be employed to diminish cravings for tobacco, alcohol, and cocaine (Boggio et al. 2008b; Conti and Nakamura-Palacios 2014; da Silva et al. 2013; Wing et al. 2013) by manipulating the activity of brain areas implicated in reward, reinforcement, and cognitive control. Hypothetically speaking, these areas might potentially be stimulated in a different way (perhaps involving opposite electrode polarity), so as to make it more difficult for subjects to resist substance cravings. However, it is not clear that even in this hypothetical scenario whether repeated tDCS administration would lead to subjects craving tDCS itself.

8.5.2 Personhood and character

Another concern related to cognitive enhancement has to do with complex questions about what aspects of a person's thoughts, emotional states, or behaviors lie at the core of who (s)he is, and whether NBS techniques like tDCS may affect those essential traits in ways we do not want. In the words of Farah and Wolpe (2004), "neuroscience-based enhancement intersects with our understanding of what it means to be a person, to be healthy and whole, to do meaningful work, and to value human life in all its imperfections." The concern is that by exogenously altering the activity of our brains, we may be intentionally or inadvertently fundamentally changing those elements of ourselves that we think are central to who we are as persons. Insofar as tDCS generally induces modest changes in behavior and that the data that support the enduring benefits of this technology are still limited (see section 8.4), it may seem a little presumptuous and premature to discuss the impact that tDCS will have on our selfhood. However, because these issues touch upon deep philosophical questions with no forthcoming easy answers, it is helpful to talk about these hypothetical problems now rather than to forestall discussion until the day when cosmetic brain stimulation becomes a reality.

One concern about personhood derives from our view that tDCS may have unintended long-term consequences on our mental experience. Earlier in this chapter we argued that stimulation of a site in the brain may have unforeseen effects related to either unanticipated effects of tDCS at that site or else downstream effects in other regions of the brain.

While most studies demonstrating the effects of tDCS on cognition have focused on functions like attention, language ability, or memory, it goes without saying that the brain also represents other features of our cognition that we tend to think of as more fundamental to our self-identity. While many of us would welcome technologies that can make us think faster, remember more, or read more efficiently, many of us have more mixed feelings when we consider devices that could alter traits like honesty, kindness, or compassion (Riis et al. 2008). Yet preliminary evidence from tDCS and other brain stimulation studies suggests that the ability to alter complex social and emotional traits is within reach (Fecteau et al. 2007; Karim et al. 2010).

A separate hypothetical concern relates to how the ability to enhance our minds might affect individual and societal values. The availability of technologies that can mitigate the difficulty of challenging intellectual tasks or painful life experiences could potentially hamper some aspects of psychological development, as well as the development of valued social attributes such as patience and determination (Chatterjee 2007), empathy, sense of accomplishment, and satisfaction in one's actions (Farah 2005; Farah and Wolpe 2004; Hamilton et al. 2011; Kramer 1997). Moreover, some have imagined that cognitive enhancement techniques could undermine intellectual diversity (Butcher 2003) by engendering greater uniformity of patterns of thought at the expense of more eccentric and potentially innovative thoughts and behaviors. As the future unfolds, it will become clearer whether tDCS and other technologies that have the capacity to augment our minds will enable or imperil our ability to realize our true selves (Riis et al. 2008).

8.5.3 Autonomy

There are at least two separate concerns related to the notion of autonomy with regards to cognitive enhancement using tDCS. The first is whether individuals should be freely allowed to pursue NBS for the purpose of self-enhancement, and the degree to which governments and regulatory bodies have the right or responsibility to interfere. The philosophical, cultural, and political traditions of Western society set a clear precedent for individuals being free to do what they choose with their own bodies, and by extension their brains. There are, of course, limits on the extent to which this freedom can be exercised in our society, and one can point to many examples in which laws restrict certain activities for the express purpose of protecting the safety of individuals. Generally speaking, however, these restrictions are imposed when either common sense or evidence indicates that the behaviors in question pose a danger to those engaging in them. As we have discussed, there is no concrete evidence to suggest that tDCS poses a significant health risk to its users. Therefore, within certain broad limits, there is reason to argue that individuals ought not be banned from attempting to enhance their intellect using tDCS, and that individual autonomy in this matter should be upheld.

A second concern related to autonomy is in some ways the converse of the first: if cognitive enhancement becomes widespread, will we be able to protect the right of individuals to avoid being coerced into receiving it if they do not want it? The answer to this question must address two different kinds of coercion. So-called "hard" coercion involves forcing individuals to undergo tDCS for the "sake of the whole" or for the greater good of the society (Chatterjee 2013; Hamilton et al. 2011). Examples of populations that have historically been affected by these forms of coercion include prisoners and persons in the military. By contrast, "soft" coercion refers to less tangible but nonetheless powerful social pressure, peer pressure to maintain standards of intelligence, performance, and productivity compared with one's peers and competitors. Given the constant pressures that society places on individuals to perform in the workplace, school, or other settings, it is easy to imagine that if cognitive enhancement techniques like tDCS become popular, they may be hard for individuals in these high-pressure environments to refuse (Chatterjee 2013; Hamilton et al. 2011).

8.5.4 Distributive justice

One of the potential problems with the introduction of technologies that augment human intellect is that they threaten to further increase the ever-growing divide in opportunity that separates "haves" in our society from "have nots." It may strike some as especially unjust that, in addition to the many other ways in which socioeconomic conditions separate the wealthy from the poor, it may now soon be the case that those who have the means will be able to modify their mental abilities in a way that will be inaccessible to those who cannot. In some ways tDCS mitigates this issue because it is relatively inexpensive. Currently, the least expensive tDCS kits on the market are sold for less than $400 on the internet (FOC.US LABS 2013). In addition, although not recommended, custom-built tDCS units can be assembled with commonly available materials for less than $50. In addition to it being cheap, because tDCS is not regulated in the same way that medications are, one does not need access to a prescribing physician or health insurance to undergo tDCS. Of note, more expensive NBS techniques like TMS are more likely to be affected by socioeconomic factors and therefore seem more liable to disparities in distribution.

Another potentially interesting unexplored issue is the role that culture may eventually play in the acceptance of these technologies (Riis et al. 2008). Differences in cultural, religious, and philosophical beliefs related to personal identity and whether one's essential being (e.g., "soul") is intimately linked to one's body and brain could impose differences in the willingness to use tDCS and other NBS techniques across different groups of people. Taking a global perspective, these differences in belief could engender disparity by giving rise to communities of "enhanced" versus "unenhanced" individuals that might be defined in part by cultural, religious, or ethnic identities.

8.6 Conclusions and future directions

In this chapter, we have tried to address two basic questions. The first is whether NBS techniques like tDCS actually work with respect to enhancing cognition. We believe that the answer to this question is probably "yes." Cognitive neuroscience has provided a large body of converging evidence in support of the notion that brain stimulation can be used to temporarily improve cognitive functions in healthy individuals, and preliminary evidence also suggests that longer lasting changes in cognition can be induced with repeated sessions of stimulation. The second and much more complicated question we have attempted to address is what concerns should we have about the emergence of NBS as a tool for cognitive enhancement. Focusing on tDCS, the current data suggest that it is safe with respect to major known health risks, and there are no well-founded data to date to indicate that it or any other NBS technique has the potential for addiction. On the other hand, there are important theoretical risks to consider, including the possibility of inadvertently manipulating cognitive abilities in unanticipated or undesirable ways, including traits that we consider important to our selfhood. Other issues, such as the capacity for coercion and the unfair distribution of beneficial technologies, are potentially concerning, but are not unique to NBS. Of course, as NBS technologies mature, unanticipated problems may arise. After all, when the internet was first developed as a novel technology, could anyone have foreseen the eventual emergence of recent problems like cyberbullying or social media depression?

In conclusion, it seems that NBS as a means of cognitive enhancement is very promising but should not be approached haphazardly. What should the scientific community and society in general do to address the opportunities and the risks posed by these potentially disruptive technologies? We believe that the first step is to engage in more systematic research to characterize the desired and inadvertent neural effects of NBS. In addition, exhaustive investigation should explore the potential safety hazards of techniques like tDCS and also allow for clearer insights into the stimulation parameters most relevant to cognitive and behavioral manipulation (e.g., intensity, electrode polarity, current flow). For now, there are at least some applications of tDCS for which benefits seem to outweigh the known risks. In light of this, respect for the autonomy of persons would suggest that individuals should not be banned from engaging in tDCS for the purposes of self-enhancement. Instead, the role of the scientific community may be to better engage and educate the public about the appropriate use of these technologies, and to take steps to promote their safe, fair, and coercion-free use (Chatterjee 2013).

References

Ambrus, G. G., H. al-Moyed, L. Chaieb, L. Sarp, A. Antal, and W. Paulus. 2012. The fade-in—short stimulation—fade out approach to sham tDCS—reliable at 1 mA for naive and experienced subjects, but not investigators. *Brain Stimulation* 5:499–504.

American Society of Addiction Medicine. 2011. *Definition of Addiction* [online]. Available at http://www.asam.org/for-the-public/definition-of-addiction.

Antal, A., K. Boros, C. Poreisz, L. Chaieb, D. Terney, and W. Paulus. 2008. Comparatively weak after-effects of transcranial alternating current stimulation (tACS) on cortical excitability in humans. *Brain Stimulation* 1:97–105.

Antal, A., and W. Paulus. 2013. Transcranial alternating current stimulation (tACS). *Frontiers in Human Neuroscience* 7:317.

Baddeley, A. D. 1986. *Working memory*. Oxford: Oxford University Press.

Barker, A. T., R. Jalinous, and I. L. Freeston. 1985. Non-invasive magnetic stimulation of human motor cortex. *Lancet* **1**:1106–7.

Batsikadze, G., V. Moliadze, W. Paulus, M. F. Kuo, and M. A. Nitsche. 2013. Partially non-linear stimulation intensity-dependent effects of direct current stimulation on motor cortex excitability in humans. *Journal of Physiology* **591**:1987–2000.

Boggio, P. S., S. P. Rigonatti, R. B. Ribeiro, M. L. Myczkowski, M. A. Nitsche, A. Pascual-Leone, and F. Fregni. 2008a. A randomized, double-blind clinical trial on the efficacy of cortical direct current stimulation for the treatment of major depression. *International Journal of Neuropsychopharmacology* 11:249–54.

Boggio, P. S., N. Sultani, N., S. Fecteau, L. Merabet, T. Mecca, A. Pascual-Leone, A. Basaglia, and F. Fregni. 2008b. Prefrontal cortex modulation using transcranial DC stimulation reduces alcohol craving: a double-blind, sham-controlled study. *Drug and Alcohol Dependence* **92**:55–60.

Boggio, P. S., S. Zaghi, and F. Fregni. 2009. Modulation of emotions associated with images of human pain using anodal transcranial direct current stimulation (tDCS). *Neuropsychologia* **47**:212–17.

Bolognini, N., F. Fregni, C. Casati, E. Olgiati, and G. Vallar. 2010. Brain polarization of parietal cortex augments training-induced improvement of visual exploratory and attentional skills. *Brain Research* 1349:76–89.

Brain Blast. 2013. DIY attempts at electrical brain stimulation to improve cognition are to get easier. *Nature* **498**:271–2.

Brunoni, A. R., R. Ferrucci, F. Fregni, P. S. Boggio, and A. Priori. 2012. Transcranial direct current stimulation for the treatment of major depressive disorder: a summary of preclinical, clinical and translational findings. *Progress in Neuropsychopharmacology and Biological Psychiatry* 39:9–16.

Butcher, J. 2003. Cognitive enhancement raises ethical concerns. Academics urge pre-emptive debate on neurotechnologies. *Lancet* **362**:132–3.

Cappelletti, M., E. Gessaroli, R. Hithersay, M. Mitolo, D. Didino, R. Kanai, R. Cohen Kadosh, and V. Walsh. 2013. Transfer of cognitive training across magnitude dimensions achieved with concurrent brain stimulation of the parietal lobe. *Journal of Neuroscience* **33**:14899–907.

Cattaneo, Z., A. Pisoni, and C. Papagno. 2011. Transcranial direct current stimulation over Broca's region improves phonemic and semantic fluency in healthy individuals. *Neuroscience* **183**:64–70.

Chatterjee, A. 2004. Cosmetic neurology: the controversy over enhancing movement, mentation, and mood. *Neurology* **63**:968–74.

Chatterjee, A. 2007. "Cosmetic neurology" and the problem of pain. *Cerebrum* [Online]. Available at http://www.dana.org/Cerebrum/2007/Cosmetic_Neurology_and_the_Problem_of_Pain/.

Chatterjee, A. 2009. Is it acceptable for people to take methylphenidate to enhance performance? No. *BMJ* **338**:b1956.

Chatterjee, A. 2013. The ethics of neuroenhancement. *Handbook of Clinical Neurology* **118**:323–34.

Chrysikou, E. G., R. H. Hamilton, H. B. Coslett, A. Datta, M. Bikson, and S. L. Thompson-Schill. 2013. Noninvasive transcranial direct current stimulation over the left prefrontal cortex facilitates cognitive flexibility in tool use. *Cognitive Neuroscience* **4**:81–9.

Clark, V. P., B. A. Coffman, A. R. Mayer, M. P. Weisend, T. D. Lane, V. D. Calhoun, E. M. Raybourn, C. M. Garcia, and E. M. Wassermann. 2012. TDCS guided using fMRI significantly accelerates learning to identify concealed objects. *Neuroimage* **59**:117–28.

Coffman, B. A., V. P. Clark, and R. Parasuraman. 2014. Battery powered thought: enhancement of attention, learning, and memory in healthy adults using transcranial direct current stimulation. *Neuroimage* **85**(Pt 3):895–908.

Cohen Kadosh, R., S. Soskic, T. Iuculano, R. Kanai, and V. Walsh. 2010. Modulating neuronal activity produces specific and long-lasting changes in numerical competence. *Current Biology* **20**:2016–20.

Conti, C. L., and E. M. Nakamura-Palacios. 2014. Bilateral transcranial direct current stimulation over dorsolateral prefrontal cortex changes the drug-cued reactivity in the anterior cingulate cortex of crack-cocaine addicts. *Brain Stimulation* 7:130–2.

Da Silva, M. C., C. L. Conti, J. Klauss, L. G. Alves, H. M. do Nascimento Cavalcante, F. Fregni, M. A. Nitsche, and E. M. Nakamura-Palacios. 2013. Behavioral effects of transcranial direct current stimulation (tDCS) induced dorsolateral prefrontal cortex plasticity in alcohol dependence. *Journal of Physiology – Paris* **107**:493–502.

Datta, A., V. Bansal, J. Diaz, J. Patel, D. Reato, and M. Bikson. 2009. Gyri-precise head model of transcranial direct current stimulation: improved spatial focality using a ring electrode versus conventional rectangular pad. *Brain Stimulation* **2**:201–7, 207 e1.

Dockery, C. A., T. Gaber, N. Birbaumer, and C. Plewnia. 2009a. Functional hemispheric lateralization of executive function by transcranial direct current stimulation. *Psychophysiology* **46**:S38.

Dockery, C. A., R. Hueckel-Weng, N. Birbaumer, and C. Plewnia. 2009b. Enhancement of planning ability by transcranial direct current stimulation. *Journal of Neuroscience* **29**:7271–7.

Durand, S., B. Fromy, P. Bouye, J. L. Saumet, and P. Abraham. 2002. Vasodilatation in response to repeated anodal current application in the human skin relies on aspirin-sensitive mechanisms. *Journal of Physiology* **540**:261–9.

Edwards, D., M. Cortes, A. Datta, P. Minhas, E. M. Wassermann, and M. Bikson. 2013. Physiological and modeling evidence for focal transcranial electrical brain stimulation in humans: a basis for high-definition tDCS. *Neuroimage* **74**:266–75.

Farah, M. J. 2005. Neuroethics: the practical and the philosophical. *Trends in Cognitive Sciences* **9**:34–40.

Farah, M. J., C. Haimm, G. Sankoorikal, M. E. Smith, and A. Chatterjee. 2009. When we enhance cognition with Adderall, do we sacrifice creativity? A preliminary study. *Psychopharmacology (Berlin)* **202**:541–7.

Farah, M. J., and P. R. Wolpe. 2004. Monitoring and manipulating brain function: new neuroscience technologies and their ethical implications. *Hastings Center Report* **34**:35–45.

Fecteau, S., D. Knoch, F. Fregni, N. Sultani, P. Boggio, and A. Pascual-Leone. 2007. Diminishing risk-taking behavior by modulating activity in the prefrontal cortex: a direct current stimulation study. *Journal of Neuroscience* **27**:12500–5.

Fertonani, A., C. Pirulli, and C. Miniussi. 2011. Random noise stimulation improves neuroplasticity in perceptual learning. *Journal of Neuroscience* **31**:15416–23.

Fertonani, A., S. Rosini, M. Cotelli, P. M. Rossini, and C. Miniussi. 2010. Naming facilitation induced by transcranial direct current stimulation. *Behavioural Brain Research* **208**:311–18.

Filmer, H. L., J. B. Mattingley, and P. E. Dux. 2013. Improved multitasking following prefrontal tDCS. *Cortex* **49**:2845–52.

Fitzgerald, P. B., S. Fountain, and Z. J. Daskalakis. 2006. A comprehensive review of the effects of rTMS on motor cortical excitability and inhibition. *Clinical Neurophysiology* **117**:2584–96.

Floel, A., N. Rosser, O. Michka, S. Knecht, and C. Breitenstein. 2008. Noninvasive brain stimulation improves language learning. *Journal of Cognitive Neuroscience* **20**:1415–22.

FOC.US LABS. 2013. *A tDCS headset for gamers* [Online]. Available at http://www.foc.us/.

Fregni, F., P. S. Boggio, M. Nitsche, F. Bermpohl, A. Antal, E. Feredoes, M. A. Marcolin, S. P. Rigonatti, M. T. Silva, W. Paulus, and A. Pascual-Leone. 2005. Anodal transcranial direct current stimulation of prefrontal cortex enhances working memory. *Experimental Brain Research* **166**:23–30.

Fregni, F., P. S. Boggio, M. A. Nitsche, S. P. Rigonatti, and A. Pascual-Leone. 2006. Cognitive effects of repeated sessions of transcranial direct current stimulation in patients with depression. *Depression and Anxiety* **23**:482–4.

Fregni, F., F. Orsati, W. Pedrosa, S. Fecteau, F. A. Tome, M. A. Nitsche, T. Mecca, E. C. Macedo, A. Pascual-Leone, and P. S. Boggio. 2008. Transcranial direct current stimulation of the prefrontal cortex modulates the desire for specific foods. *Appetite* **51**:34–41.

Galea, J. M., and P. Celnik. 2009. Brain polarization enhances the formation and retention of motor memories. *Journal of Neurophysiology* **102**:294–301.

Gandiga, P. C., F. C. Hummel, and L. G. Cohen. 2006. Transcranial DC stimulation (tDCS): a tool for double-blind sham-controlled clinical studies in brain stimulation. *Clinical Neurophysiology* 117:845–50.

Hamilton, R., S. Messing, and A. Chatterjee. 2011. Rethinking the thinking cap: ethics of neural enhancement using noninvasive brain stimulation. *Neurology* **76**:187–93.

Heinz, A., R. Kipke, H. Heimann, and U. Wiesing. 2012. Cognitive neuroenhancement: false assumptions in the ethical debate. *Journal of Medical Ethics* **38**:372–5.

Hoogendam, J. M., G. M. Ramakers, and V. di Lazzaro. 2010. Physiology of repetitive transcranial magnetic stimulation of the human brain. *Brain Stimulation* **3**:95–118.

Hyman, S. E., and R. C. Malenka. 2001. Addiction and the brain: the neurobiology of compulsion and its persistence. *Nature Reviews: Neuroscience* **2**:695–703.

Im, C. H., Jung, H. H., Choi, J. D., Lee, S. Y. and Jung, K. Y. 2008. Determination of optimal electrode positions for transcranial direct current stimulation (tDCS). *Physics and Medicine Biology*, **53**, N219–N225.

Im, C. H., J. H. Park, M. Shim, W. H. Chang, and Y. H. Kim. 2012. Evaluation of local electric fields generated by transcranial direct current stimulation with an extracephalic reference electrode based on realistic 3D body modeling. *Physics and Medicine Biology* **57**:2137–50.

Iyer, M. B., U. Mattu, J. Grafman, M. Lomarev, S. Sato, and E. M. Wassermann. 2005. Safety and cognitive effect of frontal DC brain polarization in healthy individuals. *Neurology* **64**:872–5.

Jacobson, L., M. Koslowsky, and M. Lavidor. 2012. tDCS polarity effects in motor and cognitive domains: a meta-analytical review. *Experimental Brain Research* **216**:1–10.

Jausovec, N., K. Jausovec, and A. Pahor. 2014. The influence of theta transcranial alternating current stimulation (tACS) on working memory storage and processing functions. *Acta Psychologica (Amst)* **146**:1–6.

Javadi, A. H., and P. Cheng. 2013. Transcranial direct current stimulation (tDCS) enhances reconsolidation of long-term memory. *Brain Stimulation* **6**:668–74.

Javadi, A. H., and V. Walsh. 2012. Transcranial direct current stimulation (tDCS) of the left dorsolateral prefrontal cortex modulates declarative memory. *Brain Stimulation* **5**:231–41.

Karim, A. A., M. Schneider, M. Lotze, R. Veit, P. Sauseng, C. Braun, and N. Birbaumer. 2010. The truth about lying: inhibition of the anterior prefrontal cortex improves deceptive behavior. *Cerebral Cortex* **20**:205–13.

Kessler, S. K., P. E. Turkeltaub, J. G. Benson, and R. H. Hamilton. 2012. Differences in the experience of active and sham transcranial direct current stimulation. *Brain Stimulation* **5**:155–62.

Kincses, T. Z., A. Antal, M. A. Nitsche, O. Bartfai, and W. Paulus. 2004. Facilitation of probabilistic classification learning by transcranial direct current stimulation of the prefrontal cortex in the human. *Neuropsychologia* **42**:113–37.

Klingberg, T., H. Forssberg, and H. Westerberg. 2002. Training of working memory in children with ADHD. *Journal of Clinical and Experimental Neuropsychology* **24**:781–91.

Kramer, P. D. 1997. *Listening to Prozac: The landmark book about antidepressants and the remaking of the self*, revised edition. New York: Penguin Group.

Kuo, H. I., M. Bikson, A. Datta, P. Minhas, W. Paulus, M. F. Kuo, and M. A. Nitsche. 2013. Comparing cortical plasticity induced by conventional and high-definition 4 x 1 ring tDCS: a neurophysiological study. *Brain Stimulation* **6**:644–8.

Kuo, M. F., and A. M. Nitsche. 2012. Effects of transcranial electrical stimulation on cognition. *Clinical and EEG Neuroscience* **43**:192–9.

Leite, J., S. Carvalho, F. Fregni, and O. F. Goncalves. 2011. Task-specific effects of tDCS-induced cortical excitability changes on cognitive and motor sequence set shifting performance. *PLoS One* **6**:e24140.

Marshall, L., M. Molle, M. Hallschmid, and J. Born. 2004. Transcranial direct current stimulation during sleep improves declarative memory. *Journal of Neuroscience* **24**:9985–92.

Martin, D. M., R. Liu, A. Alonzo, M. Green, M. J. Player, P. Sachdev, and C. K. Loo. 2013. Can transcranial direct current stimulation enhance outcomes from cognitive training? A randomized controlled trial in healthy participants. *International Journal of Neuropsychopharmacology* **16**:1927–36.

McCreery, D. B., W. F. Agnew, T. G. Yuen, and L. Bullara. 1990. Charge density and charge per phase as cofactors in neural injury induced by electrical stimulation. *IEEE Transactions on Biomedical Engineering* **37**:996–1001.

Meinzer, M., S. Jahnigen, D. A. Copland, R. Darkow, U. Grittner, K. Avirame, A. D. Rodriguez, R. Lindenberg, and A. Floel. 2014. Transcranial direct current stimulation over multiple days improves learning and maintenance of a novel vocabulary. *Cortex* **50**:137–47.

Meinzer, M., R. Lindenberg, D. Antonenko, T. Flaisch, and A. Floel. 2013. Anodal transcranial direct current stimulation temporarily reverses age-associated cognitive decline and functional brain activity changes. *Journal of Neuroscience* **33**:12470–8.

Melville, N. A. 2013. DIY brain stimulation raises concerns. *Medscape Medical News* July 10, 2013.

Miniussi, C., J. A. Harris, and M. Ruzzoli. 2013. Modelling non-invasive brain stimulation in cognitive neuroscience. *Neuroscience and Biobehavioral Reviews* **37**:1702–12.

Moos, K., S. Vossel, R. Weidner, R. Sparing, and G. R. Fink. 2012. Modulation of top-down control of visual attention by cathodal tDCS over right IPS. *Journal of Neuroscience* **32**:16360–8.

Mulquiney, P. G., K. E. Hoy, Z. J. Daskalakis, and P. B. Fitzgerald. 2011. Improving working memory: exploring the effect of transcranial random noise stimulation and transcranial direct current stimulation on the dorsolateral prefrontal cortex. *Clinical Neurophysiology* **122**:2384–9.

Murphy, K. 2013. Jump-starter kits for the mind. *The New York Times*, October 29, 2013.

Nitsche, M. A., J. Koschack, H. Pohlers, S. Hullemann, W. Paulus, and S. Happe. 2012. Effects of frontal transcranial direct current stimulation on emotional state and processing in healthy humans. *Frontiers in Psychiatry* **3**:58.

Nitsche, M. A., D. Liebetanz, N. Lang, A. Antal, F. Tergau, and W. Paulus. 2003. Safety criteria for transcranial direct current stimulation (tDCS) in humans. *Clinical Neurophysiology* **114**:2220–2; author reply 2222–3.

Nitsche, M. A., and W. Paulus. 2000. Excitability changes induced in the human motor cortex by weak transcranial direct current stimulation. *Journal of Physiology* **527**:633–9.

Nitsche, M. A., and W. Paulus. 2001. Sustained excitability elevations induced by transcranial DC motor cortex stimulation in humans. *Neurology* **57**:1899–1901.

O'Connell, N. E., J. Cossar, L. Marston, B. M. Wand, D. Bunce, G. L. Moseley, and L. H. de Souza. 2012. Rethinking clinical trials of transcranial direct current stimulation: participant and assessor blinding is inadequate at intensities of 2mA. *PLoS One* **7**:e47514.

Ohn, S. H., C. I. Park, W. K. Yoo, M. H. Ko, K. P. Choi, G. M. Kim, Y. T. Lee, and Y. H. Kim. 2008. Time-dependent effect of transcranial direct current stimulation on the enhancement of working memory. *Neuroreport* **19**:43–7.

Palm, U., D. Keeser, C. Schiller, Z. Fintescu, M. Nitsche, E. Reisinger, and F. Padberg. 2008. Skin lesions after treatment with transcranial direct current stimulation (tDCS). *Brain Stimulation* **1**:386–7.

Pascual-Leone, A., J. M. Tormos, J. Keenan, F. Tarazona, C. Canete, and M. D. Catala. 1998. Study and modulation of human cortical excitability with transcranial magnetic stimulation. *Journal of Clinical Neurophysiology* **15**:333–43.

Pascual-Leone, A., J. Valls-Sole, E. M. Wassermann, J. Brasil-Neto, L. G. Cohen, and M. Hallett. 1992. Effects of focal transcranial magnetic stimulation on simple reaction time to acoustic, visual and somatosensory stimuli. *Brain* **115**(Pt 4):1045–59.

Paulus, W. 2011. Transcranial electrical stimulation (tES—tDCS; tRNS, tACS) methods. *Neuropsychology Rehabilitation* **21**:602–17.

Plazier, M., K. Joos, S. Vanneste, J. Ost, and D. de Ridder. 2012. Bifrontal and bioccipital transcranial direct current stimulation (tDCS) does not induce mood changes in healthy volunteers: a placebo controlled study. *Brain Stimulation* **5**:454–61.

Pogosyan, A., L. D. Gaynor, A. Eusebio, and P. Brown. 2009. Boosting cortical activity at Beta-band frequencies slows movement in humans. *Current Biology: CB* **19**:1637–41.

Poreisz, C., K. Boros, A. Antal, and W. Paulus. 2007. Safety aspects of transcranial direct current stimulation concerning healthy subjects and patients. *Brain Research Bulletin* **72**:208–14.

Reis, J., H. M. Schambra, L. G. Cohen, E. R. Buch, B. Fritsch, E. Zarahn, P. A. Celnik, and J. W. Krakauer. 2009. Noninvasive cortical stimulation enhances motor skill acquisition over multiple days through an effect on consolidation. *Proceedings of the National Academy of Science USA* **106**:1590–5.

Riis, J., J. P. Simmons, and G. P. Goodwin. 2008. Preferences for enhancement pharmaceuticals: the reluctance to enhance fundamental traits. *Journal of Consumer Research* **35**:495–508.

Robbins, T. W., and B. J. Everitt. 1996. Neurobehavioural mechanisms of reward and motivation. *Current Opinion in Neurobiology* **6**:228–36.

Rossi, S., P. Pasqualetti, P. M. Rossini, B. Feige, M. Ulivelli, F. X. Glocker, N. Battistini, C. H. Lucking, and R. Kristeva-Feige. 2000. Effects of repetitive transcranial magnetic stimulation on movement-related cortical activity in humans. *Cerebral Cortex* **10**:802–8.

Rossini, P. M., A. T. Barker, A. Berardelli, M. D. Caramia, G. Caruso, R. Cracco, M. R. Dimitrijevic, M. Hallett, Y. Katayama, C. H. Lucking, et al. 1994. Non-invasive electrical and magnetic stimulation of the brain, spinal cord and roots: basic principles and procedures for routine clinical application. Report of an IFCN committee. *Electroencephalography and Clinical Neurophysiology* **91**:79–92.

Rothwell, J. C., P. D. Thompson, B. L. Day, S. Boyd, and C. D. Marsden. 1991. Stimulation of the human motor cortex through the scalp. *Experimental Physiology* **76**:159–200.

Russo, R., D. Wallace, P. B. Fitzgerald, and N. R. Cooper. 2013. Perception of comfort during active and sham transcranial direct current stimulation: a double blind study. *Brain Stimulation* **6**:946–51.

Sehm, B., and P. Ragert. 2013. Why non-invasive brain stimulation should not be used in military and security services. *Frontiers in Human Neuroscience* **7**:553.

Shah, P. P., J. P. Szaflarski, J. Allendorfer, and R. H. Hamilton. 2013. Induction of neuroplasticity and recovery in post-stroke aphasia by non-invasive brain stimulation. *Frontiers in Human Neuroscience* **7**:888.

Sparing, R., M. Dafotakis, I. G. Meister, N. Thirugnanasambandam, and G. R. Fink. 2008. Enhancing language performance with non-invasive brain stimulation—a transcranial direct current stimulation study in healthy humans. *Neuropsychologia* **46**:261–8.

Stagg, C. J., G. Jayaram, D. Pastor, Z. T. Kincses, P. M. Matthews, and H. Johansen-Berg. 2011. Polarity and timing-dependent effects of transcranial direct current stimulation in explicit motor learning. *Neuropsychologia* **49**:800–4.

Stagg, C. J., and M. A. Nitsche. 2011. Physiological basis of transcranial direct current stimulation. *Neuroscientist* **17**:37–53.

Terney, D., L. Chaieb, V. Moliadze, A. Antal, and W. Paulus. 2008. Increasing human brain excitability by transcranial high-frequency random noise stimulation. *Journal of Neuroscience* **28**:14147–55.

Wagner, T., F. Fregni, S. Fecteau, A. Grodzinsky, M. Zahn, and A. Pascual-Leone. 2007. Transcranial direct current stimulation: a computer-based human model study. *Neuroimage* **35**:1113–24.

Walsh, V., and A. Pascual-Leone. 2003. *Transcranial magnetic stimulation a neurochronometrics of mind.* London: MIT Press.

Weiss, M., and M. Lavidor. 2012. When less is more: evidence for a facilitative cathodal tDCS effect in attentional abilities. *Journal of Cognitive Neuroscience* **24**:1826–33.

Wing, V. C., M. S. Barr, C. E. Wass, N. Lipsman, A. M. Lozano, Z. J. Daskalakis, and T. P. George. 2013. Brain stimulation methods to treat tobacco addiction. *Brain Stimulation* **6**:221–30.

Wise, R. A. 1996. Neurobiology of addiction. *Current Opinion in Neurobiology* **6**:243–51.

Zaehle, T., S. Rach, and C. S. Herrmann. 2010. Transcranial alternating current stimulation enhances individual alpha activity in human EEG. *PLoS One* **5**:e13766.

Zaehle, T., P. Sandmann, J. D. Thorne, L. Jancke, and C. S. Herrmann. 2011. Transcranial direct current stimulation of the prefrontal cortex modulates working memory performance: combined behavioural and electrophysiological evidence. *BMC Neuroscience* **12**:2.

Zaghi, S., M. Acar, B. Hultgren, P. S. Boggio, and F. Fregni. 2010. Noninvasive brain stimulation with low-intensity electrical currents: putative mechanisms of action for direct and alternating current stimulation. *Neuroscientist* **16**:285–307.

Ziemann, U. 2001. Transcranial magnetic stimulation: new applications to measure cortical and corticospinal excitability. *Aktuelle Neurologie* **28**:249–64.

Chapter 9

The use of brain stimulation technology for cognitive enhancement and the potential for addiction

Mark Attiah

9.1 Introduction

Brain stimulation technologies are at once promising, fascinating, and terrifying. For those bedridden or otherwise crippled by movement disorders, they represent relief that pharmaceuticals can't afford them. Those suffering from psychiatric illnesses are mired in a no-man's land of changing disease classifications and even less clear therapeutic options. For them, brain stimulation offers a solution that may have life-saving potential. For biomedical researchers and a growing segment of the intellectually curious population that have never put on a white coat, brain stimulation is a window into what some consider to be the final frontier of human discovery. It is a way to not only shine light into the black box between our ears but to fashion that box to suit our needs more expediently than any pill might. Finally, for many of the lay public, the phrase "brain stimulation" conjures images of a dystopian future, devoid of the natural laws that have, up until now, held our baser natures in check. This is a valid fear. The ability to alter brain function has weighty implications given that much of what our brain does is make up our personalities—our thoughts, our emotions, our memories, and our dreams. Thus, we are at the cusp of being able to fundamentally alter *who we are* … not an endeavor to be undertaken lightly, or, some would say, to be undertaken at all.

The ability to alter brain functioning with stimulation comes fraught with ethical questions, most of which will only be able to be answered empirically as the possibilities for these technologies unfold. Such is the history of bioethics, and this will likely be no exception. Regardless, some of the more concerning issues are worthy of our attention, such as the problem of cognitive enhancement. This isn't a new subject—issues of enhancement have surfaced in athletics with performance-enhancing drugs and in pharmaceuticals with amphetamines as mental stimulants. However, brain stimulation technologies can be more invasive, and offer the potential to enhance more directly and more effectively than other methods, and so occupy a potentially unique space in this discourse.

9.2 **What exactly is disease?**

Any discussion of enhancement must necessarily begin with its definition. This definition often differentiates enhancement from therapy. What often is neglected, though, is an adequate examination of what disease actually is (Caplan et al. 2004). Functionally, a disease is anything that can be diagnosed or that can be catalogued in a book of billing codes or an authoritative handbook on a particular medical field. Our need to categorize and label illnesses as diseases belies a certain misnomer; an entity such as asthma or sickle cell may be a state or condition, but its status as a *disease* is highly variable, based on individual experience and societal consensus.

The evolution of the DSM, the handbook of psychiatry, presents a perfect case study. Homosexuality, originally regarded as a sociopathic personality disturbance in the first edition published in 1952, was later removed from the second revised edition 22 years later after a large study found that homosexuals were very well adjusted (Bieber 1988; Rubinstein 1995; Silverstein 2009). People who had Asperger's syndrome under the DSM IV are now classified as under the umbrella of autism spectrum disorder (Wing et al. 2011). In these cases, conditions aren't changing, but our classification of these conditions as illnesses is, which manifests in differing social dialogues. Whether someone wants to change their condition, to treat their illness, is their prerogative in some circumstances and not others. Using the example of the transgendered community, an individual's original sex is "normal" for all intents and purposes, but he or she may endeavor to change because it is not normal or natural to *them*. As one can see, a definition of what disease is leads us further down the rabbit hole to what "normal" actually is.

The answer is that there is no one answer. The definition of normal is a moving target. It changes across geographic landscapes, boundaries, climates, and ethnic heritages and contexts. Out of necessity, medical science has settled for a definition that is within two standard deviations of the mean for some value. However, when dealing with conditions that do not have readily quantifiable metrics like blood pressure or body mass index, the concept or normality is only relevant to how a person functions in everyday life. However, even "normal" functioning is highly variable. For global mental function, there is no benchmark number of math problems to complete or threshold amount of words per minute to type that determines what is normal. Even if there was some standard measure of normal global mental function, it would have to be normalized by societal contexts, of which there can be infinite subcategorizations. There would be some aspects of mental function that would be more variable and hard to capture, such as artistic and religious expression. Synofzik and Schlaepfer (2008) point out that such an attempt would lead to a "double dissociation" or individuals who might not have statistically normal function values but function normally and vice versa.

This isn't unique to mental function. For instance, as we age, things like eyesight and skin elasticity deteriorate and fall out of the range of "normal," although this is seen as normal by the medical establishment and wouldn't be treated. So then, mental abnormality is not adequately defined by what is treated or treatable. It also cannot be defined by what society considers as normal, because there are many different societal contexts that are also in constant flux from generation to generation (Hansson 2005). We must be content with a constantly evolving definition of disease.

9.3 The treatment/enhancement distinction

If the concept of disease is always evolving, then treatment evolves as well. This is where the discussion of the treatment/enhancement distinction invariably falters; it is always assumed that for the sake of the argument at hand, we can suspend the belief that what we think about things changes from day to day. Then a single definition of enhancement can be agreed upon, at least for now. One such statement defining enhancement as "ameliorating and broadening of human body functions and abilities using medical methods without corresponding medical indication" (Schmitz-Luhn et al. 2012) isn't helpful when we consider that much of what we do on a daily basis would be considered enhancement by this definition. Further, because what constitutes a medical indication varies especially in the case of mental states, what is enhancement must necessarily vary as well. If approached from the most minimalist of perspectives, and the natural history of human experience without modern medicine is the baseline, then all therapy can be considered as a form of enhancement (Elliott 2004). Hypocrisy is inherent in public discussions of neuroenhancement that turn a blind eye to other forms of cognitive and mood enhancement that are already parts of our everyday life and considered to be beyond or beneath discussion: the use of coffee, chocolate, exercise, and sex (Frary et al. 2005; Yeung 1996). When we speak of enhancement, it is effectively a code word for only certain *types* of enhancement, ones that have already established regulatory roadblocks or at least are unpleasant to bring up at the dinner table. These are the types that the public discourse is interested in. Altering the way the brain works by using electric current rests squarely in this category.

9.4 Current brain stimulation technologies: a primer

Deep brain stimulation (DBS) was first used by the neurosurgeon Jean Siegfried in 1976 and he performed the procedure for dyskinesia in 1982 (Siegfried 1986; Siegfried and Shulman 1987). This procedure involves inserting electrodes with multiple contacts into regions deep within the brain to administer electrical currents. Over the course of two decades, it has been successfully used to treat movement disorders like Parkinson's and essential tremor in more than 85,000 patients worldwide (Christen and Müller 2012). DBS is used in well-selected patients who are refractory to other forms of medical treatment. DBS is very effective at alleviating motor symptoms of these ailments and is advantageous over pharmacological therapies because of its reversibility and ability to adjust the amount and timing of current delivered (Benabid 2003).

Given DBS's success in treating movement disorders, attention has been expanding to treat neuropsychiatric disorders. Most notably, DBS was FDA approved in 2009 for treatment of obsessive-compulsive disorder (OCD). It has also been used to treat patients with treatment-resistant depression, Tourette's syndrome, addiction, and Alzheimer's dementia, and is even being investigated for obesity (Halpern et al. 2013; Holtzheimer and Mayberg 2011).

DBS falls under the purview of functional neurosurgery, which is described by the World Society for Stereotactic and Functional Neurosurgery as concerned with improving the "function of both the structurally normal and abnormal nervous system." This attention to the structure of the nervous system is telling. Indeed, many Parkinsonian brains look structurally unremarkable (Brooks 2010). Unlike neurosurgery that removes a tumor or corrects a misaligned spinal column, the primary goal of functional neurosurgery is to alter brain function.

Noninvasive brain stimulation techniques, namely, transcranial magnetic stimulation (TMS) and transcranial direct current stimulation (tDCS), have been FDA approved to treat refractory depression and migraine headaches. TMS involves the generation of a magnetic field via a coil of wire. The magnetic field is of sufficient strength to depolarize neurons and generate action potentials when placed near an individual's head. Repetitive TMS (rTMS) involves a series of magnetic pulses delivered in rapid succession at a predetermined frequency (Hamilton et al. 2011). The effects of stimulation from rTMS outlast the application of the stimulation. It is thought that low-frequency rTMS decreases cortical excitability, whereas higher frequencies increase excitability (Maeda and Pascual-Leone 2003). Transcranial DCS involves the delivery of small electrical currents to the brain between two electrodes applied to the scalp. It is thought that tDCS works via modulation of the resting membrane potential of neurons (Hamilton et al. 2011). The effect of stimulation via tDCS is dependent upon which electrode is applied: cathodal stimulation causes hyperpolarization of cortical neurons, while anodal stimulation leads to increased excitability. The effects last anywhere on the order of minutes to hours, depending on the intensity, polarity, and the duration of the stimulation period (Nitsche and Paulus 2000; Priori 2003). Because tDCS applies current to the entire cortex between two electrodes, TMS allows for more precise cortical targeting. However, tDCS is at present less costly and is presumably safer (George and Aston-Jones 2010). Noninvasive stimulation is currently being investigated for the treatment of Parkinson's disease and chronic pain (Lee et al. 2012; Pridmore et al. 2005; Vonloh et al. 2013).

9.5 The specter of enhancement

Both invasive and noninvasive forms of brain stimulation can be adjusted by altering the characteristics of the current or magnetic field applied. This highlights a distinction between these and pharmaceutical therapy: assuming that all that needs to be done is calibrate a stimulation parameter, these technologies will have the ability to be directly and instantly dialed up or down to achieve normal or supranormal therapeutic levels. This potential is also troublesome; there is no upper limit of normal to define a normal range of function, which is the best substitute for a disease metric in this realm. Except in extreme cases such as the famous synaesthete Solomon Shereshevsky who was unable to forget memories, it is almost never a bad thing to be hyperfunctioning. When a patient undergoes physical rehabilitation after a traumatic injury, there is no such thing as becoming "too strong" or "too coordinated." Can eyesight ever be deemed "too good?" Similarly, in the hypothetical case of major depression, if a person has DBS electrodes implanted and a certain amount of current is needed to alleviate the depressive state, there is no rule, no law, no protocol that states that more current cannot be applied to make them feel even "happier" than they were before major depression set in. This scenario is a gross simplification, but it presents a question: how happy is too happy? In mood disorders there is no quantifiable normal range of a neurotransmitter to correct to or neuroanatomical structure to repair. Because DBS can be controlled by the patient themselves, who is to say that they couldn't increase the current to give themselves a little more pep before going out on a date, for instance?

Who has control over these stimulation parameters? Should it be the patient, the doctor, or the state? Consider the scenario of a patient who undergoes brain stimulation to treat OCD (Schermer 2013). If the symptoms of the illness are not better but the patient *feels* qualitatively better, is the doctor obligated to discontinue the treatment? A doctor's bioethical duty is to uphold the principles of beneficence, nonmaleficence, autonomy, and justice (Beauchamp and Childress 2001). As for the first two, not only is the stimulation not causing any side effects, it also seems to be improving the patient's quality of life. Currently, this is an important nuance to medical administrations since patient satisfaction (and not patient outcomes) is becoming increasingly publicized and is an important factor in malpractice litigation (Stelfox et al. 2005). Autonomy is a different story, as a patient may want to continue the stimulation treatments regardless of the doctor's willingness to do so. Let us assume that the patient is not a threat to himself or anyone else. Whether treatment discontinuation is an ethical thing to do depends on the medical system's definition of treatment goals. If these include an improvement in quality of life, independent of the alleviation of obsessions and compulsions, then it would be ethically irresponsible to discontinue the treatment. Here again, we return to the theme of an ethical decision being contingent on the definition of treatment.

This scenario also illustrates a problem that is singular to brain stimulation as treatment for psychiatric purposes: the changing capacity for consent (Skuban et al. 2011). Imagine that a patient is to undergo DBS for Alzheimer's dementia, assuming that this therapy has the potential to reverse dementia and restore higher order brain function and memory. However, the disease slowly siphons a patient's ability to provide informed consent. A patient can become demented to the point where they no longer have capacity to consent. Does this mean that there is a certain cutoff in the progression of the disease beyond which the patient is irretrievable? Conventional ethics laid out by the Nuremberg code would say yes: no consent equals no procedure, unless a surrogate decision-maker is permitted to step in (Lebacqz and Levine 1977). However, perhaps the potential for brain stimulation to effectively restore decision-making capacity in a demented individual would obviate the need for surrogate decision-making, and in so doing would add a new dimension to conventional informed consent discussions.

The fact that the effects of stimulation are reversible would allow for a situation where consent could be given or denied *retroactively*. If the patient, with a clear mind, decides that he or she did not want the procedure after the fact, then the stimulation can be turned off (and removed in the case of DBS). Conversely, there is the possibility that stimulation will cause side effects that will alter the patient's executive functions, including decision-making, visuospatial memory, and even global cognitive functioning (Alegret et al. 2001; Saint-Cyr et al. 2000). In this case, the person post-stimulation may be different from the person pre-stimulation. The person who originally gave consent is no longer there. Would they then need to sign a new consent form? Would it be prudent, then, to offer each patient undergoing such stimulation advanced directives? If so, it would require making a judgement by one party or a court of law as to which mental state is the "true" state of the patient, and that any change in judgment as a result of stimulation would be null and void, regardless of what the decision is.

The decision to undergo any treatment, especially an invasive one, requires a risk–benefit analysis to guide both the patient's and physician's or researcher's decision-making. Traditionally, this analysis is predicated on a certain baseline state which a disease has compromised to some extent and the therapy is intended to ameliorate. The potential for brain stimulation to increase levels of function beyond baseline could be categorized as either a risk or a benefit, depending on whether returning to normal was seen as the gold standard. As in the case of the patient who underwent DBS for OCD, these effects might be best categorized as side effects rather than as adverse events when the procedure is explained to the patient. Such categorization involves a value judgement that may or may not be accurate in the viewpoint of the patient. Currently, because the possibility of cognitive enhancement is only hypothetical in DBS, a traditional risk–benefit analysis would not support the use of invasive electrode implantation specifically for this purpose. However, the same cannot be said for the use of TMS or tDCS which has been expressly investigated for this purpose (Woopen 2012). Because the effects of TMS and tDCS only last for a short while and there are no surgical or anesthetic risks involved, there are considerably fewer regulatory hurdles for their use in this way. As Heinrichs points out, there is a double standard in the enhancement debate that mildly demonizes pharmaceuticals and DBS in enhancement capacities, whereas noninvasive techniques are less often criticized (Heinrichs 2012). This may be due to the fact that DBS and pills have been around for much longer than TMS and tDCS, and the latter two are not as well known.

This ethical leniency toward TMS and tDCS has led to studies specifically investigating enhancement in healthy individuals. It has been shown that TMS can improve performance in motor-learning tasks and visuospatial processing (Hilgetag et al. 2001; Kim et al. 2004; Kobayashi et al. 2004; Nitsche et al. 2003). Transcranial DCS over Broca's area can improve object-naming ability and the ability to learn new grammar (de Vries et al. 2010; Flöel et al. 2008). Transcranial DCS has also been shown to improve performance on the remote association test, which requires complex problem-solving skills and is linked to other forms of executive functioning and intelligence (Cerruti and Schlaug 2009). The wealth of accrued knowledge about the enhancement of cognitive abilities in the relatively short time period that these technologies have been available is a testament to the innocuous nature ascribed to stimulating the brain when there is no need to open the skull.

The quick on- and offset, the relatively benign side effect profile, and the lack of the visceral response to the idea of a foreign body inside one's head has allowed the use of noninvasive brain stimulation by untrained individuals at home at their own discretion (Hamilton et al. 2011). While this is probably far from being a widespread reality, the use for enhancement in the home bypasses the traditional medical ethics debates of autonomy and nonmaleficence, because the technology has been taken out of the medical setting. Safety is of the utmost importance in the transition to the home setting. It might be the case that highly motivated individuals will be able to construct home brain stimulators before a governing body has a chance to institute regulations governing their use. If these technologies migrate from research laboratories to home garages with or without the blessing of regulatory oversight, the ethical questions raised are largely legal and social and beyond the scope of this chapter.

9.6 Societal viewpoints: values, judgements, and boundaries

The fate of brain stimulation technologies depends largely on the prevailing opinions of the medical establishment, politicians, and the public at large. Very little data have been gathered on what people actually think on the matter (Illes et al. 2008; Lipsman et al. 2009). Mendelsohn et al. (2010) conducted a qualitative study examining 47 neurosurgeons' views on DBS for neuropsychiatric indications. The responses to questions about elective personality alteration and cognitive enhancement were met with a visceral aversion to the idea, although later the participants reconsidered more rationally. The study found that most participants believed that surgery to alter nonpathological traits was not ethically justified. One person stated that interfering with human nature was beyond the scope of medical practice. Further, when asked to compare this "cosmetic neurosurgery" with cosmetic plastic surgery, participants either disagreed with both or argued that the idea of cosmetic neurosurgery as changing someone's personality outside of pathology was ethically untenable. Some, however, endorsed the idea of neural enhancement as long as it was safe, stating that the patient's autonomy was of primary importance. More recently, a similar survey was conducted with researchers in the field of noninvasive stimulation (Riggall et al. 2015). A total of 265 tDCS researchers of varying levels of seniority and academic focuses were asked about their attitudes toward tDCS in four realms: effectiveness, ethical issues, self use, and availability to the public. Interestingly enough, though the majority of respondents were cautiously optimistic about the effectiveness of tDCS, the vast majority also expressed concerns about its use for enhancement purposes and its availability to the public. The reasons for this were mainly safety concerns and not enough adequate data to support whether such a risk would be justified. Interestingly, little or no comment was made about the "fairness" of the use of such technology or how it should be regulated. Nonetheless, the same general aversion to brain stimulation for use in an enhancement capacity seems to hold in the scientific community regardless of invasiveness.

Autonomy, one of the stalwart pillars of medical ethics, is the main argument in favor of using brain stimulation for neural enhancement (Sententia 2004). However, the willingness to undergo such treatment may in the future lead to coercive situations when others stand to gain something from a person having quicker, sharper mind. One can imagine an individual in the military being told that he must have an electrode implantation in order to become or remain a part of a high-functioning elite unit (Heinrichs 2012), or a criminal suspect forced to undergo TMS to disinhibit their frontal lobe, making them speak more truthfully (Lo et al. 2003; Luber et al. 2009). These are extreme situations, but there will be pressures to use this technology for the betterment of a group.

Perhaps more palatable is a scenario akin to what already happens across competitive academic environments, in which individuals used brain stimulation to gain a competitive edge at school or in the workplace. Similar to Adderall and Ritalin use by students attempting to study for long periods of time (Sussman et al. 2006), there would be implicit pressure, both internally and externally, to have a better memory and sharper focus and to be faster at problem solving. Such pressures may jeopardize the true nature of "autonomy" set forth in the Belmont report, the bedrock of modern medical ethics (National Commission for the Protection of Human Subjects of Biomedical and Behavioral Research 1978).

9.7 **The personal identity crisis**

The transformation of a depressed person into a happy one or a person with OCD who is no longer driven by compulsions is a fundamental change in behavior. Whether this constitutes a change in personality is a highly individual judgement. Mathews (2011) holds that a requirement of personal identity is that one has a continuous sense of self over time. This is an issue in psychiatric disorders, the symptoms of which may wax and wane over time. There are many individuals with depression who are aware when they are in the midst of a major depressive episode and feel that they are "not themselves" (Karp 1994; Kraemer 2011). In treating these individuals with brain stimulation, one could argue that either (1) their personality is being fundamentally changed or (2) it is being freed from the snares of depression. It seems that there is no such ambiguity when it comes to applying the same therapies to healthy individuals.

Opinions on the matter were examined in a study interviewing patients who were scheduled to undergo a neurosurgical procedure (Lipsman et al. 2009). These patients found personality-altering surgery to be unconscionable; believing that we should not tamper with who we are as people and even as a species. Interestingly, patients were less likely to approve a procedure that enhanced a good personality trait than they were of one that rid them of a bad one. This highlights the uncertainty with which people conceive of fairness, probably the most troubling potential aspect of brain stimulation technologies.

A strongly held belief by the participants' doctors and patients in both Mendelsohn and Lipsman's studies is that they were strongly against altering personal identity in the absence of medical need. In fact, patients became less sure whether surgery for a psychiatric indication should be allowed when the question was rephrased as a "change in identity."

How we think about therapeutic changes to our personalities may be influenced by our views on mind–brain duality (Demertzi et al. 2009). Do we believe that our thoughts, feelings, and emotions are dictated by our brain chemistry and neuronal firing, or do we believe that the whole of who we are is greater than the sum of our brain's activities? Some people believe that some mental qualities such as honesty are essential to who we are and should not be changed, whereas concentration is a more peripheral mental faculty which doesn't fall into the same category. Indeed, in Lipsman's study, most patients believed that that mind was more than just a product of the brain.

The fact that someone can undergo a procedure to, say, attain a photographic memory touches a collective nerve in a society that values and rewards hard work and effort. If someone can essentially *buy* a competitive advantage, then one might believe that any accomplishments attained afterward would lose their value. This commodification of our traits has the potential to create social inequality based on access to the treatment (Fuchs 2006; Glannon 2006; Reichlin 2007). Those who are able to get the treatment, because they live close by a center that offers it, are more likely to have heard of it, or have more means to attain it, will be able to enjoy the benefits while others cannot. This group ostensibly could go on to become more successful in their occupations or studies, creating more advantages for themselves, which would enable themselves and their loved ones to get more "upgrades," perpetuating their advantage. Indeed, this already happens to a certain extent with individuals and families of disparate socioeconomic classes, with some able to attain advantages that others cannot. It stands to reason that without a priori legal regulation, the same would happen with brain stimulators (Caplan 2003). However, potential future scenarios featuring the growing rift between the haves and the have-nots determined by access to these particular technologies may be a little far off at present. If our worldview on performance-enhancing drugs in the sports arena and our apprehension about genetic engineering (Sandel 2004) is any indication though, there is a real, widely and deeply held anxiety that creating societal unfairness at such a fundamental level is a line in the sand that we collectively are not willing to cross.

9.8 The potential for addiction

Apart from concerns about fairness in accessing forms of neuroenhancement, there are concerns these technologies may lead to addiction. Addiction is a real concern for stimulant use and it would follow that there would be similar concerns in stimulation technologies that are able to bring about some of the same effects. Substances used to alter brain function to feel "good" are frequently addictive, including alcohol and nicotine. Ironically, some advocate brain stimulation as a treatment for addiction to heroin, cocaine, nicotine, and alcohol, and even high-fat food (Müller et al. 2013). Many of these studies, conducted in both animal models and humans, target the nucleus accumbens, a region in the brain associated with the rewarding properties of actions and objects. Although we are unsure of exactly why DBS works, some hypotheses have been put forth as its mechanism (Vitek 2002). One theory is that DBS has a net activating effect. In this case, the stimulation acts as a "substitute" for the neural activity in the area that would have otherwise been produced by smoking a cigarette or snorting cocaine. Another theory is that DBS causes a net inhibitory effect by lowering the set point for the drive for reward, i.e., that someone wouldn't want that extra drink because he or she doesn't feel the need. For instance, a case study of a man who had bilateral DBS implants in the nucleus accumbens and participated in a gambling scenario found that he exhibited better behavioral control when the stimulation was on rather than off (Heldmann et al. 2012).

The risk of addiction to brain stimulation is a faint though real one. There is one report of a patient who underwent DBS in the subthalamic nucleus (STN) to treat significant resting Parkinsonian tremor (Morgan et al. 2006). The patient then elected to increase the frequency of programming visits, and even turned his stimulator on and off against the counsel of the medical team, because the stimulator produced a "morphine-like effect." This feeling was reproduced for a few seconds under certain stimulation parameters but the patient developed tolerance after multiple programming adjustments. Though many patients report that they can feel when the stimulator is turned on, this is usually a tingling sensation or paresthesia (Deuschl et al. 2006), unlike this man's experience. Drugs of addiction have been shown to activate fibers adjacent to the STN, perhaps within the sphere of stimulation in this individual (Wise 1996).

Addiction is a very specific condition diagnosed using specific criteria, although there has been confusion about addiction and physical dependence (O'Brien 2011). This patient exhibited tolerance but it would be erroneous to say that this person was addicted. However, a more plausible case of addiction was reported in a woman who developed self-stimulating behaviors after undergoing DBS in the ventral posterolateral nucleus of the thalamus for chronic pain (Portenoy et al. 1986). After implantation, she reported feeling erotic sensations during stimulation, which could be increased by increasing the intensity of the current. She self-stimulated despite adverse events including several cardiac arrythmias. She began to neglect personal hygiene and family commitments and failed to stop after asking the family to help her. This behavior could be characterized as an addiction. It is important to note in these cases that specific areas involved with reward or pleasure were being stimulated, both of which were deep structures in the brain. Because neural circuitry is highly complex and different functions of the brain are diffusely spread throughout different regions, it is possible that surgeons run the risk of implanting DBS electrodes in a place where stimulation can produce these side effects. This may be especially true in investigational settings, where a novel structure is being targeted for a new treatment indication. The risks of these side effects should be thoroughly explained before the patient undergoes the procedure, as part of the informed consent process. It is also important to note that it is the structure being stimulated that produces these adverse effects. Noninvasive stimulation modalities have thus far not produced similar case reports, perhaps because the responsible structures were deep in the brain, beyond the reach of TMS or tDCS. Even so, the nonspecific nature of stimulation in DBS may stimulate nearby structures collaterally. Novel methods of neural stimulation using light-sensitive ion channels, termed optogenetics, would be able to bypass this problem, as only cells that are transfected with these proteins would be stimulated when light is applied to them (Fenno et al. 2011). This cellular-specific method of stimulation may represent the future of brain stimulation technologies, but currently it has only been used in research contexts in animal models.

If there is a risk of addiction from brain stimulation technologies, this doesn't stem from their potential use as neural enhancers. They should not be seen as equivalent to pharmacological stimulants like amphetamines in this respect. This risk stems solely from an area of the brain that is involved in reward being inadvertently stimulated. Indeed, the authors of the report note that the electrode may have shifted and the stimulation of outside structures was likely given the high stimulation intensities (Portenoy et al. 1986). The extreme nature of the side effect, however, is cause for concern, especially because the effect persisted after the stimulator was turned off. This is unlike most of the side effects produced by DBS, which only last for the duration of the stimulation. If the behavior change had only lasted while the electrodes were on, one's claim that "the stimulator made me do it" would be plausible to friends, colleagues, and perhaps even a jury. But a durable change in personality seems less excusable to society, and so would provide a reason for people to oppose brain stimulation, as Mendelsohn et al. (2010) point out in their study. They also suggest that while doctors believe that society is not ready for neuroenhancement via brain stimulation, what is acceptable is subject to societal definition and so "society will move in what it sees acceptable."

It may seem ludicrous at first glance to have metal electrodes fixed in our brains that can change our movements, our behaviors, and who we are. But the idea of the internet, having a phone that can fit in your pocket, or having a pacemaker in your heart seemed just as absurd 50 years ago. Although using brain stimulation technology for the mental alteration of people with a glistening bill of health is a radical departure from what many feel is permissible, it is a social value judgment whether the practice is seen as good or bad. History tells us that these judgments can change just as much as the technologies themselves.

References

Alegret, M., C. Junqué, F. Valldeoriola, P. Vendrell, M. Pilleri, J. Rumià, and E. Tolosa. 2001. Effects of bilateral subthalamic stimulation on cognitive function in Parkinson disease. *Archives of Neurology* **58**(8):1223–7.

Beauchamp, T. L., and J. F. Childress. 2001. *Principles of biomedical ethics.* Oxford: Oxford University Press.

Benabid, A. 2003. Deep brain stimulation for Parkinson's disease. *Current Opinion in Neurobiology* **13**(6):696–706.

Bieber, I. 1988. *Homosexuality: A psychoanalytic study.* Jason Aronson, Inc.

Brooks, D. J. 2010. Imaging approaches to Parkinson disease. *Journal of Nuclear Medicine* **51**(4):596–609.

Caplan, A. L. 2003. Is better best? *Scientific American* **289**(3):104–5.

Caplan, A. L., J. J. McCartney, and D. A. Sisti. 2004. *Health, disease, and illness: concepts in medicine.* Washinton, DC: Georgetown University Press.

Cerruti, C., and G. Schlaug. 2009. Anodal transcranial direct current stimulation of the prefrontal cortex enhances complex verbal associative thought. *Journal of Cognitive Neuroscience* **21**(10):1980–7.

Christen, M., and S. Müller. 2012. Current status and future challenges of deep brain stimulation in Switzerland. *Swiss Medical Weekly* **142**:w13570.

de Vries, M. H., A. C. Barth, S. Maiworm, S. Knecht, P. Zwitserlood, and A. Flöel. 2010. Electrical stimulation of Broca's area enhances implicit learning of an artificial grammar. *Journal of Cognitive Neuroscience* **22**(11):2427–36.

Demertzi, A., C. Liew, D. Ledoux, M. A. Bruno, M. Sharpe, S. Laureys, and A. Zeman. 2009. Dualism persists in the science of mind. *Annals of the New York Academy of Sciences* 1157(1):1–9.

Deuschl, G., J. Herzog, G. Kleiner-Fisman, C. Kubu, A. M. Lozano, K. E. Lyons, M. C. Rodriguez-Oroz, F. Tamma, A. I. Tröster, J. L. Vitek, J. Volkmann, and V. Voon. 2006. Deep brain stimulation: postoperative issues. *Movement Disorders* S219–S237.

Elliott, C. 2004. *Better than well: American medicine meets the American dream.* New York: **W. W. Norton.**

Fenno, L., O. Yizhar, and K. Deisseroth. 2011. The development and application of optogenetics. *Annual Review of Neuroscience* **34**:389–412.

Flöel, A., N. Rösser, O. Michka, S. Knecht, and C. Breitenstein. 2008. Noninvasive brain stimulation improves language learning. *Journal of Cognitive Neuroscience* **20**(8):1415–22.

Frary, C. D., R. K. Johnson, and M. Q. Wang. 2005. Food sources and intakes of caffeine in the diets of persons in the United States. *Journal of the American Dietetic Association* **105**(1):110–13.

Fuchs, T. 2006. Ethical issues in neuroscience. *Current Opinion in Psychiatry* **19**(6):600–7.

George, M. S., and G. Aston-Jones. 2010. Noninvasive techniques for probing neurocircuitry and treating illness: vagus nerve stimulation (VNS), transcranial magnetic stimulation (TMS) and transcranial direct current stimulation (tDCS). *Neuropsychopharmacology* **35**(1):301–16.

Glannon, W. 2006. Neuroethics. *Bioethics* **20**(1):37–52.

Halpern, C. H., A. Tekriwal, J. Santollo, J. G. Keating, J. A. Wolf, D. Daniels, and T. L. Bale. 2013. Amelioration of binge eating by nucleus accumbens shell deep brain stimulation in mice involves D2 receptor modulation. *Journal of Neuroscience* **33**(17):7122–9.

Hamilton, R., S. Messing, and A. Chatterjee. 2011. Rethinking the thinking cap: Ethics of neural enhancement using noninvasive brain stimulation. *Neurology* **76**(2):187–93.

Hansson, S. O. 2005. Implant ethics. *Journal of Medical Ethics* **31**(9):519–25.

Heinrichs, J.-H. 2012. The promises and perils of non-invasive brain stimulation. *International Journal of Law and Psychiatry* **35**(2):121–9.

Heldmann, M. G. Berding, J. Voges, B. Bogerts, I. Galazky, U. Müller, G. Baillot, H. J. Heinze, and T. F. Münte. 2012. Deep brain stimulation of nucleus accumbens region in alcoholism affects reward processing. *PloS one* 7(5):e36572.

Hilgetag, C. C., H. Theoret, and A. Pascual-Leone. 2001. Enhanced visual spatial attention ipsilateral to rTMS-induced "virtual lesions" of human parietal cortex. *Nature neuroscience* **4**(9):953–7.

Holtzheimer, P. E., and H. S. Mayberg. 2011. Deep brain stimulation for psychiatric disorders. *Annual Review of Neuroscience* **34**:289–307.

Illes, J., S. Lombera, J. Rosenberg, and B. Arnow. 2008. In the mind's eye: provider and patient attitudes on functional brain imaging. *Journal of Psychiatric Research* **43**(2):107–14.

Karp, D. A. 1994. Living with depression: illness and identity turning points. *Qualitative Health Research* **4**(1):6–30.

Kim, Y.-H., J. W. Park, M. H. Ko, S. H. Jang, and P. K. Lee. 2004. Facilitative effect of high frequency subthreshold repetitive transcranial magnetic stimulation on complex sequential motor learning in humans. *Neuroscience Letters* **367**(2):181–5.

Kobayashi, M., S. Hutchinson, H. Théoret, G. Schlaug, and A. Pascual-Leone. 2004. Repetitive TMS of the motor cortex improves ipsilateral sequential simple finger movements. *Neurology* **62**(1):91–8.

Kraemer, F. 2011. Authenticity anyone? The enhancement of emotions via neuro-psychopharmacology. *Neuroethics* **4**(1):51–64.

Lebacqz, K., and R. J. Levine. 1977. Respect for persons and informed consent to participate in research. *Clinical Research* **25**(3):101–7.

Lee, J. C., D. M. Blumberger, P. B. Fitzgerald, Z. J. Daskalakis, and A. J. Levinson. 2012. The role of transcranial magnetic stimulation in treatment-resistant depression: a review. *Current Pharmaceutical Design* **18**(36):5846–52.

Lipsman, N., R. Zener, and M. Bernstein. 2009. Personal identity, enhancement and neurosurgery: a qualitative study in applied neuroethics. *Bioethics* **23**(6):375–83.

Lo, Y. L., S. Fook-Chong, and E. K. Tan. 2003. Increased cortical excitability in human deception. *Neuroreport* **14**(7):1021–4.

Luber, B., C. Fisher, P. S. Appelbaum, M. Ploesser, S. H. Lisanby. 2009. Non-invasive brain stimulation in the detection of deception: scientific challenges and ethical consequences. *Behavioral Sciences and the Law* **27**(2):191–208.

Maeda, F., and A. Pascual-Leone. 2003. Transcranial magnetic stimulation: studying motor neurophysiology of psychiatric disorders. *Psychopharmacology* **168**(4):359–76.

Mathews, D. J. H. 2011. Deep brain stimulation, personal identity and policy. *International Review of Psychiatry (Abingdon, England)* **23**(5):486–92.

Mendelsohn, D., N. Lipsman, and M. Bernstein. 2010. Neurosurgeons' perspectives on psychosurgery and neuroenhancement: a qualitative study at one center. *Journal of Neurosurgery* **113**(6):1212–18.

Morgan, J. C., C. J. di Donato, S. S. Iyer, P. D. Jenkins, J. R. Smith, and K. D. Sethi. 2006. Self-stimulatory behavior associated with deep brain stimulation in Parkinson's disease. *Movement Disorders* **21**(2):283–5.

Müller, U. J., J. Voges, J. Steiner, I. Galazky, H. J. Heinze, M. Möller, J. Pisapia, C. Halpern, A. Caplan, B. Bogerts, and J. Kuhn 2013. Deep brain stimulation of the nucleus accumbens for the treatment of addiction. *Annals of the New York Academy of Sciences* 1282:119–28.

National Commission for the Protection of Human Subjects of Biomedical and Behavioral Research. 1978. *The Belmont Report: Ethical Principles and Guidelines for the Protection of Human Subjects of Research-the National Commission for the Protection of Human Subjects of Biomedical and Behavioral Research.* US Government Printing Office.

Nitsche, M. A., and W. Paulus. 2000. Excitability changes induced in the human motor cortex by weak transcranial direct current stimulation. *Journal of Physiology* **527**(Pt 3):633–9.

Nitsche, M. A., A. Schauenburg, N. Lang, D. Liebetanz, C. Exner, W. Paulus, and F. Tergau. 2003. Facilitation of implicit motor learning by weak transcranial direct current stimulation of the primary motor cortex in the human. *Journal of Cognitive Neuroscience* **15**(4):619–26.

O'Brien, C. 2011. Addiction and dependence in DSM-V. *Addiction* **106**(5):866–867.

Portenoy, R. K., J. O. Jarden, J. J. Sidtis, R. B. Lipton, K. M. Foley, and D. A. Rottenberg. 1986. Compulsive thalamic self-stimulation: a case with metabolic, electrophysiologic and behavioral correlates. *Pain* **27**(3):277–90.

Pridmore, S., G. Oberoi, M. Marcolin, and M. George. 2005. Transcranial magnetic stimulation and chronic pain: current status. *Australasian Psychiatry* **13**(3):258–65.

Priori, A. 2003. Brain polarization in humans: a reappraisal of an old tool for prolonged non-invasive modulation of brain excitability. *Clinical Neurophysiology* **114**(4):589–95.

Reichlin, M. 2007. The challenges of neuroethics. *Functional Neurology* **22**(4):235–42.

Riggall, K., C. Forlini, A. Carter, W. Hall, M. Weier, B. Partridge, and M. Meinzer. 2015. Researchers' perspectives on scientific and ethical issues with transcranial direct current stimulation: An international survey. *Scientific Reports* **5**:10618.

Rubinstein, G. 1995. The decision to remove homosexuality from the DSM: twenty years later. *American Journal of Psychotherapy* **49**(3):416–27.

Saint-Cyr, J. A., L. L. Trépanier, R. Kumar, A. M. Lozano, and A. E. Lang. 2000. Neuropsychological consequences of chronic bilateral stimulation of the subthalamic nucleus in Parkinson's disease. *Brain* **123**(Pt 10):2091–108.

Sandel, M. J. 2004. The case against perfection: what's wrong with designer children, bionic athletes, and genetic engineering. *Atlantic Monthly* **292**(3):50–62.

Schermer, M. 2013. Health, happiness and human enhancement-dealing with unexpected effects of deep brain stimulation. *Neuroethics* **6**:435–45.

Schmitz-Luhn, B., C. Katzenmeier, and C. Woopen. 2012. Law and ethics of deep brain stimulation. *International Journal of Law and Psychiatry* **35**(2):130–6.

Sententia, W. 2004. Neuroethical considerations: cognitive liberty and converging technologies for improving human cognition. *Annals of the New York Academy of Sciences* 1013:221–8.

Siegfried, J. 1986. Effect of stimulation of the sensory nucleus of the thalamus on dyskinesia and spasticity. *Revue neurologique* **142**(4):380–3.

Siegfried, J., and J. Shulman. 1987. Deep brain stimulation. *Pacing and Clinical Electrophysiology: PACE* **10**(1 Pt 2):271–2.

Silverstein, C. 2009. The implications of removing homosexuality from the DSM as a mental disorder. *Archives of Sexual Behavior* **38**(2):161–3.

Skuban, T., K. Hardenacke, C. Woopen, and J. Kuhn. 2011. Informed consent in deep brain stimulation—ethical considerations in a stress field of pride and prejudice. *Frontiers in Integrative Neuroscience* **5**:7.

Stelfox, H. T., T. K. Gandhi, E. J. Orav, and M. L. Gustafson. 2005. The relation of patient satisfaction with complaints against physicians and malpractice lawsuits. *American Journal of Medicine* **118**(10):1126–33.

Sussman, S., M. A. Pentz, D. Spruijt-Metz, and T. Miller. 2006. Misuse of "study drugs:" prevalence, consequences, and implications for policy. *Substance Abuse Treatment, Prevention, and Policy* **1**:15.

Synofzik, M., and T. E. Schlaepfer. 2008. Stimulating personality: ethical criteria for deep brain stimulation in psychiatric patients and for enhancement purposes. *Biotechnology Journal* **3**(12):1511–20.

Vitek, J. L. 2002. Mechanisms of deep brain stimulation: excitation or inhibition. *Movement Disorders* **17**(Suppl 3):S69–72.

Vonloh, M., R. Chen, and B. Kluger. 2013. Safety of transcranial magnetic stimulation in Parkinson's disease: a review of the literature. *Parkinsonism and Related Disorders* **19**(6):573–85.

Wing, L., J. Gould, and C. Gillberg. 2011. Autism spectrum disorders in the DSM-V: Better or worse than the DSM-IV? *Research in Developmental Disabilities* **32**(2):768–73.

Wise, R. A. 1996. Addictive drugs and brain stimulation reward. *Annual Review of Neuroscience* **19**:319–40.

Woopen, C. 2012. Ethical aspects of neuromodulation. *International Review of Neurobiology* **107**:315–32.

Yeung, R. R. 1996. The acute effects of exercise on mood state. *Journal of Psychosomatic Research* **40**(2):123–41.

Chapter 10

Debunking the ethical neuroenhancement debate

Stephan Schleim and Boris B. Quednow

10.1 Introduction

Since the end of the 1990s, scholars from a wide range of disciplines including neurobiology, law, and philosophy have been discussing the possibility of improving cognitive skills of the healthy by neuroscience technologies, in particular psychopharmacological drugs (Farah et al. 2004; Greely et al. 2008; Morris 2008; Whitehouse et al. 1997). The social and scientific urgency of this debate rested on two basic assumptions. First, the increasing understanding of the brain makes it likely that new drugs or other approaches for cognitive improvement of healthy people will be developed soon (*pharmacological optimism hypothesis*). Second, a high and/or increasing number of healthy people already make use of such means, in particular psychostimulants, for the purpose of cognitive or so-called neuroenhancement (*neuroenhancement prevalence hypothesis*). In this essay, we demonstrate that both assumptions must be reconsidered and conclude with a recommendation to give priority to clinical research that aims to improve the situation of people with neurological or psychiatric disorders instead of the healthy.

10.2 Assumption 1: the pharmacological optimism hypothesis

When the Decade of the Brain was proclaimed by the US government (Bush 1990) and the European Commission (Pandolfi 1993), a wide range of institutions as well as the public at large were called to support the efforts of neuroscience to develop treatments for neurodegenerative disorders such as Alzheimer's and Parkinson's diseases and, only for the US initiative, to fight the "war on drugs." These announcements were made in the context of an increased understanding of the brain but also the recognition that much more research is needed to face these clinical challenges of high social relevance. The current situation, more than 20 years later, is comparable. Our neuroscientific knowledge has increased much more, but in many fields the brain's complexity makes direct translations from basic research to the practice difficult. For example, the behavior of a comparatively simple model organism such as *Caenorhabditis elegans* with only 302 neurons still resists complete understanding (Mausfeld 2012; Thomas and Lockery 2000); or researchers who investigate the mechanisms of neurotransmitter release by amphetamine still face mechanistic challenges, more than 100 years after the substance was discovered (Sulzer et al. 2005). Additionally, the development of new pharmacological treatments for mental illnesses has been stagnating since decades and academic researchers as well as the pharmaceutical industry have become aware that the brain mechanisms associated with psychiatric disorders are much more complex than expected by the previously mentioned US and European brain decades as well as by the German decade of the brain 2000–10.[1] Due to this eminent lack of knowledge we are currently in a "crisis" of psychopharmacology (Amara et al. 2011).

Many of the early papers on cognitive enhancement suggested that new dementia or attention disorder drugs, once they are approved for clinical practice, can also be used to improve memory or attention performance in the healthy (Farah et al. 2004; Whitehouse et al. 1997). Yet, it turns out that the development of such drugs to later succeed in clinical trials, let alone to make them available to the general healthy public without causing serious harm, is very difficult. Recently, as an expression of the crisis, several pharmaceutical companies announced they would give up their research into disorders of the central nervous system because comparably low clinical approval success rates and comparably high development costs make these investments too risky (Miller 2010; van Gerven and Cohen 2011). Understandably, these developments caused serious concerns among clinical pharmacologists. If it is already difficult to re-establish a state of health or to prevent further neurodegeneration, despite the vast amount of research carried out, then we may assume that cognitively improving healthy brains will not be easier.

It is therefore likely that the brain of a well-rested person is already optimally tuned and can hardly be enhanced. Instead, every pharmacological intervention is likely to be fraught with several trade-offs (de Jongh et al. 2008; Quednow 2010a,b; Wood et al. 2014). These trade-offs can be described on the dimension of enhancement–impairment or more general with regard to desired effect and undesired side effects. That means acute enhancement in one aspect of cognition will be secured with an acute impairment in another cognitive, emotional, or behavioral facet. Moreover, it is likely that a chronic use of any pharmacological enhancer will also be followed by a sustained or even chronic cognitive and emotional impairment, because the brain will probably adapt to the repeated pharmacological stimulus, as shown with the stimulant drug cocaine (Vonmoos et al. 2014). In addition, these cost effectiveness trade-offs are highly individual and it is impossible to predict which individual will have a benefit or an impairment in any cognitive domain (Quednow 2010a,b; Wood et al. 2014). Therefore, the pharmacological optimism hypothesis is not plausible in the light of recent developments in neuroscience.

10.3 Assumption 2: the neuroenhancement prevalence hypothesis

With regard to the neuroenhancement prevalence hypothesis, others and we have already warned that widely communicated figures of 15, 20, and even 30% of regular neuroenhancement users may be based on selective or biased interpretations (Hall and Lucke 2010; Lucke et al. 2011a; Quednow 2010a,b; Schleim 2010, 2014b). Investigating 142 English newspaper articles on the prevalence of neuroenhancement for the period 2008–10,

Partridge and colleagues found that 94% of the articles present it as common, increasing, or both (Partridge et al. 2011). Two-thirds of them referred to scientific sources to justify these claims. If our concerns about the neuroenhancement prevalence hypothesis are correct, this finding implies that the public is systematically misled about the frequency of stimulant consumption and that some scientists—wittingly or unwittingly—contribute to this misrepresentation. In our view, a careful look at the prevalence studies does not support this second basic assumption of the neuroenhancement debate as outlined in the following.

The studies investigating larger ($N > 1000$) and more representative samples, mostly of college and university students, usually report a past-year or lifetime prevalence (i.e., at least once in the respective period) of nonmedical stimulant consumption of less than 10% (Smith and Farah 2011). However, interpretations and comparability of these figures are made difficult by different operationalizations of neuroenhancement in these studies. Some surveys asked for specific drugs such as methylphenidate or amphetamine only, others for a wide array of substances. Additionally, most studies defined nonmedical drug consumption broadly so that they included various motivations besides neuroenhancement, such as losing weight or feeling high, what is generally referred to as "recreational" or "lifestyle use." A survey investigating a nonrepresentative sample of 1500 high school pupils and university students in Germany exclusively asked about the consumption of prescription stimulants and illicit drugs for the purpose of improving vigilance, concentration, or memory (Franke et al. 2011). Lifetime prevalence for prescription stimulants was 1.3% but for illicit drugs was 2.6%. Past-year prevalence rates showed low levels ranging between 0.1 and 1.0% for the different groups and substance classes. In contrast to widespread presentations of an epidemic use of neuroenhancers particularly in students, these figures suggest the opposite (Schleim 2010, 2014b). On the basis of a more recent and representative survey, the same group estimated a much higher 12-month-prevalence of 20% for Germany, but only after including caffeine tablets, vitamins, and other dietary supplements that are legally sold over the counter in pharmacies without a prescription (Dietz et al. 2013).

Furthermore, new figures from Switzerland contradict the neuroenhancement prevalence hypothesis as well. A representative survey among 6275 students from the universities of Zurich and Basel revealed that the lifetime prevalence for the use of methylphenidate (4.1%), modafinil (0.3%), amphetamine (0.4%), and cocaine (0.2%) as neuroenhancers is rather low, while a considerable number of students used these substances as recreational drugs: methylphenidate 2.2%, modafinil 0.1%, amphetamine 3.7%, and cocaine 4.2% (Maier et al. 2013). Moreover, representative surveys of health insurance firms from Germany and Switzerland have shown that less than 1% of employees had used stimulants at least once in their lives (DAK 2009; Maier and Schaub 2014). In the DAK Study, of 3000 participants aged between 20 and 50 years, only 0.93% reported to have ever used stimulant drugs against tiredness and sleepiness, while only 0.63% have used these compounds for better concentration at least once (DAK 2009). In the SUVA study, only 0.60% of 10,000 participants between 15 and 74 years have used methylphenidate at least once for neuroenhancement purposes (Maier and Schaub 2014; see also Maier et al. 2015).[2] These numbers strongly suggest that the claimed epidemic use of stimulant drugs as neuroenhancers is a myth. With respect to the claim that the prevalence rates are increasing, an analysis of the scientific literature before the 1990s may shed more light on the validity of the neuroenhancement prevalence hypothesis.

10.4 Neuroenhancement before the 1990s

For the purpose of this paper, we searched for surveys on the prevalence of nonmedical drug consumption from the 1960s to the 1980s. It has been already noted in the cognitive enhancement debate (Quednow 2010a), that medical amphetamine consumption was common in the USA in the 1960s and that prevalence dropped until the 1980s, but then increased again since the early 1990s and reached the levels of the 1960s in mid-2000s, when combining amphetamine and the nowadays much more commonly used methylphenidate (Rasmussen 2008a). However, to our knowledge, nobody has ever investigated whether neuroenhancement might have already been common or perhaps even more frequent in earlier decades.

Interestingly, already in the 1950s, amphetamine sold as Raphetamine was advertised to increase "mental alertness" (see figure in Rasmussen 2008a, 976). Thus, the idea to improve cognitive performance by stimulants is not unique to our time. Much to our surprise, we also found several studies as well as reviews on the prevalence of drug consumption for recreational, self-treatment, and instrumental use decades before some ethicists conceived this as a new problem. For example, Smith and Blachly (1966) investigated the prevalence of amphetamine usage in a nonepresentative sample of 297 medical students. From the 208 subjects who completed the anonymous questionnaires, 92 (44%) had used amphetamine at least once, 73 (79% of the 92) more than once. When asked for the benefits of the drug consumption, 46% named reduced fatigue, 11% reduced appetite, 10% increased alertness, and 4% each increased attention span and motivation.

Wechsler and Rohman (1981) investigated drug use patterns of a larger representative sample of 7170 New England college students in 1977. Within the previous year, 59.3% of the students had used marijuana, 16.2% stimulants, 11.1% cocaine, 9.6% tranquilizers, 7.8% hallucinogens, and lower percentages a number of other drugs. Most importantly, 16% had used stimulants, 11% cocaine, and 10% tranquilizers to stay awake or study, but only 5, 2, and 3%, respectively, reported using these substances more than once a month. These authors also reviewed mean percentages of college students who have used different kinds of drugs (i.e., lifetime prevalence) for the period from 1967 to 1972. Marijuana consumption was most frequent, with 20 to 50%, followed by stimulants at about 20%, depressants/sedatives at about 15%, hallucinogens 7 to 13%, and opiates at about 4%.

Additionally, McAuliffe and colleagues investigated the prevalence of nonmedical drug use in samples of young physicians and premedical, medical, nursing, and other students in New England. They also reviewed prevalence data from 1964 until 1982 (McAuliffe et al. 1984). In a sample of 152 premedical and 211 nonpremedical students at Harvard College surveyed in 1979, 56 and 59%, respectively, had used marijuana, and 24 and 28% had used amphetamines, tranquilizers, or hallucinogens, respectively. From 257 undergraduates surveyed in 1981 at the same college, 56% had used marijuana, 14% amphetamines, 6% tranquilizers, and 9% hallucinogens. A survey of about 2000 nursing, business, medical, communications, law, social work, and counseling professional school students yielded a lifetime prevalence of marijuana use of 61 to 79%, amphetamines 14 to 26%, tranquilizers 13 to 33%, hallucinogens 9 to 30%, sedatives 5 to 17%, cocaine 14 to 29%, and heroin and other opiates 2 to 7%, respectively, for the different student groups. Confined to health professionals and students, they reviewed 21 other studies reporting similar figures of nonmedical drug use.

More generally, a survey from 1960 carried out in the city of Newcastle upon Tyne, UK, found that the equivalent of 200,000 5-mg tablets of amphetamine sulfate was prescribed to a population of 269,389 inhabitants, a quantity that would have been sufficient to supply 1.24% of the city's population with a daily dose of 10 mg. Depression and anxiety, fatigue, and obesity were the most common reasons for the prescription, while "to stay awake," to "pep up," or to delay ejaculation was also mentioned (Kiloh and Brandon 1962). Analogously, about 5% of the US population used prescribed amphetamines in 1971 (Rasmussen 2008a). Whereas such consumption would be considered as medical use in current surveys, these figures support the notion that stimulants were used for various "lifestyle" motives in earlier periods.

In summary, these figures demonstrate that the phenomenon of drug use for reasons other than treating a recognized medical disorder, including instrumental use to improve one's cognitive performance, was already present and even scientifically investigated long before the present debate on the ethics of neuroenhancement was started. In harsh contrast to the neuroenhancement prevalence hypothesis underlying the putative social and scientific urgency of the debate, this use was at least just as frequent in the 1960s–80s as it is nowadays. Most notably, already in 1966, Smith and Blachly had formulated a couple of socially relevant questions derived from their research on amphetamine use of medical students:

> Several questions are raised by this survey which deserves further investigation. (a) Do individuals who feel that their performance is improved by the use of amphetamines actually perform better? (b) What is the class standing of those individuals who use the drug to improve their performance? (c) If increased learning occurs, is it transient or permanent? (d) If students who frequently use the drug to facilitate learning do derive benefit from its use, why don't they use the drug continuously? (e) What other drugs are used and to what extent do medical students undertake self-medication? How is this affected by drug salesmen with their samples? (Smith and Blachly 1966, 170)

These questions have lost none of their relevance 50 years later and reflect some recent debates on the facts and fallacies of neuroenhancement; e.g., whether stimulants are rather secondary enhancers by improving vigilance and motivation instead of cognitive functions directly (Quednow 2010a,b). A major difference between the 1960s–80s and the present seems to be a lack of ethical reflection in scholarly journals in the past, while drug regulation was an important point on the political agenda at the time (Rasmussen

2008b). One notable exception was the psychiatrist Gerald Klerman, who discussed drugs in the light of social values (Klerman 1970, 1972).

10.5 **Drugs and social values**

Initially, two main positions on the ethical aspects of neuroenhancement can be identified. Proponents such as Henry Greely and colleagues have stated,

> "We should welcome new methods of improving our brain function. In a world in which human workspans and lifespans are increasing, cognitive enhancement tools—including the pharmacological—will be increasingly useful for improved quality of life and extended work productivity, as well as to stave off normal and pathological age-related cognitive declines. Safe and effective cognitive enhancers will benefit both the individual and society" (Greely et al. 2008, 705).

In contrast, bioconservatives such as Leon Kass has cautioned,

> "The greatest moral challenges headed our way do not in fact come from hate-filled fanatics threatening death and destruction. They come rather from well meaning scientists and technologists offering life, pleasure, and enhancement" (Kass 2008, 5).

These antithetic ethical positions toward neuroenhancement seem to resemble a social antagonism that was originally coined by Gerald Klerman in order to explain different attitudes regarding the drug culture of the 1960s and 1970s: *psychotropic hedonism* versus *pharmacological Calvinism*[3] (Klerman 1970, 1972). As the term "hedonism" is less useful with respect to a putatively performance-enhancing technology, we would like to slightly adapt this antagonism to the ethical neuroenhancement debate as follows: *pharmacological utilitarianism* versus *pharmacological Calvinism*. The former focus on the reward results (or utility) expected by neuroenhancement and they believe that cognitive enhancement is deeply human, in essence not different from self-betterment through education, culture, sports, sleep, diet, and any information technology. For the welfare of mankind, such interventions will increase the quality of life, can potentially prevent cognitive decline with aging, and are particularly desirable in specific professions (e.g., pilots, air-traffic controllers, surgeons). Quite the opposite is assumed by the pharmacological Calvinists. They emphasize the reward effort and suppose that neuroenhancement of the healthy is absolutely inhuman, unnatural, and artificial, is nothing more than deception, will likely promote unfairness and social inequalities, is the same as drug abuse, is a form of biocapitalism and consequently a coercion of freedom, and in the end is only another form of disease mongering by the pharmaceutical (for review of the different positions see Bostrom and Sandberg 2009; Dees 2007; Greely et al. 2008; Hughes 2009; Kass 2008; Mehlman 2004; Morris 2008; Reiner 2013; Sandberg and Bostrom 2006; Whitehouse et al. 1997).

Here, we would like to propose an additional axis on which attitudes to neuroenhancement can be described, distinguishing different beliefs about the degree of novel knowledge science will provide in the future in order to develop improved pharmacological interventions: *pharmacological optimism* versus *pharmacological skepticism*. With these two axes for ethical and knowledge attitudes, we aim to present a two-dimensional matrix enabling the location of different stances on neuroenhancement. We hope that this

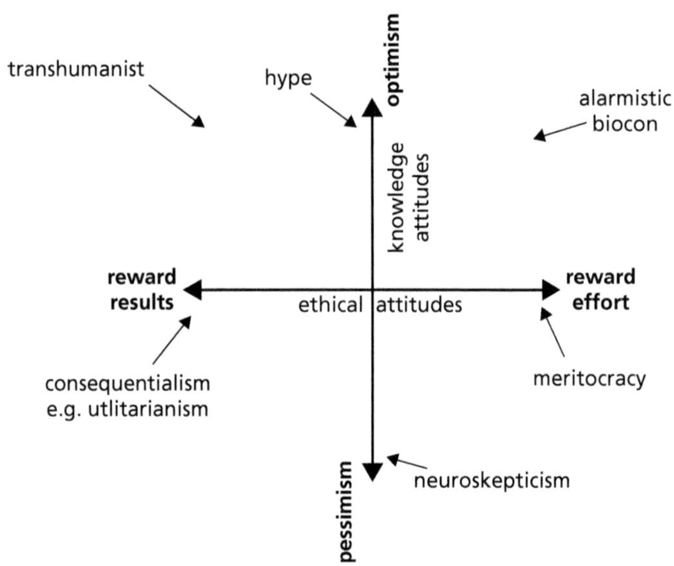

Figure 10.1 The two axes of knowledge and ethical attitudes define a two-dimensional matrix that allows the location of several positions within the ethical neuroenhancement debate. For example, both transhumanists and alarmistic bioconservatives are pharmacological optimists, as they believe that pharmacological neuroenhancement is possible; yet they take opposing ethical views, with the former rewarding results or the utility of an intervention and the latter rewarding merit or effort spent to achieve a certain aim.

provides a useful theoretical tool to better evaluate the ethical, legal, and social aspects of drug use for enhancement purposes. This framework properly illustrates that, e.g., proponents and bioconservatives share an optimistic attitude as both suppose that neuroenhancement actually works (Figure 10.1). An extremely affirmative position is taken by transhumanists, who are aiming at limitless human enhancement, therefore combining an extreme position on the knowledge and ethics dimension (Bostrom and Sandberg 2009; Sandberg and Bostrom 2006).

As stated earlier, the present authors would like to argue for their more cautious position as they have demonstrated that the ethical significance of neuroenhancement is currently over-rated because it is unsafe and less efficient (see section 10.2), most likely addictive (e.g., Bell et al. 2012; Mohamed 2012; Morton and Stockton 2000), much less prevalent than often assumed (see section 10.3), and only desired by a minority (Quednow 2010a, Maier et al. 2015). A balanced position between optimism and skepticism seems to be more appropriate to us. Without suggesting that we know "the truth," we would like to coin an ideal position: *pharmacological realism*. However, when considering the vast complexity of the brain, the development of safe and efficient compounds for use in healthy subjects seems implausible at least in the medium term. This perspective might even be described as pharmacological skepticism and ranges into the Calvinistic dimension (see Figure 10.1). Finally we would agree with our Brazilian colleague João R. Oliveira that "myriad personality traits are just as important as memory or 'intelligence' in the overall scheme of a successful life" (Oliveira 2009, 532).

10.6 Conclusion

The discussion of the pharmacological optimism and neuroenhancement prevalence hypotheses in sections 10.2 and 10.3 strongly suggests that neuroenhancement is not epidemic and neither a new nor an increasing phenomenon, putting these important presuppositions of the ethical debate into question. Thus, we again propose that this discussion is mainly a *phantom debate* (Quednow 2010a, Schleim 2014a) or a *bubble* (Lucke et al. 2011b), which was primarily created and is still kept alive in order to generate public research funding for a few bioethicists, psychiatrists, and neuroscientists. It seems that in particular the newly established neuroethicists strongly benefited from framing what may just have been common drug consumption by a few students as a completely new and urgent problem in the light of the decades of the brain. In correspondence with our conclusion, the US Presidential Commission for the Study of Bioethical Issues recently warned about the detrimental social and scientific effects of the hype on neuroenhancement (Allen and Strand 2015; Presidential Commission for the Study of Bioethical Issues 2015).

Somewhat ironically, the exponential growth in stimulant production, particularly in the USA (Figure 10.2), hardly received attention by neuroethicists as they understood that as a reflection of medical treatment and therefore not as a form of enhancement and, consequentially, not as an important ethical problem. The enhancement/treatment distinction is a very basic one in ethical discourse (US President's Council on Bioethics 2003) and may explain why many neuroethicists fail to see the actual social relevance of exponentially increasing figures of stimulant prescriptions for attention disorders, which is discussed by medical sociologists under the label of medicalization or pharmaceuticalization (Abraham 2010; Bell and Figert 2012; Rose 2006). To put this idea in a provocative manner, many neuroethicists created a new problem and then presented themselves as the experts to regulate the legal, ethical, and social aspects of it (Schleim 2014b). Indeed, scholarly publications on neuroenhancement literally exploded right after the institutionalization of neuroethics in 2004 (Figure 10.2).

It is common to compare benefits and risks for the ethical evaluation of a practice such as neuroenhancement. However, we would like to point out that in addition to the individuals' and societies' benefits and risks that have been frequently discussed—such as competitive advantages, distributive justice, coercion to conform—additional opportunity costs are present which were not sufficiently considered previously. Scientific means and resources are scarce, yet there are many people suffering from psychiatric disorders for whom the development of new psychopharmacological drugs might lead to a serious decrease of suffering and increase in quality of life. Particularly in current times when many pharmacological companies already have dropped or have announced to drop psychopharmacological projects and, thus, means and resources have become even scarcer,

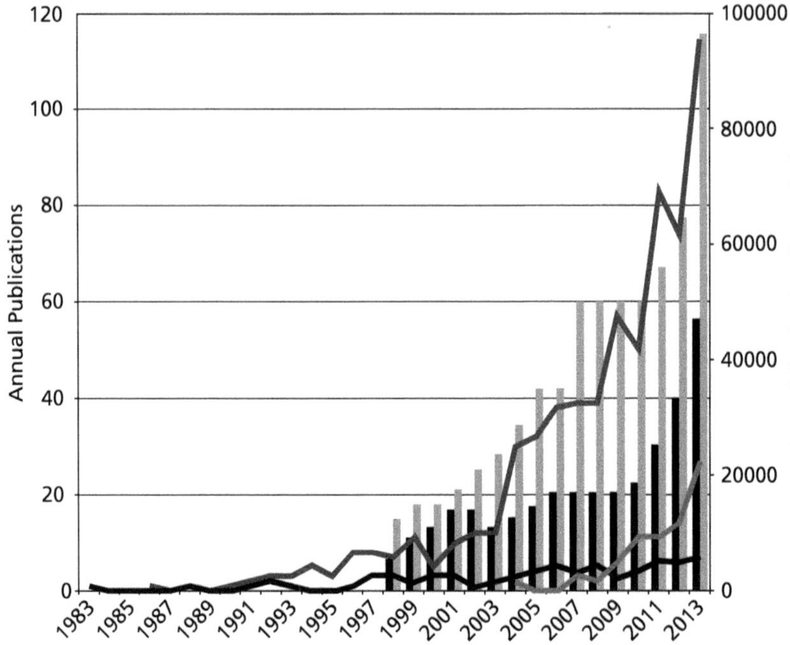

Figure 10.2 Number of publications (left axis) listed in the ISI Web of Science per year for the topic searches "cognitive enhancement" (dark gray line), "mood enhancement" (black), and "neuroenhancement" (light gray) and annual production quotas of methylphenidate (light gray bars) and amphetamine (excluding amphetamine produced for conversion; dark gray bars) in kilograms (right axis), based on official figures published by the US Drug Enforcement Agency and the Federal Register. Publications on enhancement topics exploded right after the institutionalization of neuroethics in 2004. Aggregate publication numbers (all three kinds of enhancement) and production quotas (methylphenidate plus amphetamine) correlate at $r = 0.97$. We refrain from an attempt to explain this extremely high correlation here, but speculate that both increases may reflect the medicalization/pharmaceuticalization of social problems.

we would like to encourage the investigation of treatments for psychiatric disorders instead of the development of neuroenhancement drugs for the healthy (Schleim and Quednow 2015). What applies to pharmacologists applies to theoreticians and ethicists competing for research grants, too. Our discussion suggests that important perspectives are neglected when only one kind of experts such as neuroethicists analyzes these problems. As Raymond De Vries emphasized some years ago, it may be important to guard the self-appointed guardians of neuroscience (De Vries 2007).

10.7 Acknowledgments

We would like to thank Felix Hasler and Nicolas Langlitz for helpful discussions before and during the preparation of the present paper. The research visit of Stephan Schleim at the Division of Experimental and Clinical Pharmacopsychology (Head: Professor Dr. Boris B. Quednow), Department of Psychiatry, Psychotherapy, and Psychosomatics,

Psychiatric Hospital, University of Zurich was supported by a grant from the Barbara Wengeler Foundation.

Notes

1. See this German press release of November 5, 1999, http://idw-online.de/de/news15426 (accessed December 23, 2014). In our perception, not many activities of the German decade of the brain were accomplished.
2. It should be noted that we were approved to contact Swiss participants who reported a neuroenhancement use of methylphenidate in the online surveys of Maier et al. (2013) and Maier and Schaub (2014) in order to investigate their neuropsychological performance and personality profile at the Psychiatric Hospital of the University of Zurich (see Maier et al. 2015). However, in personal interviews it turned out that most of these participants tried methylphenidate only once or that the compound was often prescribed for an underlying attention deficit hyperactivity disorder, which was not declared in the online survey. Therefore, regular use of methylphenidate for neuroenhancement purposes in mentally healthy individuals was very rare and we will show in a future publication that the personality profile of these users is highly specific.
3. It was noted by Louis Menand in an article in *The New Yorker* published on March 1, 2010 that Klerman may have misunderstood Calvinist theology when he coined that term, but we would like to leave that discussion aside for our purposes and use the concept since it has become established in the scholarly debate (http://www.newyorker.com/magazine/2010/03/01/head-case-2, accessed December 23, 2014).

References

Abraham, J. 2010. Pharmaceuticalization of society in context: Theoretical, empirical and health dimensions. *Sociology-the Journal of the British Sociological Association* **44**(4):603–22.

Allen, A. L., and N. K. Strand. 2015. Cognitive enhancement and beyond: recommendations from the bioethics commission. *Trends in Cognitive Sciences* **19**(10):549–51.

Amara, S. G., S. Grillner, T. Insel, D. Nutt, and T. Tsumoto. 2011. Neuroscience in recession? *Nature Reviews: Neuroscience* **12**(5):297–302.

Bell, S. E., and A. E. Figert. 2012. Medicalization and pharmaceuticalization at the intersections: Looking backward, sideways and forward. *Social Science and Medicine* **75**(5):775–83.

Bell, S. K., J. C. Lucke, and W. D. Hall. 2012. Lessons for enhancement from the history of cocaine and amphetamine use. *AJOB Neuroscience* **3**(2):24–9.

Bostrom, N., and A. Sandberg. 2009. Cognitive enhancement: Methods, ethics, regulatory challenges. *Science and Engineering Ethics* **15**(3):311–41.

Bush, G. H. W. 1990. *Presidential Proclamation 6158: Project on the decade of the brain.* Washington, DC: Library of Congress.

DAK. 2009. *Gesundheitsreport 2009: Analyse der Arbeitsunfähigkeitsdaten. Schwerpunktthema Doping am Arbeitsplatz.* Hamburg: DAK Forschung.

de Jongh, R., I. Bolt, M. Schermer, and B. Olivier. 2008. Botox for the brain: enhancement of cognition, mood and pro-social behavior and blunting of unwanted memories. *Neuroscience and Biobehavioral Reviews* **32**(4):760–76.

De Vries, R. 2007. Who will guard the guardians of neuroscience?—Firing the neuroethical imagination. *EMBO Reports* **8**(1S):S65–9.

Dees, R. H. 2007. Better brains, better selves? The ethics of neuroenhancements. *Kennedy Institute of Ethics Journal* **17**(4):371–95.

Dietz, P., H. Striegel, A. G. Franke, K. Lieb, P. Simon, and R. Ulrich. 2013. Randomized response estimates for the 12-month prevalence of cognitive-enhancing drug use in university students. *Pharmacotherapy* **33**(1):44–50.

Farah, M. J., J. Illes, R. Cook-Deegan, H. Gardner, E. Kandel, P. King, E. Parens, B. Sahakian, and P. R. Wolpe. 2004. Neurocognitive enhancement: what can we do and what should we do? *Nature Reviews: Neuroscience* **5**(5):421–5.

Franke, A. G., C. Bonertz, M. Christmann, M. Huss, A. Fellgiebel, E. Hildt, and K. Lieb. 2011. Non-medical use of prescription stimulants and illicit use of stimulants for cognitive enhancement in pupils and students in Germany. *Pharmacopsychiatry* **44**(2):60–6.

Greely, H., B. Sahakian, J. Harris, R. C. Kessler, M. Gazzaniga, P. Campbell, and M. J. Farah. 2008. Towards responsible use of cognitive-enhancing drugs by the healthy. *Nature* **456**(7223):702–5.

Hall, W. D., and J. C. Lucke. 2010. The enhancement use of neuropharmaceuticals: more scepticism and caution needed. *Addiction* **105**(12):2041–3.

Hughes, J. 2009. Technoprogressive biopolitics and human enhancement. In *Progress in Bioethics*, ed. J. Moreno and S. Berger, 163–188. Cambridge, MA: MIT Press.

Kass, L. 2008. How brave a new world? *Society* **45**(1):5–8.

Kiloh, L. G., and S. Brandon. 1962. Habituation and addiction to amphetamines. *BMJ* **2**(5296):40–3.

Klerman, G. L. 1970. Drugs and social values. *International Journal of the Addictions* **5**(2):313–19.

Klerman, G. L. 1972. Psychotropic hedonism vs. pharmacological Calvinism. *The Hastings Center Report* **2**(4):1–3.

Lucke, J. C., S. K. Bell, B. J. Partridge, and W. D. Hall. 2011a. Academic doping or Viagra for the brain? The history of recreational drug use and pharmacological enhancement can provide insight into these uses of neuropharmaceuticals. *EMBO Reports* **12**(3):197–201.

Lucke, J. C., S. K. Bell, B. J. Partridge, and W. D. Hall. 2011b. Deflating the neuroenhancement bubble. *AJOB Neuroscience* **2**(4):38–43.

McAuliffe, W. E., M. Rohman, P. Fishman, R. Friedman, H. Wechsler, S. H. Soboroff, and D. Toth. 1984. Psychoactive drug use by young and future physicians. *Journal of Health and Social Behavior* **25**(1):34–54.

Maier, L. J., M. E. Liechti, F. Herzig, and M. P. Schaub. 2013. To dope or not to dope: Neuroenhancement with prescription drugs and drugs of abuse among Swiss university students. *PLOS One* **8**(11):e77967.

Maier, L. J., and M. P. Schaub. 2014. *"Doping" am Arbeitsplatz und in der Bildung in der Schweiz.* Zürich: Schweizer Institut für Sucht- und Gesundheitsforschung (ISGF).

Maier, L. J., M. D. Wunderli, M. Vonmoos, A. T. Rommelt, M. R. Baumgartner, E. Seifritz, M. P. Schaub, and B. B. Quednow. 2015. Pharmacological cognitive enhancement in healthy individuals: a compensation for cognitive deficits or a question of personality? *PLOS One* **10**(6):e0129805.

Mausfeld, R. 2012. On some unwarranted tacit assumptions in cognitive neuroscience. *Frontiers in Psychology* **3**:67.

Mehlman, M. J. 2004. Cognition-enhancing drugs. *Milbank Quarterly* **82**(3):483–506.

Miller, G. 2010. Is pharma running out of brainy ideas? *Science* **329**(5991):502–4.

Mohamed, A. D. 2012. Modafinil has the potential for addiction. *AJOB Neuroscience* **3**(2):36–8.

Morris, K. 2008. Experts urge smart thinking on cognitive enhancers. *Lancet Neurology* **7**(6):476–7.

Morton, W. A., and G. G. Stockton. 2000. Methylphenidate abuse and psychiatric side effects. *Journal of Clinical Psychiatry* **2**(5):159–64.

Oliveira, J. R. 2009. Much ado about cognitive enhancement. *Nature* **457**(7229): 532.

Pandolfi, F. M. 1993. European decade of brain research. *Italian Journal of Neurological Sciences* **14**(5):395–7.

Partridge, B. J., S. K. Bell, J. C. Lucke, S. Yeates, and W. D. Hall. 2011. Smart drugs "as common as coffee": media hype about neuroenhancement. *PLOS One* **6**(11):e28416.

Presidential Commission for the Study of Bioethical Issues. 2015. *Gray Matters; Topics at the intersection of neuroscience, ethics, and society,* Volume 2. Washington, DC.

Quednow, B. B. 2010a. Ethics of neuroenhancement: A phantom debate. *BioSocieties* **5**(1):153–6.

Quednow, B. B. 2010b. Neurophysiologie des Neuroenhancements: Möglichkeiten und Grenzen. *SuchtMagazin* **2**:19–26.

Rasmussen, N. 2008a. America's first amphetamine epidemic 1929–1971: A quantitative and qualitative retrospective with implications for the present. *American Journal of Public Health* **98**(6):974–85.

Rasmussen, N. 2008b. *On speed; the many lives of amphetamine* New York: New York University Press.

Reiner, P. B. 2013. The biopolitics of cognitive enhancement. In *Cognitive enhancement: An interdisciplinary perspective,* ed. E. Hildt and A. Franke. Heidelberg: Springer. 189–200

Rose, N. 2006. Disorders without borders? The expanding scope of psychiatric practice. *BioSocieties* 1:465–84.

Sandberg, A., and N. Bostrom. 2006. Converging cognitive enhancements. *Annals of the New York Academy of Sciences* 1093:201–27.

Schleim, S. 2010. Second thoughts on the prevalence of enhancement. *BioSocieties* **5**(4):484–5.

Schleim, S. 2014a. Critical neuroscience—or critical science? A perspective on the perceived normative significance of neuroscience. *Frontiers in Human Neuroscience* **8**:336.

Schleim, S. 2014b. Whose well-being? Common conceptions and misconceptions in the enhancement debate. *Frontiers in Systems Neuroscience* **8**:148.

Schleim, S., and B. B. Quednow. 2015. Cognitive Enhancement; Ein kritischer Rückblick auf fünfzehn Jahre Forschung. *InFo Neurologie and Psychiatrie* **13**(5):2–5.

Smith, M. E., and M. J. Farah. 2011. Are prescription stimulants "smart pills"? The epidemiology and cognitive neuroscience of prescription stimulant use by normal healthy individuals. *Psychological Bulletin* **137**(5):717–41.

Smith, S. N., and P. H. Blachly. 1966. Amphetamine usage by medical students. *Journal of Medical Education* **41**(2):167–70.

Sulzer, D., M. S. Sonders, N. W. Poulsen, and A. Galli. 2005. Mechanisms of neurotransmitter release by amphetamines: a review. *Progress in Neurobiology* **75**(6):406–33.

Thomas, J. H., and S. R. Lockery. 2000. Neurobiology. In *C. elegans: A practical approach,* ed. I. A. Hope, 143–80. Oxford: Oxford University Press.

US President's Council on Bioethics. 2003. *Beyond therapy: Biotechnology and the pursuit of happiness.* New York: ReganBooks.

van Gerven, J., and A. Cohen. 2011. Vanishing clinical psychopharmacology. *British Journal of Clinical Pharmacology* **72**(1):1–5.

Vonmoos, M., L. M. Hulka, K. H. Preller, F. Minder, M. R. Baumgartner, and B. B. Quednow. 2014. Cognitive impairment in cocaine users is drug-induced but partially reversible: evidence from a longitudinal study. *Neuropsychopharmacology* **39**(9):2200–10.

Wechsler, H., and M. E. Rohman. 1981. Patterns of drug-use among New England college-students. *American Journal of Drug and Alcohol Abuse* **8**(1):27–37.

Whitehouse, P. J., E. Juengst, M. Mehlman, and T. H. Murray. 1997. Enhancing cognition in the intellectually intact. *Hastings Center Report* **27**(3):14–22.

Wood, S., J. R. Sage, T. Shuman, and S. G. Anagnostaras. 2014. Psychostimulants and cognition: a continuum of behavioral and cognitive activation. *Pharmacological Reviews* **66**(1): 193–221.

Part III

Ethical, philosophical, legal, and policy issues of cognitive enhancement

Chapter 11

The evolutionary limits to neuroenhancement

Ralph Hertwig and Thomas Hills

11.1 An evolutionary paradox

According to a recent, but possibly already outdated estimate, "more than 100 drugs are currently being developed, tested or used for cognitive enhancement" (Förstl 2009, 841). The psychoactive substances under investigation include cholinesterase inhibitors, memantine, dimebon, ampakines, fluoxetine and other antidepressants, methylphenidate, and modafinil. All of these substances are being considered for use as neuroenhancers in healthy individuals (Förstl 2009), with potential advantages for attention, memory, motivation, creativity, and alertness. It is also speculated that psychoactive substances may be able to improve moral decision-making, relationships, and longevity in healthy people (Earp et al. 2012; Finkel and Holbrook 2000; Persson and Savulescu 2012). On the surface of things, neuroenhancers thus seem to offer vast potential for improving human life.

According to media coverage, this message has not fallen on deaf ears. Numerous reports have suggested that the use of prescription drugs to improve cognitive and psychological functions—neuroenhancement—in healthy people is "common, increasing in prevalence, or both" (Partridge et al. 2011, 4). Beyond the media hype, however, representative surveys suggest that neuroenhancement is not (yet) the norm. For instance, German surveys of the general population, student populations, and employees gauge the prevalence of neuroenhancer use in healthy individuals to be about 1.5 to 5% (Kowalski 2013). Yet while these numbers are much smaller than those broadcast by the media, they still translate into an estimated two million employees in Germany using psychoactive drugs for nonmedical purposes.

This purported rise in pharmaceutical self-enhancement represents a potential sea change in the way the general public and, to some extent, scientists view cognition. In the past, cognitive ability was conceptualized as a nearly fixed quantity, one that people essentially have to live with until it inevitably declines with age. To the extent that there is plasticity, it was assumed that cognitive abilities could be boosted only through diligent practice involving great effort (e.g., Kliegl et al. 1989). Neuroenhancement, in contrast, bears the promise of made-to-order, off-the-shelf optimized cognition. It promises to make people smarter, more alert, more focused, or more persistent—and that these gains come without pain. As the old adage goes, however, "there's no such thing as a free lunch." In this chapter, we explore the hidden cognitive costs of neuroenhancement. By hidden costs, we do not mean the potentially serious and unanticipated medium- and long-term health consequences of extended use of psychoactive substances (e.g., Förstl 2009), neither do we mean the many ethical issues associated with the rise of neuroenhancement (e.g., Farah 2005). Our starting point is a different one: If enhancers are truly beneficial, why have the respective cognitive capacities not been developed through evolution?

Adaptationist accounts of evolution posit that, in proportion to the selective pressures from the environment, evolution will optimize over constraints and produce phenotypes that are close to optimal (Pigliucci and Kaplan 2000). The bolder the claims made about enhancement, the more likely it would seem that the capacities being targeted should already exist. If they offer competitive advantages to the organisms that have them, then natural selection should have favored them. With the potential of pharmacology to improve school performance, reduce age-related cognitive decline, improve human performance in combat, and even enhance scientific productivity (e.g., Greely et al. 2008), it is almost too easy to overlook the question of why these capacities have not been developed through evolution (see, e.g., Humphrey 2002). If these enhancements are so promising and powerful, it seems doubly important to understand why humans are not yet endowed with them.

One possible answer from an evolutionary perspective is that the notion that cognitive capacities can be enhanced at no cost rests on a naïve "more-is-better" assumption. More memory is better; more focus is better; more happiness is better; more self-control and willpower are better; and so on. Just as people cherish faster processing speed and larger memory in their digital electronics, they may assume that boosting a particular cognitive trait will bring better mental performance and affective well-being. As we discuss in this chapter, however, the empirical evidence indicates that in many domains the more-is-better assumption is incorrect. Unfortunately, both users of neuroenhancers and some of their scholarly advocates seem to buy into this assumption. Claims that "the drugs … should be viewed in the same general category as education, good health habits, and information technology" (Greely et al. 2008, 702), for example, appear to rest on this naïve and erroneous view of cognition.

Admittedly, numerous technological additions to modern life can be viewed as enhancers of cognition—pencils, books, and computers can all help to improve decision-making and quality of life. The same applies to enhancers of cognition that target the representation of information and knowledge in the world: presenting information in ways that are more easily processed can help to improve the statistical reasoning and decision-making of patients, doctors, lawyers, and the public more generally (e.g., Hoffrage et al. 2000). How is a pharmacological nudge any different? From an evolutionary perspective, the difference is profound. Education, health habits, computers, and information presentation formats are subject to horizontal (i.e., cultural) transmission. They have not been around long enough, nor has the selection pressure been strong enough, to lead to evolutionary changes at the level of the genome.

Cultural changes do affect cognitive faculties and can improve cognitive performance. Indeed, systematic gains in human intelligence—both crystallized and fluid—have been observed since the middle of the past century in most of the developed world (Flynn

2012). The proposed explanations for these changes are cultural in origin; predominantly knowledge-based accounts explain the gains in terms of improvements in, for example, abstract reasoning and analogy (Flynn 2012; Fox and Mitchum 2013). The possibility that the gains could be evolutionary in origin is flatly discounted, because the time scale over which the changes have taken place is simply too short.

However, the targets of neuroenhancement—e.g., attention and memory—*have been* subject to selection. Evolution has had thousands, if not millions, of years to fine-tune these cognitive traits. If evolution has not adapted them to the environment, the question is why. One may argue that today's environment is sufficiently different from the ancestral environment to warrant additional tuning. We briefly return to this caveat in section 11.3. Yet it is deeply curious that some single-gene mutations appear to enhance memory and cognition, but do not appear to have been selected for in wild-type populations. Seemingly trivial neuromolecular changes, such as the overexpression of *N*-methyl-D-aspartic acid (NMDA) receptors in the hippocampus, appear to enhance memory (Tang et al. 1999). Mutant mice lacking the nociceptin receptor also show improved learning and memory (Manabe et al. 1998), as do mice lacking the α5 subunit of the $GABA_A$ receptor (Collinson et al. 2002). Given the likely ease with which evolution could have accessed these mutations, their absence in wild-type populations is difficult to reconcile with their apparent added value. Something is amiss.

The resolution of this evolutionary paradox is the focus of this chapter. In particular, we consider why cognitive capacities have evolved to their currently observable state. Our explanation requires an understanding of the constraints (or trade-offs) under which cognitive traits have evolved. Without such trade-offs, selective forces associated with improved performance (and thus fitness) would drive the performance capacities of cognitive traits ever upward. As we explain later, however, all known evolutionary trajectories inevitably run up against constraints that prevent such runaway selection. The costs eventually outweigh the benefits. So which evolutionary trade-offs are built into the cognitive system and where do they come from? In what follows, we discuss two kinds of evolutionary trade-offs that must be considered in the context of neuroenhancement. We then briefly provide theoretical and empirical evidence for an alternative to the more-is-better view of human cognition and decision making. Finally, we outline what follows from the evolutionary view that we propose in this chapter.

11.2 Evolutionary cognitive trade-offs

The evolution of any living system is the result of trade-offs over multiple constraints. Consider the human female pelvis. Because its dimensions are small relative to a baby's head, obstetric complications during labor are common. Why hasn't evolution improved the survival chances of both mother and baby by selecting for a larger female pelvis? The widely accepted explanation is that the optimal pelvis for bipedal locomotion and the optimal pelvis for encephalization (the progressive increase in the baby's brain size) place competing demands on the human pelvis. Bipedal locomotion requires substantial skeletal changes, including alterations in the pelvic architecture (Wittman and Wall 2007), and such changes must compete (in an evolutionary sense) with the obstetric demands of human babies' relatively large brains. Recent evidence suggests that this limit may be still more complicated, involving metabolic demands on the mother, who is unable to cope with babies above a certain size (Dunsworth et al. 2012).

Cognition is the product of similar trade-offs over multiple constraints. Notably, these constraints originate from two different sources: (1) the kinds of problems a flexible intelligence has evolved to solve and (2) the underlying biology. These two kinds of constraints generate within-domain and between-domain trade-offs, respectively. These trade-offs have typically been confounded in the literature on enhancement, but their distinctness is evident in the ubiquity of *inverted U-shaped performance functions* (henceforth ∩-shaped performance functions) and the *side effects* of enhancement (Cools and Robbins 2004; Husain and Mehta 2011), respectively. These two kinds of trade-offs explain why we are not smarter than we are.

11.2.1 Trade-offs within domains: ∩-shaped performance functions

Performance functions that are ∩-shaped are often observed for optimal-control problems in which the goal is to maximize benefits subject to some cost function (i.e., specified constraints). Such problems often require a decision on when to stop taking one action and switch to another. Specific examples include when to accept a particular job candidate or mating partner rather than to continue looking for another; when to leave one resource patch to move to another; and when to stop collecting information and make a decision. Cognition must solve these kinds of attention-switching problems in countless domains that combine poorly defined completion criteria (i.e., in which it is unclear when to stop) with opportunity costs (Hills et al. 2010). Furthermore, even in problems that provide well-defined completion criteria, there may be multiple possible trajectories to a solution, and finding a solution may involve abandoning an approach that is not working.

Myriad cognitive problems of organisms from nematodes to office workers have been formulated as optimal-control problems (e.g., Pirolli 2007). The ecological regularity of such problems has led to the proposal that regulating goal maintenance and abandonment (i.e., persistence in action) is one of the key evolutionary building blocks leading to the flexible intelligence associated with executive cognition (Hills and Dukas 2012). For support, this proposal points to the shared structure of control problems across domains and the shared neural correlates and cognitive function of the processes that solve these problems across species. As one example, domain-general cognitive-control processes related to working memory span are governed by attentional control and updating (see Unsworth and Engle 2007). These control processes, by their nature, influence how attention is distributed over potential goals in an environment.

Mathematically, optimizing the control of attention over opportunity costs can be reduced to a search problem, in which cognition attempts to maximize its payoff by choosing how long, t^*, to perseverate on one action or goal state before switching to another.

With the realistic assumption that any course of action is associated with reduced payoffs over time, $F(t^*)' < 0$, and that it costs some amount of time, T, to switch between actions, the optimal t^* solves the equation

$$F(t^*)' = \frac{\overline{F}(t^*)}{t^* + T}$$

where $\overline{F}(t^*)$ is the mean payoff associated with all other action payoff functions. In words, the giving-up (or abandonment) rate, $F(t^*)'$, associated with one action should be related to the opportunity costs associated with switching to other possible actions. For any problem that fits this basic framework, an ∩-shaped performance curve arises naturally over the values of t^* (see Figure 11.1). Given that problems of optimizing over subgoals make up a significant portion of real and laboratory cognitive tasks (e.g., Tower of London, category fluency, and operation span), one should expect to find ∩-shaped dose-response curves among people who have taken pharmaceuticals designed to increase the duration of focused attention, t, with respect to a given task.

For illustration, consider the following accuracy-effort trade-off faced by a chess player planning her next move. The time to investigate future moves is limited (e.g., at tournaments, each player often has not more than 2 hours to make the first 40 moves). Constrained by the capacity to investigate only one sequence of future moves at a time, the player has to solve the problem of how long to investigate any one sequence of moves. She may choose to move her pawn first, which she anticipates will be met with the opponent

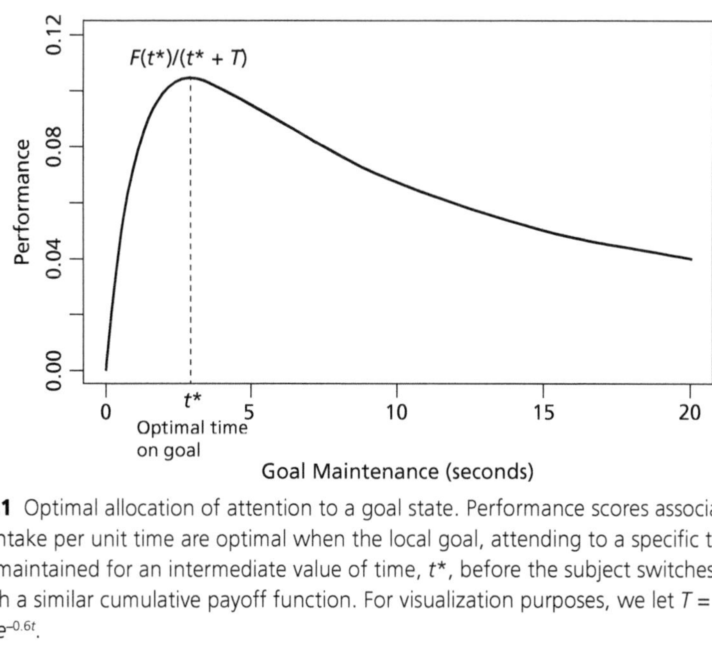

Figure 11.1 Optimal allocation of attention to a goal state. Performance scores associated with resource intake per unit time are optimal when the local goal, attending to a specific target or action, is maintained for an intermediate value of time, t^*, before the subject switches to a new action with a similar cumulative payoff function. For visualization purposes, we let $T = 5$ s and $F(t) = 1 - e^{-0.6t}$.

playing their rook, followed by her playing her knight, and so on. But how long should she continue down this route? The longer she investigates this sequence, the less time she will have to investigate alternatives. If she investigates many alternatives, she will know little about the long-term consequences of any of them. Yet the cost of thoroughly investigating only a few alternatives is that she may overlook better ones. There are many facets to being a good chess player, but a key one is the capacity to optimize the trade-off between too little and too much exploration of a given sequence of moves. As in Figure 11.1, the best solution lies somewhere in between.

The importance of striking a balance between too much and too little is not just theoretical, but relates directly to research on neuroenhancement. Consider amphetamines, Ritalin, and modafinil, all of which have been proposed as neuroenhancers of attention. These drugs exhibit some positive effects on cognition, especially among individuals with lower baseline abilities. However, individuals of normal or above-average cognitive ability often show negligible improvements or even decrements in performance following drug treatment (for details, see de Jongh et al. 2008). For instance, Randall et al. (2005) found that modafinil (a drug licensed for the treatment of narcolepsy and sleep apnea) improved performance only among individuals with lower IQ, but not among those with higher IQ. Similarly, Mohamed (2014) showed that, after taking modafinil, participants low in the personality trait of creativity scored higher on a convergent thinking task of creativity than did those scoring high on this trait. In addition, modafinil appeared to reduce divergent creative thinking in the same participants. Similarly, Farah et al. (2009) found a non-linear relationship of dose to response for amphetamines in a remote-associates task, with low-performing individuals showing enhanced performance, but high-performing individuals showing reduced performance. Such ∩-shaped dose–response curves are common (see Cools and Robbins 2004). More importantly, they demonstrate that improving attention is much more complicated than simply giving individuals the capacity to increase their focus.

11.2.2 Trade-offs between domains: cognitive side effects

Within-task trade-offs represent a key constraint on both the evolution and the pharmacological enhancement of cognition. However, even when within-task trade-offs can be avoided, between-task trade-offs may remain. The reason is interdependencies across domains. Recall our example of the female pelvis. Expanding the birth canal would reduce the likelihood of obstetric complications, but at the expense of efficient bipedal locomotion. Such interdependencies apply across cognitive domains as well as between cognition and other more general domains, such as mental and physical health. One common example is the rise in anxiety and loss of fine motor control often found following high doses of caffeine (Smith 2002). By the same token, the enhancements to memory and attention associated with nicotine (Warburton 1992) have well-documented long-term effects on mortality and health (Rostron 2013).

It has been controversially argued that the Ashkenazi Jew population provides a less well-known, but more dramatic example of between-domain trade-offs (for all arguments in this paragraph, see Cochran et al. 2005; Cochran and Harpending 2009; and also Pinker 2006 for a critical review). Among Ashkenazi Jews, the average IQ is approximately 0.7 to 1 standard deviations above that of the general European population. According to Cochran and colleagues, this rise in IQ may have been the consequence of evolutionary selection for greater intelligence among Ashkenazi Jews in medieval Europe. However, this greater cognitive capacity appears to have exacted costs on other dimensions. In particular, although Ashkenazi Jews perform better on verbal and mathematical tests, they perform roughly half a standard deviation below the European average on visuospatial abilities. Perhaps more importantly, sphingolipid disorders such as Tay-Sachs, Niemann Pick, Gaucher, and mucolipidosis are more prevalent in this population. These diseases are correlated with the same neural causes that rendered increased IQ possible, such as increased dendrite development.

Cases of individuals with superior memory provide additional examples of cognitive trade-offs. Parker et al. (2006) reported on the case of a woman, AJ, who described her memory as "non-stop, uncontrollable, and totally-exhausting" (p. 35). Tests showed her to achieve perfect or near-ceiling performance on visual memory, visual paired associates, word recognition, autobiographical memory, and digit span. However, AJ's performance on tasks of executive function and reasoning were impaired. For example, on the Wisconsin Card Sorting Test—a task that requires adaptive switching to new contexts—AJ showed abnormal levels of perseveration on past rules. She also showed impaired performance on a face recognition test. Moreover, AJ's enhanced memory performance did not translate into outstanding school grades. She explained that her memory—although prodigious—was "automatic, not strategic," and therefore of limited use in willful memorization.

Similar accounts of memory and other cognitive trade-offs are found for people with synaesthesia, which is the experience of cross-modal associations—for example, seeing the color yellow when hearing the number 4. Individuals with synaesthesia often show enhanced memory for numbers and words (Yaro and Ward 2007). However, they also often have deficits in mathematical and spatial reasoning (Cytowic 1995; Rothen et al. 2012). Luria's (1987) famous examination of a man named Shereshevskii, whose memory appeared to have "no distinct limits" (p. 11) and in whom Luria diagnosed an extremely strong version of synaesthesia, offers another illustration of the potential costs of perfect memory. For instance, the experience of cross-modal associations caused Shereshevskii difficulties in daily life:

> One time I went to buy some ice cream ... I walked over to the vendor and asked her what kind of ice cream she had. "Fruit ice cream," she said. But she answered in such a tone that a whole pile of coals, of black cinders, came bursting out of her mouth, and I couldn't bring myself to buy any ice cream after she had answered in that way. (Luria 1987, 82)

Even more profoundly, he had problems extracting the meaning of simple passages of text and especially of poetry or metaphors. In Luria's (1987) words, he

> was unable to grasp an idea unless he could actually see it, and so he tried to visualize the idea of "nothing", to find an image with which to depict "infinity". And he persisted in these agonizing

> attempts all his life, forever coping with a basically adolescent conflict that made it impossible for him to cross that "accursed" threshold to a higher level of thought. (Luria 1987, 133)

The paradox of near-perfect memory paired with comprehension deficits bedeviled Shereshevskii. Similar paradoxes are often experienced by savants, who offer perhaps the clearest natural evidence for between-domain trade-offs in performance across tasks. Their spectacular skills in one domain go hand-in-hand with poor performance in other domains (Treffert 2009). This is no coincidence. Savant-like skills can be induced in healthy participants by *turning off particular functional areas of the brain*—for example, via repetitive transcranial magnetic stimulation (Snyder 2009).

11.3 More capacity is not invariably better

What is the origin of the more-is-better assumption that is the leitmotif of the enhancement venture? One can only speculate, but one source may be psychology's portrayal of the human mind as profoundly limited—a view that emerged in the second half of the twentieth century. Our brain has been weighed, measured, and found wanting. In research on memory, for instance, the list of documented limits includes the now classic thesis that the number of "chunks" of information that can be stored in short-term memory is restricted to the magical number of seven, plus or minus two (Miller 1956); more recently, this number has been trimmed to just four chunks (Cowan 2001). In research on attention, the influential filter theory (Broadbent 1958) envisioned a limited transmission capacity that requires a filter to protect the human information processing machinery from overload. Kahneman (1973) replaced the filter by a supply metaphor, in which attention is an unspecified capacity ("energy") that, due to its limited nature, must be strategically allocated. More recently, Pashler (1998) revived the filter theory and argued that such a filter acts as a central bottleneck in decision-making and memory retrieval. Still others have proposed that the human budget of strategic processing capacity is restricted (e.g., Schweickert and Boggs 1984), and that the ability to pursue multiple intentional goals at any one time is therefore limited. This list of limitations goes on and on.

The view that the human capacity to process information is limited is often accompanied by another assumption, namely, that these limitations pose a liability (see Hertwig and Todd 2003). They restrict people's cognitive potential, barring them from performing feats such as computing the square roots of large numbers in their heads or reciting all of Shakespeare's sonnets by heart. More sinister still, these cognitive limitations are suspected not only of undermining performance, but also of thwarting the ability to reason, judge, and decide rationally. The close link between cognitive limitations and suboptimal reasoning is made in such varied psychological research programs as Piaget's theory of cognitive development in children (e.g., Flavell 1985), Johnson-Laird's mental model theory (1983, 2001), and Kahneman and Tversky's influential heuristics-and-biases program (e.g., Kahneman et al. 1982). The latter, for instance, assumes that people, weighed down by their cognitive limitations and pressed to make decisions in a complex and uncertain world, are often unable to reason and decide rationally (i.e., in agreement with the norms of probability theory, rational choice theory, and statistics). Instead, they have no choice but to recruit mental shortcuts—heuristics—that are assumed to make them vulnerable to committing systematic and consequential reasoning errors.

If the human mind is indeed a profoundly limited information processing machine, people need all the help they can get to surmount their deficiencies. And perhaps neuroenhancers are just what the doctored ordered. Yet a new conception of human decision-making suggests that simple heuristics are in fact indispensable and powerful mental strategies that enable people to reckon with uncertainty and complexity. This research demonstrates that there is no general trade-off between accuracy and effort. Simple heuristics can be as accurate as or more accurate than strategies requiring much more computation and information (e.g., Gigerenzer et al. 2011; Hertwig et al. 2013).

Relatedly, work on language acquisition has demonstrated that limiting "memory span" can facilitate language acquisition (Cochran et al. 1999; Elman 1993; Kersten and Earles 2001; Newport 1990; see also Plunkett et al. 1997). Memory researchers have emphasized the importance of forgetting for an adaptive memory system (Altmann and Gray 2002; Benjamin 2011), mental health (de Quervain 2008), and the accuracy of simple heuristics (Schooler and Hertwig 2005). Still other researchers have demonstrated that reliance on small samples enables surprisingly accurate inferences (e.g., Hertwig and Pleskac 2010; Vul et al. 2014), and so on.

The crucial implication of this and related research is that endowing the human mind with increased cognitive capacities, and thus boosting its ability to encode, store, retrieve, and process information, is not a silver bullet to improve performance. Not infrequently, leaner computations, less information, and the forgetting of unneeded data *enable* good or even superior performance—less is often more. This regularity is likely to hold across different environments, ancestral and modern alike. We believe that these empirical findings and theoretical considerations profoundly challenge the more-is-better premise of neuroenhancement.

11.4 Why aren't humans smarter already?

We believe that asking this question can provide much-needed direction in determining where neuroenhancers may be truly beneficial and where their success is likely to be compromised by trade-offs. The following points are crucial to an evidence-based approach to neuroenhancement. The prevalence of ∩-shaped performance functions (Figure 11.1) suggests that investigations of enhancers need to describe the performance functions associated with the tasks for which they are intended to produce optimal behavior; these task-specific performance functions then need to be combined with dose–response curves. Such investigations should report performance expectations relative to individual differences in baseline performance in the task. Between-subject designs that overlook such differences are almost guaranteed to over- or underestimate actual effects and invite improper generalizations of their usefulness to people of different abilities. Furthermore, the possibility of performance trade-offs between domains suggests that researchers need to cast a wide net for potential side effects until principled methods for predicting prospective side effects are developed.

Identifying cognitive side effects is crucial, because optimization over multiple constraints implies a gain–loss asymmetry. Figure 11.2 illustrates this asymmetry. Assuming that the values of a cognitive trait (and related performance scores) follow decelerating functions (i.e., gains in functionality have diminishing returns), then beyond the point of the optimal trade-off, t^*, between two traits *A* and *B*, shifting the values of trait *A* upward (through neuroenhancers) yields a gain ($\Delta T1$) on performance scores correlated with trait *A*. At the same time, however, there is a loss ($\Delta T2$) in performance of larger magnitude on trait *B*. Such an asymmetry is an evolutionary necessity for any trait that has reached an evolutionarily stable state. That is, the asymmetry of gains and losses stabilizes selective forces around an optimal trade-off (for examples related to mental disorders, see Keller and Miller 2006).

Figure 11.2 is concerned with potential declines in performance in dimensions that are tied to the dimension to be enhanced through the evolution of cognitive processes. In reality, of course, the prospect of between-domain trade-offs is much more complex. Many drugs with the potential to enhance cognition are highly addictive, and their extended use is associated with the risk of long-term health consequences (e.g., nicotine, cocaine, and amphetamines). Understanding these adverse effects—especially in the still-developing brains of children and adolescents—is central to any evaluation of the long-term viability of neuroenhancers. Indeed, if the human brain permanently adapts to the chemical imbalance produced by the continual use of neuroenhancers and does not

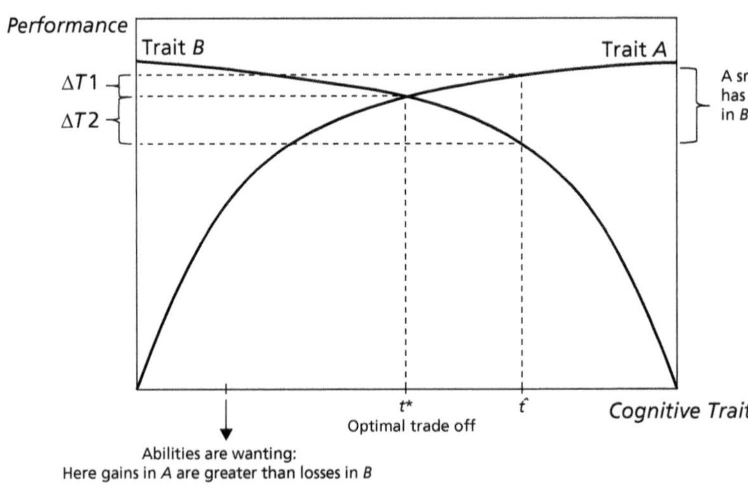

Figure 11.2 A gain–loss asymmetry. Performance scores associated with traits *A* and *B* follow decelerating functions (i.e., gains in functionality have diminishing returns); t^* represents the point of the optimal trade-off between both traits. Shifting the values of trait *A* upward (through neuroenhancers) yields a performance gain ($\Delta T1$) associated with trait *A* that is smaller in magnitude than the corresponding performance loss associated with trait *B* ($\Delta T2$).

revert to its normal state when intake is stopped, then adverse effects including cognitive impairment may be persistent. Various psychoactive drugs appear to have such severe long-term effects (e.g., Berke and Hyman 2000; Whitaker 2010).

Even if there were an ethically acceptable way to investigate these potential adverse effects in healthy people, it may be some time before the effects of drugs with cumulative and delayed negative consequences are fully realized. For instance, it was only recently possible to study the long-term health effects of smoking on women born around 1940—the first generation in which many women smoked substantial numbers of cigarettes throughout their adult lives. The observed long-term health consequences include an approximately tripled all-causes death rate and at least 10 years' reduction in life expectancy relative to nonsmokers (Pirie et al. 2013).

To conclude, the current hype about neuroenhancment in media reports that tend to emphasize the benefits of neuroenhancers and overlook their potential risks (Partridge et al. 2011) endorses a simplistic view of human cognition—namely, that more of a cognitive resource, available at demand, must invariably be a good thing. Taking an evolutionary perspective, we questioned the foundations of this view. Finally, let us stress that there are numerous nonpharmacological means to enhance cognitive capabilities throughout the life span, including physical exercise, sleep, meditation, and mnemonic strategies. Several of these strategies "seem to be more efficacious compared to currently available pharmaceuticals" (Dresler et al. 2013, 529), and they come without severe adverse effects. What they do require is investment in terms of time and/or effort. There is no such thing as a free lunch.

11.5 Acknowledgments

Parts of this chapter are drawn, with permission, from Thomas Hills and Ralph Hertwig (2011), Why aren't we smarter already: Evolutionary trade-offs and cognitive enhancements. *Current Directions in Psychological Science* 20:373-7. We are grateful to Susannah Goss for editing the manuscript and we thank the German Research Foundation and the Swiss National Science Foundation for grants to the first author (HE 2768/7-2; 100014-130397).

References

Altmann, E. M., and W. D. Gray. 2002. Forgetting to remember: The functional relationship of decay and interference. *Psychological Science* **13**(1):27–33.

Benjamin, A. S. 2011. *Successful remembering and successful forgetting: A festschrift in honor of Robert A. Bjork.* New York: Psychology Press.

Berke, J. D., and S. E. Hyman, S. E. (2000). Addiction, dopamine, and the molecular mechanisms of memory. *Neuron* **25**(3):515–32.

Broadbent, D. E. 1958. *Perception and communication.* Oxford: Pergamon Press.

Cochran, B. P., J. L. McDonald, and S. J. Parault. 1999. Too smart for their own good: The disadvantage of a superior processing capacity for adult language learners. *Journal of Memory and Language* **41**(1):30–58.

Cochran, G., J. Hardy, and H. Harpending. 2005. Natural history of Ashkenazi intelligence. *Journal of Biosocial Science* **38**:659–93.

Cochran, G., and H. Harpending. 2009. *The 10,000 year explosion: How civilization accelerated human evolution*. New York: Basic Books.

Collinson, N., F. M. Kuenzi, W. Jarolimek, K. A. Maubach, R. Cothliff, C. Sur, A. Smith, F. M. Otu, O. Howell, J. R. Atack, R. M. McKernan, G. R. Seabrook, G. R. Dawson, P. J. Whiting, and T. W. Rosahl. 2002. Enhanced learning and memory and altered GABAergic synaptic transmission in mice lacking the α5 subunit of the GABAAReceptor. *Journal of Neuroscience* **22**(13):5572–80.

Cools, R., and T. W. Robbins. 2004. Chemistry of the adaptive mind. *Philosophical Transactions of the Royal Society London A* **362**:2871–88.

Cowan, N. 2001. The magical number 4 in short-term memory: A reconsideration of mental storage capacity. *Behavioral and Brain Sciences* **24**:87–114.

Cytowic, R. E. 1995. Synesthesia: Phenomenology and neuropsychology. *Psyche* **2**(10):2–10.

de Jongh, R., I. Bolt, M. Schermer, and B. Olivier. 2008. Botox for the brain: Enhancement of cognition, mood and pro-social behavior and blunting of unwanted memories. *Neuroscience and Biobehavioral Reviews* **32**:760–76.

de Quervain, D. J. 2008. Glucocorticoid-induced reduction of traumatic memories: Implications for the treatment of PTSD. *Progress in Brain Research* **167**:239–47.

Dresler, M., A. Sandberg, K. Ohla, C. Bublitz, C. Trenado, A. Mroczko-Wasowicz, S. Kühn, and D. Repantis. 2013. Non-pharmacological cognitive enhancement. *Neuropharmacology* **64**:529–43.

Dunsworth, H. M., A. G. Warrener, T. Deacon, P. T. Ellison, and H. Pontzer. 2012. Metabolic hypothesis for human altriciality. *Proceedings of the National Academy of Sciences* **109**(38):15212–16.

Earp, B. D., A. Sandberg, and J. Savulescu. 2012. Natural selection, childrearing, and the ethics of marriage (and divorce): Building a case for the neuroenhancement of human relationships. *Philosophy and Technology* **25**:561–87.

Elman, J. L. 1993. Learning and development in neural networks: The importance of starting small. *Cognition* **48**:71–99.

Farah, M. J. 2005. Neuroethics: The practical and the philosophical. *Trends in Cognitive Sciences* **9**:34–40.

Farah, M. J., C. Haimm, G. Sankoorikal, and A. Chatterjee. 2009. When we enhance cognition with Adderall, do we sacrifice creativity? A preliminary study. *Psychopharmacology* **202**:541–7.

Finkel, T., and N. J. Holbrook. 2000. Oxidants, oxidative stress and the biology of ageing. *Nature* **408**:239–47.

Flavell, J. H. 1985. *Cognitive development*, 2nd edn. Englewood Cliffs, NJ: Prentice Hall.

Flynn, J. R. 2012. *Are we getting smarter? Rising IQ in the twenty-first century*. New York: Cambridge University Press.

Förstl, H. 2009. Neuro-Enhancement [Cognitive enhancement]. *Der Nervenarzt* **80**:840–6.

Fox, M. C., and A. L. Mitchum. 2013. A knowledge-based theory of rising scores on "culture-free" tests. *Journal of Experimental Psychology: General* **142**(3):979–1000.

Gigerenzer, G., R. Hertwig, and T. Pachur, eds. 2011. *Heuristics: The foundations of adaptive behavior*. Oxford: Oxford University Press.

Greely, H., B. Sahakian, J. Harris, R. Kessler, M. Gazzaniga, P. Campbell, and M. Farah. 2008. Towards responsible use of cognitive-enhancing drugs by the healthy. *Nature* **456**:702–5.

Hertwig, R., U. Hoffrage, and the ABC Research Group. 2013. *Simple heuristics in a social world*. New York: Oxford University Press.

Hertwig, R., and T. J. Pleskac. 2010. Decisions from experience: Why small samples? *Cognition* **115**:225–37.

Hertwig, R., and P. M. Todd. 2003. More is not always better: The benefits of cognitive limits. In *Thinking: Psychological perspectives on reasoning, judgment and decision making*, ed. D. Hardman and L. Macchi, 213–31. Chichester, UK: Wiley.

Hills, T., and R. Dukas. 2012. The evolution of cognitive search. In *Cognitive search: Evolution, algorithms, and the brain*, ed. P. M. Todd, T. Hills, and T. Robbins, 11–24. Cambridge, MA: MIT Press.

Hills, T., and R. Hertwig. 2011. Why aren't we smarter already: Evolutionary trade-offs and cognitive enhancements. *Current Directions in Psychological Science*, **20**:373–7.

Hills, T., P. M. Todd, and R. L. Goldstone. 2010. The central executive as a search process: Exploration and exploitation in generalized cognitive search processes. *Journal of Experimental Psychology: General* **139**:590–609.

Hoffrage, U., S. Lindsey, R. Hertwig, and G. Gigerenzer. 2000. Communicating statistical information. *Science* **290**:2261–2.

Humphrey, N. 2002. *The mind made flesh: Essays from the frontiers of psychology and evolution*. Oxford: Oxford University Press.

Husain, M., and M. A. Mehta. 2011. Cognitive enhancement by drugs in health and disease. *Trends in Cognitive Sciences* **15**:28–36.

Johnson-Laird, P. N. 1983. *Mental models*. Cambridge, UK: Cambridge University Press.

Johnson-Laird, P. N. 2001. Mental models and deduction. *Trends in Cognitive Sciences* **5**:434–42.

Kahneman, D. 1973. *Attention and effort*. Englewood Cliffs, NJ: Prentice Hall.

Kahneman, D., P. Slovic, and A. Tversky, eds. 1982. *Judgment under uncertainty: Heuristics and biases*. New York: Cambridge University Press.

Keller, M. C., and G. Miller. 2006. Resolving the paradox of common, harmful, heritable mental disorders: Which evolutionary genetic models work best? *Behavioral and Brain Sciences* **29**:385–452.

Kersten, A. W., and J. L. Earles. 2001. Less really is more for adults learning a miniature artificial language. *Journal of Memory and Language* **44**(2):250–73.

Kliegl, R., J. Smith, and P. B. Baltes. 1989. Testing-the-limits and the study of adult age differences in cognitive plasticity of a mnemonic skill. *Developmental Psychology* **25**:247–56.

Kowalski, H. 2013. Neuroenhancement—Gehirndoping am Arbeitsplatz [Cognitive enhancement—Brain doping at the workplace]. In *Fehlzeiten-Report 2013: Verdammt zum Erfolg—die süchtige Arbeitsgesellschaft?*, ed. B. Badura et al., 27–34. Berlin: Springer.

Luria, A. R. 1987. *The mind of a mnemonist: A little book about a vast memory*. New York: Basic Books (originally published 1968).

Manabe, T., Y. Noda, T. Mamiya, H. Katagiri, T. Houtani, M. Nishi, T. Noda, T. Takahashi, T. Sugimoto, T. Nabeshima, and H. Takeshima. 1998. Facilitation of long-term potentiation and memory in mice lacking nociceptin receptors. *Nature* **394**(6693):577–81.

Miller, G. A. 1956. The magical number seven, plus or minus two: Some limits on our capacity for processing information. *Psychological Review* **63**:81–97.

Mohamed, A. D. 2014. The effects of modafinil on convergent and divergent thinking of creativity: A randomized controlled trial. *Journal of Creative Behavior*, doi: 10.1002/jocb.73.

Newport, E. L. 1990. Maturational constraints on language learning. *Cognitive Science* **14**:11–28.

Parker, E. S., L. Cahill, and J. L. McGaugh. 2006. A case of unusual autobiographical remembering. *Neurocase* **12**(1):35–49.

Partridge, B. J., S. K. Bell, J. C. Lucke, S. Yeates, and W. D. Hall. 2011. Smart drugs "as common as coffee": Media hype about neuroenhancement. *PLoS One* **6**:e28416.

Pashler, H. E. 1998. *The psychology of attention*. Cambridge, MA: MIT Press.

Persson, I., and Savulescu, J. 2012. *Unfit for the future: The need for moral enhancement*. New York: Oxford University Press.

Pigliucci, M., and J. Kaplan. 2000. The fall and rise of Dr Pangloss: Adaptationism and the Spandrels paper 20 years later. *Trends in Ecology and Evolution* **15**(2):66–70.

Pinker, S. 2006. The lessons of the Ashkenazim: Groups and genes. *The New Republic* online. Posted June 17, 2006. Retrieved December 10, 2014.

Pirie, K., R. Peto, G. K. Reeves, J. Green, and V. Beral. 2013. The 21st century hazards of smoking and benefits of stopping: A prospective study of one million women in the UK. *The Lancet* 381(9861):133–41.

Pirolli, P. 2007. *Information foraging theory: Adaptive interaction with information.* New York: Oxford University Press.

Plunkett, K., A. Karmiloff-Smith, E. Bates, J. L. Elman, and M. H. Johnson. 1997. Connectionism and developmental psychology. *Journal of Child Psychology and Psychiatry* **38**(1):53–80.

Randall, D. C., J. M. Shneerson, and S. E. File. 2005. Cognitive effects of modafinil in student volunteers may depend on IQ. *Pharmacology Biochemistry and Behavior* **82**:133–9.

Rostron, B. 2013. Smoking-attributable mortality by cause in the United States: Revising the CDC's data and estimates. *Nicotine and Tobacco Research* **15**(1):238–46.

Rothen, N., B. Meier, and J. Ward. 2012. Enhanced memory ability: Insights from synaesthesia. *Neuroscience and Biobehavioral Reviews* **36**(8):1952–63.

Schooler, L. J., and R. Hertwig. 2005. How forgetting aids heuristic inference. *Psychological Review* 112:610–28.

Schweickert, R., and G. J. Boggs. 1984. Models of central capacity and concurrency. *Journal of Mathematical Psychology* **28**:223–81.

Smith, A. 2002. Effects of caffeine on human behavior. *Food and Chemical Toxicology* **40**:1243–55.

Snyder, A. 2009. Explaining and inducing savant skills: Privileged access to lower level, less-processed information. *Philosophical Transactions of the Royal Society, B* **364**:1399–1405.

Tang, Y. P., E. Shimizu, G. R. Dube, C. Rampon, G. A. Kerchner, M. Zhuo, and J. Z. Tsien. 1999. Enhancement of learning and memory in mice. *Nature* **401**:63–9.

Treffert, D. A. 2009. The savant syndrome: An extraordinary condition. A synopsis: Past, present, future. *Philosophical Transactions of the Royal Society B: Biological Sciences* **364**(1522):1351–7.

Unsworth, N., and R. W. Engle. 2007. The nature of individual differences in working memory capacity: Active maintenance in primary memory and controlled search from secondary memory. *Psychological Review* **114**:104–32.

Vul, E., N. Goodman, T. L. Griffiths, and J. B. Tenenbaum. 2014. One and done? Optimal decisions from very few samples. *Cognitive Science* **38**(4):599–637.

Warburton, D. M. 1992. Nicotine as a cognitive enhancer. *Progress in Neuro-Psychopharmacology and Biological Psychiatry* **16**(2):181–92.

Whitaker, R. 2010. *Anatomy of an epidemic: Magic bullets, psychiatric drugs, and the astonishing rise of mental illness in America.* New York: Broadway Paperbacks.

Wittman, A. B., and L. L. Wall. 2007. The evolutionary origins of obstructed labor: Bipedalism, encephalization, and the human obstetric dilemma. *Obstetrical and Gynecological Survey* **62**:739–48.

Yaro, C., and J. Ward. 2007. Searching for Shereshevskii: What is superior about the memory of synaesthetes? *Quarterly Journal of Experimental Psychology* **60**(5):681–95.

Chapter 12

Enhancement and therapy: is it possible to draw a line?

Alex McKeown

12.1 Introduction

In this chapter I analyze the therapy/enhancement distinction primarily as it relates to contemporary pharmaceutical methods of cognitive enhancement such as modafinil, Adderall, and Ritalin, all of which have been argued to improve cognitive capacities such as concentration and short-term memory retention. I will also briefly consider it in relation to potential genetic cognitive enhancements, should they become available in future.

I will advance three interconnected arguments. Firstly, consistent with a view found both within the pro-enhancement literature (Bostrom and Roache 2008; Bostrom and Savulescu, 2009; Sandberg and Savulescu 2011; Savulescu 2006) and amongst those who are more circumspect about its benefits (President's Council on Bioethics 2003; Scully and Rehmann-Sutter 2001; Shickle 2000) I argue that the distinction between therapy and enhancement is ambiguous and logically unstable. Secondly, that despite the potential threat that this instability appears to pose to identifying clear criteria for ensuring an ethical allocation of medical resources anchored by an account of "normal health" (Buchanan et al. 2000; Capps 2011; Koch 2010; Pellegrino 2004), there is a relatively simple theoretical solution. This solution could, if it were possible to implement, negotiate the difficulties raised concerning the separability of the two concepts, and thus continue to protect the just allocation of resources according to need.

However, and thirdly, despite the simplicity of this solution, contemporary medicine in the UK is not institutionally ready to implement it because of its adherence to "normality" as the boundary of appropriate medical practice. This assumption upholds the apparent nominal distinction between therapy and enhancement, despite the fact that no clear nonarbitrary line can be drawn between them.

Consequently, despite possible significant benefits available from the use of cognitive enhancements by the healthy (Bostrom and Sandberg 2009a; Nam 2015; Sandberg and Savulescu 2011), I argue that the realization of these benefits remains relatively distant. I therefore conclude that we should limit our expectations about what can practically be achieved via the widespread use of cognitive enhancement drugs until the institutional assumptions of healthcare, and the training of medical professionals informed by them, have undergone substantial reorientation.

To support these arguments I will draw on a qualitative study that I carried out in relation to enhancement via recombinant human erythropoietin (EPO). This is a drug for treating renal anemia caused by chronic kidney disease, and also a well-known and controversial illegal performance enhancer in professional sport (Fisher 2010; Mayes 2010; Mignon 2003), for example in the high profile cases of the cyclists Lance Armstrong[1] and David Millar.[2] In this study I gathered perspectives on enhancement amongst NHS nephrologists and renal scientists whose clinical or research work involves EPO. Twenty-five participants were interviewed about their views on ethical, philosophical, and conceptual issues relating to the therapy/enhancement distinction in general, and the context of EPO in particular.

Similar empirical research has been carried out into drugs that have been appropriated for nonstandard use, and argued to enhance normal cognitive capacities, such as modafinil, Ritalin, and Adderall. These have been the subject of qualitative studies of professional, user, and public perceptions (Aikins 2011; Bell et al. 2013; Fitz et al. 2013; Forlini and Racine 2009; Vrecko 2013). While of contemporary interest due to their off-label use by students for the enhancement of concentration, however, they have only begun to be appropriated for this purpose relatively recently (Cakic 2009; Coveney et al. 2009; Greely et al. 2008; Tannenbaum 2014; Williams et al. 2008). Consequently the extent to which significant social and ethical implications are discernible is limited. EPO, by contrast, is a product with over two well-documented decades of both standard clinical and "off-label" enhancement use (Diamanti-Kandirakis et al. 2005; Joyner 2003; Mignon 2003; Robinson et al. 2006). For this reason the data bring a relevant historical perspective to the ethical debate which is less available for drugs now increasingly considered for purposes of cognitive enhancement.

There is, of course, a reasonably clear difference between these two kinds of drug and the enhancement purposes to which they are put. EPO is used for enhancing physical endurance and exercise tolerance, whereas drugs such as modafinil, Ritalin, and Adderall are used in the hope that they will enhance cognitive ability. Given this difference one might ask what can be learnt from the debate about the former that is relevant to the debate about the latter. Although the two are not categorically equivalent in terms of their effects, however, I will go on to show that they nevertheless share features that are ethically relevant with respect to identifying the boundaries of appropriate medical practice. Furthermore, to the extent that the therapy/enhancement problem *as such* is a conceptual challenge that underlies medical practice in general, and not just one or two specific fields within it, there are insights available from the EPO case that are equally pertinent in the case of drugs such as modafinil, Adderall, and Ritalin.

12.2 **Separating therapy and enhancement**

There is a basic logicosemantic difficulty in clearly separating therapy from enhancement, since it is a condition of a therapy's effectiveness that it improves health and/or functioning. Insofar as a therapy works, it enhances, since if it did not do so it could not successfully be described as "therapeutic," hence Savulescu's (2006, 324) observation that "a doctor may enhance his patient's chance of survival by giving the patient a drug." Indeed Pellegrino (2004, 1) is unequivocal that there is "no question that the cure or amelioration of a disease process will also result secondarily in enhancement of the patient's life."

Nevertheless, therapy and enhancement can often be separated empirically with ease. Potentially life-saving chemotherapy for the treatment of cancer is demonstrably different from using drugs EPO for a performance advantage in professional sport. The chemotherapy case is also different from a healthy individual using a cognitive enhancer to improve their concentration or memory retention, for example while studying. All of these share a goal of enhancing health and/or functioning; however, the starting points are so different that it would be misleading to say that there is *no* important difference between them.

The most significant difference is that in examples such as these, life itself is threatened in only one of the cases. This is important because, as Toulmin (1975, 61) observes, "The chief vital functions of the human body ... are preconditions for almost any other imaginable human good." Since without assistance life is threatened in the chemotherapy case and not in the EPO or modafinil enhancement cases, it would be misleading to say that the difference in health makes no difference to how the interventions should be understood.

The therapy–enhancement distinction is therefore heuristically useful for determining the severity of needs. Although the putative account of normal health anchoring it is contestable, the distinction nevertheless implies a real gradation of needs, and provides an approximately just way of carving up medical resources according to their distribution across a population. As Daniels (2000, 315) writes:

> Within a society, relative to its normal range of opportunities, some diseases and disabilities are more important to treat than others, and this will affect our decisions about which treatments to offer when we cannot provide all the ones people need.

The cancer patient, relative to a normal range of opportunities, does have clear, severe, immediate health needs that require and legitimate medical assistance. The athlete using EPO or student using modafinil has a much more limited need because the health consequences of not intervening would be less serious. To the extent that the difference between these consequences is that "normal health" would persist in the latter two but not the first, the two forms of assistance can be nominally identified as instances of therapy and enhancement *if* one accepts the reality of the distinction.

In this respect the therapy–enhancement distinction is cogent. To the extent that it maps onto a hierarchy in which health states are ranked according to the deleteriousness of their effects, it is effective for distinguishing between different levels of need. Beyond this, its effectiveness decreases, however, following Boyd's (2000, 12) observation that "a common core of ideas about what disease is or what health is ... But beyond that ... our conceptions of disease and health begin to get fuzzy." Whilst there is a clear difference in need between a cyclist seeking EPO and a cancer patient seeking chemotherapy, it is not clear where between these two extremities the provision of medical assistance ceases to be therapeutic and becomes "enhancement." As Savulescu (2006, 334) writes, it is *we* who decide the calibration of this threshold, according to a judgment about how severe a need must be for it to be unjust that policies are not put in place to ensure its remediation:

> what IQ is necessary for a decent chance of a decent life? Perhaps, in a technologically sophisticated society, people would significantly benefit from a higher IQ. An IQ of 120 is needed to be able to complete tertiary education ... justice requires enhancement. It is on these grounds that we choose to treat those currently with an IQ less than 70. But where we set the minimum threshold for treatment or enhancement is up to us.

I will develop this line of analysis by drawing on data from the EPO study. An analogous example is available from this case in relation to changes in hemoglobin targeting for treating renal anemia.

12.3 Hemoglobin targeting, optimizing health, and functioning

A "normal" hematocrit in an adult male is reported to be within a range of approximately 13–17 g/dl (Lacson et al. 2003). EPO is typically administered when hemoglobin content falls below approximately 10 g/dl (Will 2001; Will et al. 2007), since below this threshold patients tend to become symptomatic. It is administered to restore hemoglobin to a level that allows the recipient to carry out the tasks of "normal living" as they did before the onset of anemia. One might assume, therefore, that an appropriate hemoglobin replacement target would be within the statistically normal range, as Paoletti and Cannella (2006) and Szczech et al. (2008) suggest.

However, the CHOIR and CREATE studies, which attempted to more accurately determine the optimum safe hemoglobin replacement threshold (Levin 2007), indicated that the therapeutic effect on the quality of patients' lives through hemoglobin restoration tailed off at around 10 g/dl (Besarab et al. 2012; Singh et al. 2007). Above this, EPO dosing was associated with an increased risk of thrombosis that was unnecessarily high when balanced against the positive effects of restored functioning (Fishbane and Wish 2012; Remuzzi and Ingelfinger 2006). In fact, optimal hemoglobin replacement appears to be at the upper end of the anemic—and thus *pathological*—range. It is therefore possible that a statistically normal hematocrit and the functional state that is optimally adequate for normal contemporary living are not isomorphic.

The reason for this mismatch is unclear (Fishbane and Wish 2012), and several hypotheses have been advanced. It could be, for example, that synthetic EPO, while a nearly exact copy of endogenous erythropoietin, does not mimic it perfectly and has a toxic effect for reasons not yet fully understood (Fishbane 2009). Alternatively, it may be that EPO receptors exist elsewhere in the body alongside the bone marrow receptors (Winter et al. 2005) which are the target of EPO replacement therapy. These, in addition to the bone marrow receptors, may detect the presence of erythropoietin when it is injected, causing an unanticipated spike in red cell production which thickens the blood, raising the risk of thrombosis. Until further research confirms or eliminates these possible explanations for the discrepancy between a naturally occurring normal hematocrit and a pharmaceutically induced one, the reason for the discrepancy will remain unknown.

An alternative hypothesis advanced in my data was that the normal hematocrit may be a biological "hangover" from an earlier evolutionary time in which threats to life involving blood loss were more immediate and severe than at present. These evolutionary pressures have receded but the hemoglobin content of the blood has remained the same, such that we may now be "maladapted" to our present environment (Powell and Buchanan 2011; Warnock 2003). Bostrom and Sandberg (2009b, 381) summarize the radical difference between the "environment of evolutionary adaptedness" (EEA) and the present day:

> Hunting, gathering of fruits and nuts, courtship, parasites, and hand-to-hand combat with wild animals and enemy tribes were elements of the EEA; speeding cars, high levels of trans fats, concrete ghettos, and tax return forms were not.

If this hypothesis is correct, the statistically normal hematocrit may not represent an optimal trade-off relative to the attendant threats to life in the UK in the twenty-first century. If this is true there are grounds for arguing that most healthy people would hypothetically benefit from an intervention that would reduce it by 1 or 2 g/dl, irrespective of the fact that the goal would deviate from a biostatic norm. Enhancement is frequently assumed to be additive; however, Earp et al. (2014, 4) suggest this is misleading. They recommend moving the emphasis away from how a capacity is being modified to "the overall normative goal of the modification itself … it does not matter if the capacity itself is being modified 'up' or being modified 'down.' "

Indeed, analogous preventive public health measures of this kind are already encouraged. This can be seen in recommendations for the use of statins, in combination with improvements to lifestyle, for lowering cholesterol and reducing the risk of heart disease (Armitage 2007; Davignon 2004; LaRosa 1999), and aspirin to reduce the risk of bowel cancer (Burn et al. 2012; Cuzick et al. 2009). Both interventions are biomedical modulations of biostatic norms that contribute to increasing healthy longevity. Although administered under the auspices of therapy for this reason, they exist at the ambiguous "boundary" between therapy and enhancement because, following Juengst's (1997, 126) argument in relation to genetic enhancements:

> to the extent that disease prevention is a proper goal of medicine, and the use of gene transfer techniques to strengthen or enhance human health maintenance capacities will help achieve that goal, then the treatment/enhancement distinction cannot confine or define the limits of the properly medical use of gene transfer techniques.

Furthermore, even if there *were* agreement that these interventions, or related preventive public health interventions such as vaccination, are clearly enhancements rather than therapies, their utility is such that they would not cease to be legally provided or administered under the auspices of conventional medical practice. This is significant in the case of cognitive enhancers because similar arguments have been made in their favor in view of the widespread advantages that cognitive ability in particular confers. Bostrom and Sandberg (2009a, 328) characterize such capacities as "intrinsically desirable" because having, for example, "a good memory ... is normally valuable in its own right." Moreover, pushing the analogy with vaccination, the "herd immunity" benefit derived by an entire population from a sufficient number of its members being vaccinated is redolent of Bostrom and Sandberg's (2009a, 328) justification for the enhancement of functions such as memory, since:

> An enhancement that enables an individual to solve some of society's problems would produce a positive externality: in addition to benefits for the enhanced individual, there would be spillover benefits for other members of society.

12.4 De-stigmatizing need

Implicit in a judgment that the use of a drug such as modafinil by the healthy constitutes "enhancement" is the assumption that no need is present. There are no doubt instances in which this may be truer than others. For example, an already high-flying student may seek new ways to improve their cognitive capacity to achieve more highly still and may believe that a drug such as modafinil offers such a possibility, although it must be conceded that in high-performing individuals the potential for further improvement may be more limited.

However, Sandberg and Savulescu (2011, 97) ask, "What use is enhancing the top part of cognitive distribution?" The implication of this question is that the amount of potential benefit available from enhancement is correlative with one's position within the range of cognitive ability: the high-achieving "normal" student has less to gain from using cognitive enhancements than the more cognitively limited "normal" student at the lower end of the range. The difference in capacity between those at the top and bottom ends of the range entails the judgment that the needs of those at the lower end are greater than those at the top relative to a task for which the relevant cognitive capacities are required, and thus stand to benefit more from cognitive-enhancing drugs.

This is, in fact, an unremarkable claim redolent of Hare's (1986, 178) straightforward rationale for seeking medical assistance that "If I have bad eyesight ... I ought to go to the oculist and he ought to prescribe spectacles if they will make my eyesight better." The term "bad" does not by itself map on to any criteria for determining medical need, since I would, for example, no doubt be "bad" at playing professional basketball because I am insufficiently tall for it despite being of "normal" height. Everybody will perform "badly" relative to some goal, because different goals require different skills, abilities, or capacities, and these are distributed unevenly. The worse one is at something, the more one stands to gain from the relevant capacity being enhanced. It is this insight that Powell and Buchanan (2011, 10) capture in observing "the ubiquity of suboptimal design."

An objection would be to say that the fact that the person at the lower end of the normal cognitive range has a greater need than the person at the top does not of itself provide a justification for why medicine should institutionally sanction the supply of cognitive enhancements to them.[3] Despite this need they are, after all, *not* considered *ab*normal. This objection has only limited effectiveness, however, because what counts as "normal health" is not determined a priori, and is vulnerable to technological (Stempsey 2006), geographical (Gunga et al. 2007), historical (Bullough 1987; Rayner et al. 2010), and sociocultural (Kleinman and Sung 1979; Press 1980) change. The less cognitively able of the pair is considered ineligible for assistance because of a convention in the placing of the threshold. However, if the threshold were slightly higher, the less able student would be judged cognitively "abnormal" and thus eligible for assistance, even though this change in judgment would be an arbitrary one.

A further objection is that it would be undesirable to raise the threshold in any case, because if it were, then the *de facto* reclassification of the user as cognitively "abnormal" could have a stigmatizing effect by virtue of their moving into the purview of the medical profession (Conrad 1992; Fox 1977; Lowenberg and Davis 1994). However, the response to this is simply to argue for a deliberate reduction of the significance of normality in decisions about whether to assist, and instead simply allocate resources as far as they will go. It is this move, in essence, which is at the center of the theoretical solution that I propose to the challenge of completely disaggregating therapy from enhancement.

12.5 **Institutional obstacles**

Difficulties arise, however, not in theory itself but in its practical application. Regrettably, it is the practical difficulty of applying this theory which is likely to preclude the widespread realization of any possible benefits of cognitive enhancement in the near to medium term. Sandberg and Savulescu (2011, 106) note that the present system for licensing drugs and medical assistance "was created to deal with traditional medicine which aims to prevent, diagnose, cure, or alleviate disease," and hence, that "In this framework there is no place for enhancing medicine." Consequently, as Sandberg (2010) points out, a question mark exists over whether the political will exists to enact policies that could bring about the realization of these benefits. If this is the case we may need to rein in expectations that we might have about the widespread availability of cognitive enhancement being just around the corner.

However, a difficulty for someone who *defends* the status quo is that there are already anomalous instances in which the enhancement use of medical products is legal and accepted. I have mentioned statins, aspirin, and vaccination as three kinds of standard intervention which can arguably be construed as enhancements, and there are others. Two participants in the EPO study attested to claims made within the literature that beta-blockers are regularly prescribed to individuals for dealing with a normal nervous response to certain stressful situations, for example to help musicians cope with the pressure of live performance (Neftel et al. 1982; Wesner et al. 1990). The cosmetic plastic surgery industry is a second example of enhancement on a much larger scale (Delinsky 2005; Holm 2008). In both of these cases assistance is granted in the absence of "disorder," and are thus enhancements *by definition* according to present criteria.

What is clear about these anomalous examples is precisely that they are anomalies—they are the exception rather than the rule. Their contestability, however, follows from their being located closer to the putative "boundary" between normality and abnormality than the extreme examples given earlier for a cancer patient receiving chemotherapy, or a professional cyclist using EPO. These inconsistencies indicate a confusion concerning the definition of therapies and enhancements, and how their inclusion within the corpus of medicine and health policy should be managed.

12.6 Justice and changing medical responsibility

To illustrate why I am inclined to argue that the possibility of realizing the benefits of cognitive enhancement is distant I will draw again on data from the EPO study. The prevailing view of the participants was that there are two key reasons why the prescription of EPO for enhancement purposes would be unacceptable. Firstly, it was unanimously considered too dangerous because of the increased risk of thrombosis associated with boosting the hematocrit to supranormal levels. Secondly, due to the relatively limited nature of the advantages available using EPO, i.e., for positional advantage in high-level sport, it is likely that most users are seeking only a positional advantage over unenhanced competing nonusers. These two justifications underwrite EPO's illegality for enhancement purposes, and in view of the safety and justice considerations at stake they are both legitimate prudential reasons.

Crucially, however, no principled objection to the use of enhancements *as such* was offered in the study. Assuming that a drug could be used safely and in ways that would not entail injustices, these medical professionals did not consider it ethically problematic. If Harris (2009), Bostrom and Sandberg (2009a), Sandberg and Savulescu (2011), and Nam (2015) are correct that cognitive enhancers offer intrinsic benefits and that their distribution could be used to *reduce* inequalities and natural injustice, then there are grounds for suggesting that this would not conflict with standard conditions of good medical practice. While it is therefore possible that medical professionals might be open to moves for supplying cognitive enhancers to the cognitively normal, the reality is likely to be more complex.

Although the professionals interviewed did not object to enhancement *as such*, what was also clear was that they considered decisions about any widespread and subsidized provision of enhancements to be sufficiently distant from standard medical practice that they did not have to consider the implications of such policies in detail. Decisions such as this were considered as only tangential when compared to the pressing day-to-day matter of providing assistance to sick people with severe and immediate health needs. While philosophically and conceptually interesting to them, they perceived their role as being relatively clearly defined: they are motivated to help the sick, not the well, and hence close scrutiny of the practical ethical ramifications for practice was not viewed as a threat to the conventional institutional structure of medicine. Indeed, one of the participants' (S6[4]) response was to say that "we're not ethically trained for that" with respect to decisions about enhancement. He viewed such decisions as ones in which there is no pressing need for doctors to engage at present.

This is significant because it represents a "catch-22." Unless enhancement—cognitive or otherwise—is viewed as a legitimate goal of medical practice by medical practitioners, its implementation in policy will not be seriously considered. However, it will not be considered as a proper goal of medicine, and thus something directly relevant to medical practice, until some move is made toward opening a serious discussion about the possibility of implementing policies that would enable their provision. Despite the presence of anomalies such as cosmetic surgery, beta-blockers, statins, and vaccinations for enhancing normal human capacities or increasing healthy longevity, "enhancement" remains conventionally understood as distant and distinct from standard medical practice, if the data from the EPO study are also more widely held. If these conventions were altered in view of the compelling arguments for the continuity between the goals of medicine and certain kinds of enhancement, such as those which could improve cognition, the impediments to their provision would be reduced.

However, it is of course not surprising that doctors might see enhancement as so distant from standard practice that decisions about their provision and distribution are "somebody else's" responsibility. It is uncontroversial to point out, as Nordenfelt (2001, 11) does, that "the traditional core activity of the doctor" is "to cure diseases and … alleviate the consequences of diseases" with the goal of restoring normality (Boorse 1977; Fleischhauer and Hermeren 2006; Holtug 1999. This tenet is foundational to the medical infrastructure and the principles according to which professionals are trained (Freidson 1970).

Chatterjee (2004, 973) talks of "the evolving role of the physician" as a distinct possibility if we become increasingly able to prevent illness and move from "the familiar moorings of disease" as individuals are able to live more healthily for longer. To alter this infrastructure, reorientating it in the way set out here, is not something that can be achieved rapidly. This would require a reorientation of one of medicine's basic normative assumptions (Goffette 2006; Parens 1998), and it is not yet clear that the will for such a profound change exists sufficiently generally across the medical profession. As Greely et al. (2008, 704) suggest in the context of healthcare in the USA:

> Physicians who view medicine as devoted to healing will view such prescribing as inappropriate, whereas those who view medicine more broadly as helping patients live better or achieve their goals would be open to considering such a request.

There is also a further complication. As long as health needs differ in severity, and until all life-threatening conditions have been eradicated, there will always be people whose needs are immediate or severe. Correspondingly, there will always be people in poor(er) health whose claim on resources is *prima facie* greater than those of "normally" healthy individuals seeking to achieve what would be for them supranormal functioning. It is also important to consider the connected issues of the inevitable finitude of resources, and the complexities of effectively subsidizing an enhancement such that it offsets any advantages accruing to those people who can afford to buy it privately.

Criticisms of enhancement have been leveled that once an enhancement has been legalized it will be impossible to prevent those with sufficient economic resources from accessing it privately and entrenching the advantage that they already have. This is a function of economic power since, as Buchanan et al. (2000, 340) write, "Markets permit the lucky, less vulnerable participants to detach their fortunes from those of the unlucky fellow citizens." There would also undoubtedly be doctors, or "schmoctors" (Parens 1998), willing to provide them because, as Pellegrino (2004, 3) writes, "Enhancement will also appeal to the physician's self-interest ... Physicians can say they are doing 'good' for their patients even while doing well for themselves."

Consequently not only would there be a need to ensure that a sufficiently large supply could be provided via subsidization, but moreover it would be of crucial importance that appropriate rather than inappropriate candidates for legalization were identified. The combination of these factors, therefore, means that despite the relative simplicity of the theoretical solution to reduce the emphasis on achieving only a threshold of normality and instead use resources as far as they will go, the will to take on this risk may not yet exist. Allhoff (2005, 44) provides a persuasive response to some of these concerns in relation to genetic enhancements; however, it is not one which is entirely free of complications:

> enhancement procedures alone will not lead to unjust results; there would have to be an unjust distributive scheme to enable the injustice to come about. If we can determine what constitutes a just distributive scheme, then genetic enhancement, as a good or service, can be distributed according to the principles of that scheme.

As valid as this argument is, it underplays the practical difficulties associated with constructing a "just distributive scheme" that functions effectively. Fukuyama (2002, 188) has voiced doubts that any truly "leak-proof" policy could be designed which would achieve a satisfactory ethical outcome. Worries of this kind were reflected frequently by the participants in the EPO study, for example over the coercive power of commercial factors which would come to bear in an enhancement market.

For example, one participant believed that health policy would, ultimately, be unable to constrain an enhancement market, because the increased revenue which would be available from a market in legal enhancements would create a situation in which "the temptation for the exchequer would be to move more medicines into pot B and have fewer in pot A" (C8). Still more directly, speaking in relation to the economic pressure that the pharmaceutical industry is able to exert on cash-limited healthcare systems, another participant made the bald assertion that "no-one's clean in this business" (C5).

If one takes seriously the ethical primacy of a commitment to the sick, then caution amongst medical professionals is understandable, irrespective of any possible widespread advantages available from the improvement of normal cognitive ability. Views such as these are comprehensible even if one personally takes the view that, on balance, enhancements such as those which are thought to improve cognitive ability would be worthwhile and sufficiently valuable for consideration as candidates for legalization and subsidization. Lin and Allhoff (2008, 262) remind us that "it is important to keep in mind that the human enhancement debate is not just a theoretical discussion about ethics ... it has a bearing on the real world." It is the tension between the ready availability of simple theoretical solutions, on one hand, and the difficulty of applying them, on the other, which may constitute an obstacle to realizing any potentially widespread benefits of cognitive enhancements.

12.7 Negotiating changing norms

Central to the difficulty of delineating therapy from enhancement is the challenge of providing an unambiguous and persistent account of normality. Theories of health are numerous and divergent, and it is not obvious which of the competing accounts is the most acceptable. However, *even if* it were possible to agree on any one characterization of "the normal" in particular, changing externalities would eventually require us to reconsider it. Changes in the treatment threshold for hemoglobin replacement via EPO are pertinent here again since, as we saw earlier, the falling cost of EPO has meant that more people can be treated and no longer have to become severely anemic before they receive assistance (Remak et al. 2003; Winearls 2006).

This is, in turn, bringing the initial treatment threshold ever closer to the enhancement baseline of normality, as participants C7 and S1 indicated. C7 referred to EPO rationing in the early 1990s that caused situations in which "ten people got EPO and if you were the eleventh you had to wait for someone to be transplanted or die before you could have it." He described this as "tangibly an unfair situation." S1 stated that EPO is now sufficiently cheap in the UK that anemia patients receive EPO before they become symptomatic and thus frequently need not "feel ill" before assistance is provided. He reported that "we have a whole anaemia service ... they have huge block contracts with manufacturers ... nowadays we don't allow anyone to become anaemic."

Given this network of developments it is not clear, by analogy, why the distribution of cognitive enhancers should be objected to on the basis that to do so would necessarily transgress the boundary of "appropriate" medical practice. What falls in and outside of this boundary is not fixed, since the example above indicates that determinants of "appropriateness" include not only the question of whether to assist "treats" or "enhances," but whether there is an acceptable equilibrium among the available resources, the cost of assisting, and what other health needs also need to be met. Here, again, we can consider again the earlier example of the legal prescription of beta-blockers to healthy individuals in the absence of an identifiable disorder. In instances such as these, in which the aim is to maintain healthy normal living rather than to restore it, a different strategy may be more appropriate. Rather than simply asking "is this application of an improvement in cognition a therapy or an enhancement?" it would be preferable to arrive at a balanced judgment about the net benefit of doing so, irrespective of how the intervention is characterized.

In view of all of this it is therefore not obvious either that "normality" necessarily needs to be a goal that must constrain all medical decision-making. It is also not obvious, given the ambiguity concerning the clear identification of "the normal," how this could be appealed to with any degree of permanence. It is not clear, therefore, why cognitive enhancement should be objected to if the appeal is grounded in an appeal to normality as something fundamentally significant.

The simplicity of this solution is that all one needs in order to legitimate the principled extension of cognitive enhancers to people whose health and/or functioning is normal is to accept the claim that someone can have medically improvable needs irrespective of their relative health. If, instead of allocating and apportioning medical resources by appeal to a norm, resources were simply allocated according to need as far as they would go regardless of whether assistance was considered "therapeutic" or "enhancing," then two difficulties could be avoided. Firstly, the conceptual issues associated with the therapy–enhancement distinction would be neutralized, since the distinction would cease to be of central importance to medical decision-making. Secondly, a greater range of needs could be met, and it would be possible to address areas of potential injustice that are propagated and maintained by the assumption that one only has significant needs if one is ill, diseased, or abnormal.

This runs contrary to the prevailing distributive model. However, as we have seen, the medical model already generates borderline cases whose legitimate interpretation as enhancements has no negative bearing on the judgment that it is reasonable for a doctor to provide them. Therefore, we may stick with this model, but the price we pay for doing so is an inconsistency of definition in these borderline cases. These cases create unnecessary confusion in terms of mapping applications of technology onto definitions of therapy and enhancement, and yet this confusion is avoidable if a different approach were adopted. It remains to be seen, however, whether the advantages of this approach in terms of theoretical simplicity are outweighed by the practical obstacles associated with trying to implement it.

12.8 **Comparing cases: EPO and cognitive enhancers**

Before concluding it is important to insert two qualification to the arguments laid out, in order that the picture presented is not oversimplified.

Firstly, the hypotheses put forward on the basis of my empirical data are valid beyond the EPO case only to the extent that the views expressed are held more generally across the medical profession, and of course they may not be. The data relate to EPO, and therefore may be anomalous and unrepresentative of views held about other enhancement products by specialists in fields outside nephrology. The views expressed imply conditions under which they would hypothetically find the use of enhancements ethically acceptable or unacceptable, consistent or inconsistent with the goals of medicine. However, larger scale empirical research would have to be carried out to ascertain this, and whether these views are more widely held. The data may therefore be limited, but they do nevertheless provide a starting point for framing future investigations. Secondly, although cognitive enhancers nominally fall into a different category from EPO because they extend a different set of capabilities, there have been instances which may cast doubt on the veracity of an absolute categorization based on a clear difference between mental and physical effects.

For example, in 2003 the American sprinter Kelli White tested positive for modafinil after winning the 100 m in the World Track and Field Championship (Honour 2004; Kaufman 2005). White informed the International Association for Athletics Federation

(IAAF) that she had been prescribed modafinil because of a positive family history of narcolepsy. However, since modafinil increases wakefulness and energy, and since she had not previously informed these regulators of this and applied for an exemption, it was judged that the modafinil may have conferred an unfair performance enhancement. Her case was subsequently referred to the USA Track and Field Federation (USATF) for arbitration. The ethically relevant issues here are (a) whether the cognitive effect of modafinil conferred a secondary enhancement of physical performance by reducing fatigue; and if so, (b) whether this secondary enhancement of performance is judged to be fair if it is the case that the reason for using modafinil was to remediate for the effects of her narcolepsy. As Kaufman (2005, 242) writes:

> if the athlete is using modafinil to treat a medical condition, then disqualification for having taken such would not appear, on initial evaluation, to be ethical. To not allow the use of this agent, would mean that the athlete is functioning at a handicap, for modafinil has been shown to improve fatigue associated with narcolepsy.

Such cases must be kept in mind when trying to arrive at a balanced judgment about what kinds of enhancements are fair in different competitive scenarios. Modafinil *may* indeed confer *some* competitive advantage in cases such as this. Nevertheless, the extent to which in this case it restored deficient functioning caused by narcolepsy—and thus functioned as a "therapy" rather than an "enhancement"—is unclear. It is therefore not possible to say for certain that this case undermines the drawing of at least a nominal and heuristically useful distinction between drugs such as modafinil and EPO in terms of the nature of their respective effects.

12.9 **Conclusions**

The arguments put forward here argue for the removal or significant reduction of "normality" as a constraint to the limit of standard medical practice, the advantage of this being that it would be a better way to allocate resources relative to all the needs of an entire population. However, a criterion of "normality" contributes significantly to the orientation of the entire medical infrastructure, and unless this is changed it is unlikely that the kind of system that I advocate will be able to replace it. The recommendations here are therefore to some extent inconsistent with conventional medical norms, and if so it is for this reason I draw my conclusion that we should be more circumspect about what it may *actually* be possible to achieve via cognitive enhancements in the short to medium term.

Despite the obvious weakness of the therapy–enhancement distinction and the benefits of dispensing with it, or at least attenuating its significance in medical decision-making, the changes and reorientation required to achieve this are significant. A doctor may practice for up to five or six decades and must go through lengthy training, and in the first instance the nature of this training would have to change significantly.

In addition the changes which follow from rethinking the therapy–enhancement distinction in the way that I have laid out would have to be ones that society as a whole deems desirable. As Capps (2011, 123) notes in relation to the likely practical difficulties associated with the effective implementation of cognitive enhancement policies in this way, for such policies to be introduced at all "they require legitimation in the sense that they facilitate a widespread belief that they benefit social and political stability."

Medical practice should be institutionally grounded in the same values and beliefs as the society that it serves, and healthcare policy must adopt "a perspective that is communal and social rather than focused upon the atomistic individual" (Koch 2010, 697) in order that it does not act in ways contrary to the interests and desires of its members. Indeed, other critics such as Fukuyama (2002) and Sandel (2004) have expressed concern over whether it will be possible to ensure that any benefits of enhancement contribute to justice in society, rather than being corrosive of it.

On the other hand, Bostrom and Sandberg (2009a), Sandberg and Savulescu (2011), and Nam (2015) have argued that it is not inevitable that the wider interests of society and the introduction of cognitive enhancement interventions are mutually exclusive. If they are correct, then focus should be placed not on banning enhancements because they are enhancements, but developing policies according to which they can be distributed and accessed fairly.

If the data from the EPO case represent any more generally held view, they suggest that medicine is not institutionally ready for the changes that would be required to realize any widespread potential benefits of enhancements. The therapy–enhancement distinction can be upheld, but only at the price of holding an account of "normal health" that may become increasingly obsolete. Stempsey (2006, 234) argues that if medicotechnological development accelerates, it will drive the development of new norms increasingly rapidly:

> because technological development forces us to expand our thinking about the very nature of the body, its limitations, and what it might possibly be able to do. Technology creates the possibility of new ends that can only be met by means of technology.

This observation points to the fact that the threshold of normality which apparently delineates therapy from enhancement, and medical assistance that is needed from that which is not, is a rationalization of the human biological range according to categories whose boundaries are not clearly defined. Sadegh-Zadeh (2000, 607) notes that "it may well be that a person is healthy and has a disease nonetheless, or that she is not healthy without having any disease," concluding that since the meaning of these terms are not a priori definable, "the concepts of health, illness, and disease are not amenable to classical logic."

In addition to this conceptual difficulty, the recategorization of one kind of state as another does nothing by itself to reduce or remove the presence of the need in question. It is misleading to characterize, as participants in the empirical study frequently did, normal health as an absence of needs. Just because I am judged "normal" does not mean that I have no needs. If I wished to be a basketball player, I would *need* to be taller than I am. Whether my height is or is not considered normal is irrelevant to whether I am actually tall enough to play professional basketball.

Clearly, this is not a need which is likely to be seriously deleterious to my overall health or successful functioning, and for this reason there would be no reason to prioritize enhancement of my height over more severe or limiting disease or illness states unless sufficient resources were available to do so. Certainly, it would be misleading to say that there is no difference between the need for life-saving medical assistance caused by kidney failure and the need to be above a certain height if one wishes to play professional basketball. These needs are different because in the first instance life itself may be threatened. Returning briefly to the observation by Toulmin (1975) mentioned earlier, to function sufficiently well that one is alive is a prerequisite for the attainment of any further goods. For this reason it is more important to prioritize meeting the former need than the latter.

However, recognition that need does not evaporate at the threshold of normality does not have to threaten the just allocation of medical resources. To return to another earlier observation, what Powell and Buchanan (2011, 10) have called "the ubiquity of suboptimal design," it is possible to accept both that even the healthy have needs *and* that resources should be allocated in such a way that they are titrated according to severity, or the extent to which they threaten the attainment of the goods of health. The key to the theoretical solution I propose is to attenuate the clinical significance of the restoration of normality as a threshold whose transgression would be antithetical to the appropriate goals of medical practice. Thus, if cognitive enhancers were permitted for use by people judged to be within the normal range, it would be appropriate to allocate them according to the relative position of individuals within this range.

As simple a theoretical solution as this is, its implementation would be beset with obstacles. It is not a simple matter to reduce the significance of normality as the proper goal of medical practice. However much it might be the case that normality has a tyrannical aspect in implying that one *ought* to be normal, it is not an easy task to divorce this from aggregate judgments about the benefits of normality. To be normal, after all, is simply to be similar to the majority of the other members of a class. Since "most people"—by definition—are normal, it is extremely difficult to imagine a scenario in which normality was no longer considered important or indeed central to the way in which we understand ourselves and interact. Nevertheless, even the reference class "most people" is arbitrary because, as Holm (2008) notes, one's relative normality—statistical or otherwise—will be different according to which reference class one compares oneself against.

The biomedical model of health underpinning the structure of western medicine in terms of its institutional structure, diagnostic criteria, and resource allocation criteria is reflected in its training. In order to realize a situation in which cognitive enhancements were readily available and their users not stigmatized, this model would have to be reorientated. Only then could potential doctors undertake medical training in a way that reflected this change. Enhancement would have to be understood as something that is not pernicious or coercive, and social policy would have to be able to manage the products in a way that was congruent with this. Mechanisms for offsetting the ability of the rich to buy enhancements (i.e., in a way similar to the "two-tier" structure of state-funded and private medicine at present) would have to be in place and this, again, would require a fundamentally different perspective on medical practice and its goals.

More could be achieved via cognitive enhancement if the normative pressure of normality were deliberately reduced in medical decision-making. Although this may be desirable for the reasons given here, I conclude that there is only a remote possibility of achieving it at present. To the extent that the views expressed in the EPO example represent those more widely held across the medical profession (and this would need to be ascertained by further empirical research), there is little perceived need to assist people in achieving supranormal functioning: medicine *ought* to meet those needs which restore normality before decisions about going further than this can be considered.

This principle may be reasonably clear-cut in the EPO case, because of the disabling and ultimately life-threatening nature of chronic kidney disease and the severity of its impact upon everyday life. However, in the case of cognitive enhancement it is less clear-cut. The "normal" range of cognitive function is wide, and one does not have to be considered educationally subnormal to nevertheless function significantly less well than somebody at the top of the normal range. A person who is cognitively "normal" but who could derive a benefit—both absolute and relative to others—from being free to enhance their ability to concentrate would not be eligible for this kind of assistance, because they would not be judged to have a need for it. This is regrettable because, unlike EPO which only confers a useful performance advantage in circumstances of extreme physical exertion, faculties such as concentration or memory retention are generally valuable to many in everyday life.

Finally, whatever one's view about the potential value of widening access to cognitive enhancers, the EPO case reminds us that unless we reach a point where we have managed to successfully neutralize all threats to health from illness and disease, there will always be people with needs who are more serious than those of someone in good health. In terms of a *prima facie* entitlement to medical resources, there will always be someone whose health needs take priority. This is not to say that a situation can *never* be envisioned in which there were sufficient funds available to justify the subsidization of assistance for meeting needs not caused by disease or illness, but just that it is a distant one. If one is committed to the view that it is ethically more important to treat chronic kidney disease, for example, than to provide cognitive enhancements for cognitively "normal" people, because the former poses a more serious threat to health than the latter, then it is hard to see how any widespread subsidization of cognitive enhancers could be incorporated into the health budget in a way that would be considered socially and politically acceptable.

Given the profound changes in medicine and society that would be needed in order to realize any potential large-scale benefits from cognitive-enhancing interventions, we should limit our expectations of what they can *actually* be used to achieved in the near future in terms of channeling their benefits to those who need them the most. The conclusion to be drawn from this is, for the time being at least, a pessimistic one: although there is a simple theoretical solution to certain ethical problems in cognitive enhancement, this solution is hamstrung by the importance awarded to "normality" as representing the end of medicine. Until professional perceptions about this change, there are so many practical obstacles to its implementation that it is hard to see how it could be realized in the near future.

Notes

1. http://www.bbc.co.uk/news/world-europe-20026838
2. http://news.bbc.co.uk/1/hi/programmes/hardtalk/9571648.stm
3. Here I am leaving aside the question concerning the relation between fact and value. I am assuming a broadly ethical naturalist position here in which impediments to health are judged to be bad in themselves to the extent that they have a negative impact on the person affected.
4. Numbers in brackets identify study participants, firstly by specialism and secondly by position in the series of interviews, e.g. C1 = renal clinician 1, S6 = renal scientist 6.

References

Aikins, R. D. 2011. Academic performance enhancement: A qualitative study of the perceptions and habits of prescription stimulant-using college students. *Journal of College Student Development* **52**(5):560–76.

Allhoff, F. 2005. Germ-line genetic enhancement and Rawlsian primary goods. *Kennedy Institute of Ethics Journal* **15**(1):39–56.

Armitage, J. 2007. The safety of statins in clinical practice. *The Lancet* **370**(9601):1781–90.

Bell, S., B. Partridge, J. Lucke, and W. Hall. 2013. Australian University students' attitudes towards the acceptability and regulation of pharmaceuticals to improve academic performance. *Neuroethics* **6**(1):197–205.

Besarab, A., W. K. Bolton, A. R. Nissenson, and S. J. Schwab. 2012. The normal HCT trial re-revisited: what were the actual findings? *Kidney international 82*(2):242.

Boorse, C. 1977. Health as a theoretical concept. *Philosophy of Science* 44(4):542–73.

Bostrom, N., and R. Roache. 2008. Ethical issues in human enhancement. *New Waves in Applied Ethics*, ed. J. Ryberg, T. Peterson and C. Wolf, 120–52.

Bostrom, N., and A. Sandberg. 2009a. Cognitive enhancement: methods, ethics, regulatory challenges. *Science and Engineering Ethics* **15**(3):311–41.

Bostrom, N., and A. Sandberg. 2009b. The wisdom of nature: An evolutionary heuristic for human enhancement. In *Human enhancement*, ed. J. Savulescu and N. Bostrom, 375–416. Oxford: Oxford University Press.

Bostrom, N., and J. Savulescu. 2009. Human enhancement ethics: the state of the debate. In *Human enhancement*, ed. J. Savulescu and N. Bostrom, 1–22. Oxford: Oxford University Press.

Boyd, K. M. 2000. Disease, illness, sickness, health, healing and wholeness: exploring some elusive concepts. *Medical Humanities* **26**:9–17.

Buchanan, A., D. Brock, N. Daniels, and D. Wikler. 2000. *From chance to choice: Genetics and justice.* New York: Cambridge University Press.

Bullough, V. L. 1987. Technology for the prevention of "Les Maladies Produites par la Masturbation." *Technology and Culture* **28**(4):828–32.

Burn, J., A. M. Gerdes, F. Macrae, J. P. Mecklin, G. Moeslein, S. Olschwang, and D. T. Bishop. 2012. Long-term effect of aspirin on cancer risk in carriers of hereditary colorectal cancer: An analysis from the CAPP2 randomised controlled trial. *The Lancet* **378**(9809):2081–7.

Cakic, V. 2009. Smart drugs for cognitive enhancement: Ethical and pragmatic considerations in the era of cosmetic neurology. *Journal of Medical Ethics* **35**(10):611–15.

Capps, B. 2011. Libertarianism, legitimation, and the problems of regulating cognition-enhancing drugs. *Neuroethics* **4**:119–28.

Chatterjee, A. 2004. Cosmetic neurology: The controversy over enhancing movement, mentation, and mood. *Neurology* **63**:968–74.

Conrad, P. 1992. Medicalisation and social control. *Annual Review of Sociology* **18**:209–32.

Coveney, C. M., B. Nerlich, and P. Martin. 2009. Modafinil in the media: Metaphors, medicalisation and the body. Social Science & and Medicine, **68**(3): 487–495.

Cuzick, J., F. Otto, J. A. Baron, P. H. Brown, J. Burn, P. Greenwald, and M. Thun. 2009. Aspirin and non-steroidal anti-inflammatory drugs for cancer prevention: an international consensus statement. *The Lancet Oncology* **10**(5):501–7.

Daniels, N. 2000. Normal functioning and the treatment-enhancement distinction. *Cambridge Quarterly of Healthcare Ethics* **9**:309–22.

Davignon, J. 2004. Beneficial cardiovascular pleiotropic effects of statins. *Circulation* **109**(23 suppl 1): III-39.

Delinsky, S. S. 2005. Cosmetic surgery: A common and accepted form of self-improvement? *Journal of Applied Social Psychology* **35**(10):2012–28.

Diamanti-Kandarakis, E., P. A. Konstantinopoulos, J. Papailiou, S. A. Kandarakis, A. Andreopoulos, and G. P. Sykiotis. 2005. Erythropoietin abuse and erythropoietin gene doping. *Sports Medicine 35*(10):831–40.

Fishbane, S. 2009. Erythropoiesis-stimulating agent treatment with full anemia correction: a new perspective. *Kidney International* **75**(4):358–65.

Fishbane, S., and J. B. Wish. 2012. A physician's perseverance uncovers problems in a key nephrology study. *Kidney International* **82**(2):135–7.

Fisher, J. W. 2010. Landmark advances in the development of erythropoietin. *Experimental Biology and Medicine* **235**(12):1398–1411.

Fleischhauer, K., and G. Hermerén. 2006. *Goals of medicine in the course of history and today.* Stockholm: Kungl. Vitterhets Historie och Antikvitets Akademien.

Fukuyama, F. 2002. *Our posthuman future: Consequences of the biotechnology revolution.* New York: Picador.

Fitz, N. S., R. Nadler, P. Manogaran, E. W. J. Chong, and P. B. Reiner. 2013. Public attitudes toward cognitive enhancement. *Neuroethics* 7(2):173–88.

Forlini, C., and E. Racine. 2009. Autonomy and coercion in academic "cognitive enhancement" using methylphenidate: Perspectives of key stakeholders. *Neuroethics* **2**(3):163–77.

Fox, R. C. 1977. The medicalization and demedicalization of American society. *Daedalus* **106**(1):9–22.

Freidson, E. L. 1970. *Profession of medicine: A study of the sociology of applied knowledge.* New York: Dodd, Mead.

Goffette, J. 2006. Naissance de l'anthropotechnie: de la médecine au modelage de l'humain. Paris: Vrin.

Greely, H., B. Sahakian, J. Harris, R. C. Kessler, M. Gazzaniga, P. Campbell, and M. J. Farah. 2008. Towards responsible use of cognitive-enhancing drugs by the healthy. *Nature* **456**(7223): 702–5.

Gunga, H. C., K. Kirsch, A. L. Roecker, E. Kohlberg, J. Tiedemann, M. Steinach, and W. Schobersberger. 2007. Erythropoietin regulations in humans under different environmental and experimental conditions. *Respiratory Physiology and Neurobiology* **158**(2–3):287–97.

Harris, J. 2009. "Enhancements are a moral obligation." In *Human Enhancement*, ed. J. Savulescu and N. Bostrom, 131–55. Oxford: Oxford University Press.

Holm, S. 2008. *Cosmetic surgery and the concept of medical necessity.* [Online] Available at http://www.ccels.cf.ac.uk/archives/issues/2008/holm.pdf.

Honour, J. W. 2004. The fight for fair play. *Nature* **430**(6996):143–4.

Holtug, N. 1999. Does justice require genetic enhancements? *Journal of Medical Ethics* **25**(2):137–43.

Joyner, M. J. 2003. VO2MAX, blood doping, and erythropoietin. *British Journal of Sports Medicine* 37(3):190–1.

Juengst, E. T. 1997. Can enhancement be distinguished from prevention in genetic medicine? *Journal of Medicine and Philosophy* **22**(2):125–42.

Kaufman, K. R. 2005. Modafinil in sports: ethical considerations. *British Journal of Sports Medicine* **39**(4):241–4.

Kleinman, A., and L. H. Sung. 1979. Why do indigenous practitioners successfully heal? *Social Science and Medicine. Part B: Medical Anthropology* **13**(1):7–26.

Koch, T. 2010. Enhancing who? Enhancing what? Ethics, bioethics, and transhumanism. *Journal of Medicine and Philosophy* **35**(6):685–99.

Lacson, Jr., E., N. Ofsthun, and J. M. Lazarus. 2003. Effect of variability in anemia management on hemoglobin outcomes in ESRD. *American Journal of Kidney Diseases* **41**(1):111–24.

LaRosa, J. C., J. He, and S. Vupputuri. 1999. Effect of statins on risk of coronary disease: a meta-analysis of randomized controlled trials. *JAMA* **282**(24):2340–6.

Levin, A. 2007. Understanding recent haemoglobin trials in CKD: methods and lesson learned from CREATE and CHOIR. *Nephrology, Dialysis, Transplantation* **22**(2):309–12.

Lin, P., and F. Allhoff. 2008. Untangling the debate: The ethics of human enhancement. *NanoEthics* **2**(3):251–64.

Lowenberg, J. S., and F. Davis. 1994. Beyond medicalisation–demedicalisation: the case of holistic health. *Sociology of Health and Illness* **16**(5):579–99.

Mayes, R. 2010. The modern olympics and post-modern athletics: A clash in values. *Journal of Philosophy, Science and Law* **10**:1–17.

Mignon, P. 2003. The Tour de France and the doping issue. *International Journal of the History of Sport* **20**(2):227–45.

Nam, J. 2015. Biomedical enhancements as justice: Biomedical enhancements as justice. *Bioethics* **29**(2):126–32.

Neftel, K. A., R. H. Adler, L. Käppeli, M. Rossi, M. Dolder, H. E. Käser, and H. Vorkauf. 1982. Stage fright in musicians: a model illustrating the effect of beta blockers. *Psychosomatic Medicine* **44**(5):461–9.

Nordenfelt, L. 2001. On the goals of medicine, health enhancement and social welfare. *Health Care Analysis* **9**(1):15–23.

Parens, E. 1998. Is better always good?: The enhancement project. Hastings Center Report, **28**(1):s1–s17.

Pellegrino, E. D. 2004. Biotechnology, human enhancement, and the ends of medicine. *Dignitas* **10**(4):1–5.

Paoletti, E., and G. Cannella. 2006. Update on erythropoietin treatment: should hemoglobin be normalized in patients with chronic kidney disease? *Journal of the American Society of Nephrology* **17**(4 Suppl 2):S74–7.

Powell, R., and A. Buchanan. 2011. Breaking evolution's chains: the prospect of deliberate genetic modification in humans. *Journal of Medicine and Philosophy* **36**(1):6–27.

President's Council on Bioethics. 2003. *Beyond Therapy: Biotechnology and the Pursuit of Happiness.* New York: Regan Books.

Press, I. 1980. Problems in the definition and classification of medical systems. *Social Science and Medicine. Part B: Medical Anthropology* **14**(1):45–57.

Rayner, J., P. Pyett, and J. Astbury. 2010. The medicalisation of "tall" girls: A discourse analysis of medical literature on the use of synthetic oestrogen to reduce female height. *Social Science and Medicine* **71**(6):1076–83.

Remak, E., J. Hutton, M. Jones, and M. Zagari. 2003. Changes in cost-effectiveness over time. *European Journal of Health Economics* **4**(2):115–21.

Remuzzi, G., and J. R. Ingelfinger. 2006. Correction of anaemia—payoffs and problems. *New England Journal of Medicine* **355**(20):2144–6.

Robinson, N., S. Giraud, C. Saudan, N. Baume, L. Avois, P. Mangin, and M. Saugy. 2006. Erythropoietin and blood doping. *British Journal of Sports Medicine* **40**(suppl 1):i30–i34.

Sadegh-Zadeh, K. 2000. Fuzzy health, illness, and disease. *Journal of Medicine and Philosophy* **25**(5):605–38.

Sandberg, A., and J. Savulescu, J. 2011. The social and economic impacts of cognitive enhancement. Enhancing Human Capacities, 92–112.

Sandel, M. J. 2004. The case against perfection. *Atlantic Monthly* **293**(3):51–62.

Savulescu, J. 2006. Justice, fairness, and enhancement. *Annals of the New York Academy of Sciences* 1093(1):321–38.

Scully, J. L., and C. Rehmann-Sutter. 2001. When norms normalise: The case of genetics enhancement. *Human Gene Therapy* **12**(1):87–95.

Shickle, D. 2000. Are "genetic enhancements" really enhancements? *Cambridge Quarterly of Healthcare Ethics* **9**(03):342–52.

Singh, A. K., L. Szczech, K. L. Tang, H. Barnhart, S. Sapp, M. Wolfson, and D. Reddan. 2007. Anaemia of CKD—the CHOIR study revisited. *Nephrology Dialysis Transplantation* **22**(7):1806–10.

Stempsey, W. E. 2006. Emerging medical technologies and emerging conceptions of health. *Theoretical Medicine and Bioethics* **27**(3):227–43.

Szczech, L. A., H. X. Barnhart, J. K. Inrig, D. N. Reddan, S. Sapp, R. M. Califf, U. D. Patel, and A. K. Singh. 2008. Secondary analysis of the CHOIR trial epoetin-alpha dose and achieved hemoglobin outcomes. *Kidney International* **74**(6):791–8.

Tannenbaum, J. 2014. The promise and peril of the pharmacological enhancer modafinil. *Bioethics* **28**(8):436–45.

Toulmin, S. 1975. Concepts of function and mechanism in medicine and medical science. In *Evaluation and explanation in the biomedical science: Proceedings of the First Trans-Disciplinary Symposium on Philosophy and Medicine, Held at Galveston, May 9–11, 1974*, ed. H. T. Engelhardt, Jr, and S. Spicker, 51–66. Boston: D. Reidel.

Vrecko, S. 2013. Just how cognitive is "cognitive enhancement"? On the significance of emotions in university students' experiences with study drugs. *AJOB Neuroscience* **4**(1):4–12.

Warnock, M. 2003. What Is natural? And should we care? *Philosophy* **78**(4):445–59.

Wesner, R. B., R. Noyes Jr, and T. L. Davis. 1990. The occurence of performance anxiety among musicians. *Journal of Affective Disorders* **18**(3):177–85.

Will, E. 2001. Targets and targeting. *American Journal of Kidney Diseases* **38**(2):411–14.

Will, E. J., D. Richardson, C. Tolman, and C. Bartlett. 2007. Development and exploitation of a clinical decision support system for the management of renal anaemia. *Nephrology, Dialysis, Transplantation* **22**(Suppl 4):iv31–iv36.

Williams, S. J., C. Seale, S. Boden, P. Lowe, and D. L. Steinberg. 2008. Waking up to sleepiness: Modafinil, the media and the pharmaceuticalisation of everyday/night life. *Sociology of Health and Illness* **30**(6):839–55.

Winearls, C. G. 2006. In the wake of progress-ethical problems of renal failure treated by dialysis. *Clinical Medicine* **6**(1):76–80.

Winter, S. C., K. A. Shah, L. Campo, H. Turley, R. Leek, R. J. Corbridge, and A. L. Harris. 2005. Relation of erythropoietin and erythropoietin receptor expression to hypoxia and anemia in head and neck squamous cell carcinoma. *Clinical Cancer Research*, **11**(21):7614–7620.

Chapter 13

On the argument that enhancement is "cheating"[1]

Maartje Schermer

A frequently invoked argument against the enhancement of human capacities through biotechnological innovations is that this would amount to cheating. The paradigmatic example is doping in sports, but recently the argument has also been raised against cognitive enhancement. Neuroscientist Stephen Rose expresses the common intuition as follows: "with the cognition enhancing drugs—as with the use of steroids by athletes, their use, at least in a competitive context, is seen as a form of cheating, of bypassing the use for hard work and study" (Rose 2005). And Michael Gazzaniga asks, "Why do we resist changes in our cognitive skills through drugs? It seems to me that it is because we think cognitive enhancement is cheating. If, somehow, someone gets better through hard work, that's okay.... But popping a pill and mastering the information after having read it only once seems like cheating" (Gazzaniga 2005).

In this chapter, I will compare sports and education, two practices in which medical means are increasingly being used for enhancement purposes (Chatterjee 2006), and analyze how the enhancement-is-cheating argument can be interpreted in both contexts. I will argue that although using enhancements can be a form of cheating, this is relatively easy to remedy by either changing the rules or instituting controls and sanctions. This does not, therefore, constitute a categorical objection to enhancement. However, there are some remaining worries that concern the specific goals and goods internal to both sports and education understood as "practices." These worries are less easy to articulate, but also less easy to dismiss.

13.1 Cheating

Given the *prima facie* moral wrongness of cheating, it is remarkable that so little has been written on the subject in ethics. Perhaps this is partly caused by the lack of conceptual clarity. The term cheating is used, in common language, as indicating many different forms of deception and fraud that are intended to gain some benefit for oneself. Following one of the most thorough analyses available, cheating can be described as "the intentional violation of a rule, in order to gain an unfair advantage over others" (Green 2004). Cheating usually involves deception in order to hide that one is getting an unfair advantage, but I agree with Green that deception is not the moral core of cheating. Cheating is primarily a matter of fairness. The rules that are violated by the cheater can be explicit, as in games and sports, but they can also be implicit, as in codes of social conduct. In many cases, one does not explicitly subscribe to the set of rules that determines what is fair and what is not, but one accepts them tacitly by joining a certain practice or by entering in a game (Loland 1992). Actions that involve the breaking of implicit or unwritten rules and offer an unfair advantage can also be considered cheating, according to this definition.

13.2 Enhancement and cheating in sports

The paradigmatic example of cheating as gaining an unfair advantage over others by breaking the rules comes from competitive sports (Loland 2005). Unsurprisingly, most discussion regarding the unfairness of enhancement technologies has also taken place in this field, since performance enhancement is a natural part—if not the goal—of sports. Some forms of biotechnological enhancement are against established sports rules: the International Olympic Committee issues lists of forbidden substances (doping), and doping checks are in place in all major games and contests to keep these rules and expel trespassers. Global governance is established by the World Anti-Doping Agency (http://www.wada-ama.org). The question of whether existing enhancement technologies constitute cheating can be answered simply by referring to the rules of the sport in question: do they forbid a specific technology? If so, using the technology to gain an—by definition unfair—advantage is a form of cheating; if not, there is no moral problem.

When new enhancement technologies are introduced, however, the question of cheating cannot be answered with a simple appeal to existing rules. The question is whether the old rules still suffice to deal with the new technological possibilities. If the new technology offers some unfair advantage, it may have to be banned by new rules. With each new opportunity for enhancement, this question must be considered again—for example, see a recent discussion on the use of hypoxic air machines in the training of Australian footballers (Spriggs 2005; Tamburrini 2005; Tännsjö 2005).

Obviously, it is unfair if some competitors can use an effective performance enhancer while others cannot. But this is no reason to ban the technology altogether, because this unfairness can be solved relatively easily by allowing the technology and by ensuring equal access for all. Another important set of reasons for banning enhancement technologies has to do with health and safety. Enhancements that are dangerous to the athletes' health may be banned on those—paternalistic—grounds.[2] But apart from these, what other reasons might there be to issue rules against particular new enhancement technologies, thereby making their use a form of cheating? In many cases intuitions about the unfairness of enhancement are strong. Even if there are no rules against it—yet—we may have the intuition that enhancement is a form of cheating because it violates some unwritten or inarticulate rule. We should look further to explain and assess this intuition. Are there good arguments to support it, and claim that enhancement should be forbidden because it is cheating?

One frequently invoked argument rests on the distinction between constitutive and regulative rules. Some rules are constitutive for a specific sport—for example, in running the marathon it is essential that there is running involved. Someone who covers the 42,195 metres on roller skates is not cheating on his fellow marathon men but is simply joining another game (Whitehouse et al. 1997). This notion of constitutive rules implies that some enhancements cannot or should not be allowed because they would change the sport in question in such a way that it would not be the same sport anymore. This is not a matter of fairness or cheating—it is the defining characteristics of a sport that are at stake here. However, many enhancement technologies do not touch upon the constitutive elements of specific sports. Using special shoes, erythropoietin (EPO), isotonic drinks, or steroids during a marathon is not disruptive in the same sense that using roller skates is. Still, it can be asked whether the advantage they give is unfair, a violation of some implicit rule of justice, and whether this should prompt a change in the explicit regulative rules.

Another line of argument thus appeals to the unwritten rules, the "ethos," or the "local justice," of sports (Loland 1999). In Walzer's terms, justice in sports is a matter of desert (Walzer 1983). Victory, praise, and prizes in sports are distributed fairly if they are distributed according to desert, which means one should get these goods because they are deserved due to a certain merit. But there is no consensus on what exactly constitutes merit in the case of sports. Victory in a sports contest depends on natural abilities, training efforts, equipment, sheer luck, and probably a number of other factors. Opinions vary on what exactly should, ideally, determine victory. With regard to the hypoxic air machines one commentator is of the opinion that the essence of sport is that the person with the best natural abilities (the winner of the "genetic lottery") should win the competition (Tännsjö 2005). Another commentator, however, claims that this idea is becoming obsolete and that there is no reason to let the genetic lottery decide the outcome of the sport competition (Tamburrini 2005). Technologically enhanced performance could also be seen as merit, the combination of natural luck and smart use of technology (Savulescu et al. 2004). In order to determine what is fair or just we need to think about the merit that we value in sports. What exactly do we want to praise or celebrate?

According to the President's Council on Bioethics (2003), what matters most is not even the competitive results but the specific type of human excellence displayed in a sport: "Keeping scores is meant to honour and promote a given type of human excellence, whose meaning is in the doing, not simply in the scored results." This view refers not so much to a notion of fairness or justice in the distribution of victory, but to something like the "internal goods" and internal standards of excellence of sport (MacIntyre 1985). It is telling and instructive that MacIntyre introduces his notion of internal goods in contrast to the rationale of cheating. He asks us to imagine a child whom he wants to learn how to play chess (chess being the paradigmatic example of a practice). Since the child has no particular desire to learn, he motivates the child by promising him or her candy for playing chess once-a-week, and some extra candy if the child wins the game. As long as it is only the candy that motivates the child to play chess, the child has no reason not to cheat and every reason to cheat, according to MacIntyre. However, once the child has learned to appreciate the internal goods of chess, like "the achievement of a highly particular kind of analytic skill, strategic imagination and competitive intensity," he or she will have discovered a new kind of motivation, not just for winning on a particular occasion, but for trying to excel in whatever way the game of chess demands. "Now if the child cheats," says MacIntyre (1985), "he or she will not be defeating me, but him or herself."

So, as long as one plays for the sake of winning external goods—money, prizes, status—cheating in sport may be rational. However, cheating makes no sense if what one wants to achieve are the internal goods of sport.[3] Every sport, being a practice, has its own internal standards of excellence to which each sportsperson should aim to live up to. Moreover, doing so is intimately connected with virtues such as courage, honesty, and justice. Enhancement technologies might disrupt these forms of excellence and thus bar the achievement of internal goods, while still allowing the attainment of external goods like prizes. This would constitute a different form of "cheating," namely on the practice itself. As the President's Council on Bioethics (2003) expresses it, sportsmen who would use biotechnological enhancements would be bad sportsmen—"not simply because they cheated their opponents, but because they also cheated, undermined or corrupted themselves and the very athletic activity in which they seem to excel."

Whether biotechnological enhancements would really pervert specific sports practices depends on the understanding of the specific internal goods and standards of excellence of those practices. This is something only participants in a practice can judge, according to MacIntyre (1985). One example of such a discussion is the one on hypoxic air machines mentioned earlier. Likewise, the prospect of genetically enhanced athletes has prompted discussion on the view that effort and natural giftedness are prerequisites for excellence in sport, versus the view that performance enhancement is in the spirit of sport (Sandel 2004; Savulescu et al. 2004; Walzer 1983).[4]

I believe discussions like these should primarily be conducted among sportsmen themselves, but, like Brown (1990), I see no reason why at least some biomedical enhancement technologies should not be accepted as contributing to excellence and as innovations of the practice in question.

13.3 Enhancement and cheating in education

I turn now from sports to a practice in which cognitive enhancers play or might play a role. In the discussion about cognitive enhancement the cheating-argument is used especially with regard to education, and the parallel with sports offers useful insights to discuss the argument in this context.

Recently, Turner and Sahakian (2006) considered what effects the widespread use of smart drugs would have on our educational system: "Could children in the future face blood or urine tests when sitting their A-levels or GCSE exams?" The image of students on Ritalin cramming for exams is frequently invoked in the enhancement debate; more powerful future enhancers—of memory, concentration, and other learning abilities—could indeed prompt the question of whether students using them would be cheating.[5] The case is similar to the doping-in-sports case, but not the same. Unlike in sports, the rules of the "game" are not as clear and explicit in education. With regard to exams the rules are put down in the exam regulations. These forbid cribbing or fraud, but they usually do not forbid drinking coffee beforehand, or taking a beta-blocker or other substance to calm one's nerves. More specifically, exam regulations are silent with regard to the way a student is supposed to study. So, as long as schools or universities, or other supervising authorities, do not issue rules against the use of modafinil, Ritalin, or any other cognitive enhancer, students can use whatever they want without it being cheating.

This does not end the discussion, however, because rules can be changed and the availability (or prospect) of cognitive enhancers may prompt us to do so, just like the availability of new enhancement drugs in sports prompts new anti-doping rules. Perhaps using a memory enhancer *should* be considered cheating, just like using a crib is! As empirical research shows, many have the intuition that there is something unfair or wrong with such enhancers that requires adjustment of the official rules (Sabini and Monterosso 2005).

Suppose that cognitive enhancers would make it much easier to learn the stuff needed for an exam—for example, by boosting memory so that reading a text once would be enough to reproduce it. This would be unfair when the other students had no access to the same enhancers, but, like in sports, we could simply change the rules and allow cognitive enhancers for everyone, even make them available as a school supply.[6] This would solve the problem of unfair advantage and thus of cheating. This is the line that Mehlman (2004) takes. Likewise, Whitehouse et al. (1997) states that when everyone would be allowed enhancers, this would not be different from allowing everyone to use an electronic calculator.

I believe this conclusion is too quick. First of all, exams are supposed to signify something—if one wants to test the student's performance on mental arithmetic, it makes no sense to allow the use of an electronic calculator, just like it makes no sense to run the marathon on roller skates.[7] There is an analogy here with the constitutive rules of a game or sport. If one wants to test memory capacities it makes no sense to allow cribs or memory enhancers. It depends on the exact way in which an enhancer would work, whether it would undercut the rationale of a specific exam. If the goal is to test understanding and insight, the use of cribs or open-book exams poses no problem and memory enhancers, or drugs improving mental stamina and decreasing the need for sleep, like modafinil, could be seen as legitimate study aids.

One special purpose of exams is to select the best students for scarce university places. Such entrance exams measure the applicant's abilities relative to others, and since there is a competitive advantage to be gained, it makes sense to cheat. There is also an implicit idea of local justice at stake here: the winning candidates should really merit the university admission; they should have whatever it is that the universities look for in their students. But what exactly are entrance exams supposed to indicate—talent, effort, endurance, knowledge, competences, ambition, motivation, or the willingness to take enhancing drugs to get in? Who do we want to enter the best universities?[8]

Second, in education in general, cognitive performances are not only valued as such, but are also valued for the manner in which they are achieved. There may be some inarticulate rule broken by the use of cognitive enhancers, which would make their use a form of cheating. Like in the case of sport, one might ask what kind of merit makes one fairly deserve praise for a good performance in school. A good school performance is in part determined by inborn abilities and a dose of luck, but also by effort and dedication. Praise is given for the effort and endurance put into the performance, as well as for the final outcome. A weak student may deserve a high amount of praise for an average achievement, while a brilliant student deserves blame for the same performance. Likewise, a student who improves his performance through tutoring and extra practice may, arguably, deserve more praise than one who owes his improved performance to an enhancing drug. The use of cognitive enhancers could disrupt this idea of fairly earned praise. As Juengst (1998) asks, to what extent can you take credit for accomplishments if they are not achieved through the socially valued practices, like study and effort that have traditionally produced them?

The issue of cognitive enhancement and cheating is mostly discussed with reference to competitive exam situations, but the previous considerations indicate it should also be viewed in a broader context, that of education as a whole. Just like sports matches are not all there is to sports, exams are not all there is to education. Like sport, education can be understood as a practice, or as part of the specific practices one is educated into.[9] Education and studying have internal goods next to their more instrumental goals. Such internal goods may be the attained appreciation of the internal goods of the practices one is educated in, knowledge and truth, the activity of studying with its character-building side effects,[10] or the general (moral) self-development it effectuates. According to MacIntyre, these are the ends of education, but these are increasingly substituted by a rat race in which only exams and test results count (MacIntyre and Dunne 2002). The use of cognitive enhancers might make studying easier, but it is not clear how much it would add to the ends of education, conceived in this broader way. According to Juengst (1998), for example, the use of Ritalin would undermine the disciplined study and active learning that the practice of being a student is supposed to involve. Moreover, the use of cognitive enhancers might improve memory or attention, but it does not necessarily engender more insight and understanding, or a better attitude. Generalized use of cognitive enhancers may well lead to an overemphasizing of cognitive performance and test results, at the cost of a broader idea of development. Or, in MacIntyrian terms, it may promote concentration on external goods and thus threaten the achievement of internal goods (MacIntyre and Dunne 2002).

On the other hand, there seem to be no *prima facie* reasons why cognitive enhancers could not be fitted into the practice of education and learning. They could well help increase the general level of development by helping students to read and study more and remember better. If used in the right context and with the right intentions it might contribute to attaining internal goods and living up to standards of excellence in education and the various practices that students are educated into.

13.4 Conclusions

The argument that a certain new enhancement technique is wrong because it constitutes cheating can be interpreted in a simple and straightforward way by making reference to existing rules: only if rules are broken and there is an unfair advantage does the use of enhancers constitutes cheating. This does not account for the intuitive idea—often expressed in terms of "cheating"—that there is something unfair about the use of certain enhancement techniques, perhaps because they break some unwritten rules. For new enhancement techniques, new formal rules may have to be established. With regard to sports, one reason to do so, next to safety and health-related concerns, is that a specific enhancement may violate the constitutive rules of the sport in question—and such rules cannot be changed randomly. Another reason is that even though regulative rules might be changed without changing the defining characteristics of the sport in question, such changes might still somehow violate the ethos or local justice of sport, or might disrupt a specific sports practice with its associated internal goods. This line of argument thus leads to a discussion about what sport *is*, and why we value it. In other words, it leads us away from the straightforward notion of cheating-as-rule-breaking, toward the complex meaning and value of sports as a practice, and toward an appraisal of its own standards of excellence and internal goods.

Along these same lines, the argument that the use of cognitive enhancers in the context of education constitutes cheating can be assessed. The argument that using cognitive enhancers in examinations would constitute cheating is, in general, not very convincing: making enhancers available for everyone would solve this issue. However, depending on the goal or purpose of specific exams and the specific capacities that would be enhanced, some enhancers might justifiably be banned as forms of cheating. If we consider the issue from a broader perspective and understand education in terms of practices, the use of cognitive enhancers may raise more puzzling questions with less straightforward answers. This perspective leads us away from the question of cheating and examination rules, into discussions on the values, ends, and internal goods of education. Seen from this perspective, the important question is how cognitive enhancers might be embedded in education (understood broadly)—or how they might pervert it. I will not try to conduct that discussion here, but it seems clear that it should be conducted by those involved in the practice of education—teachers, educators, students, philosophers of education—to help answer moral questions about the desirability of allowing cognition-enhancing drugs in our schools. Since the drugs involved come from the medical domain, and may also involve health effects, medical professionals in school—and youth healthcare—like those in sports medicine—also ought to think about these issues.

13.5 Acknowledgments

The research for this paper was funded by the Netherlands Organization for Scientific Research (NWO). The author would like to thank two anonymous reviewers for their helpful comments and suggestions. There are no competing interests.

Notes

1. This chapter previous appeared as Schermer, M. 2008. On the argument that enhancement is "cheating." *Journal of Medical Ethics* 34:85–8. It is reprinted here with permission from BMJ Publishing Group Limited.
2. Issues of health and safety are of course very important with regard to the regulation of doping in sports. While top sport itself is a rather unhealthy activity, the use of various kinds of doping has side effects and health risks of its own. Due to the competitive setting in which it is used, the voluntary nature of its use can also be called into question. For the sake of the present argument, however, I will set aside those concerns and focus on the issues of cheating and fairness.
3. As Brown (1990) and McNamee (1995) have argued, the distinction between external and internal goods may not be as clear as MacIntyre suggests. However, I believe this is an analytic distinction, which holds heuristic power and enables us to express a certain type of value that is specific for practices.
4. Another example, which I owe to an anonymous reviewer, is the discussion about Casey Martin, a professional golfer who, because of a leg disorder, claimed the right to use a golf cart. This prompted debate on whether walking is part of the excellence of playing golf.
5. See Farah et al. (2004). Another example of a situation where the issue of "cheating" might come up would be that of an employee who gets a job promotion because the use of modafinil has enabled him to work much longer hours than his competitors. Would he be cheating or playing unfair?
6. Disregarding here, again, for the sake of the argument, other concerns such as health risks or social pressure.
7. A point shared between educational tests and sports is that the most efficient means to a certain goal are often ruled out—I owe this point to an anonymous reviewer.
8. These issue were also at stake in a recent discussion in the Netherlands about entrance exams, for instance, for medical school. While traditionally there were no such exams in the Netherlands, there have been some universities experimenting with selection of the best students. As it turned out, exam results did not predict success in the study. Moreover, it is another question altogether whether grades, test results, or success in cognitive performance predicts very much about the ability to become a good doctor.
9. For extensive discussion on the question of whether teaching and education are practices in themselves, or part of other practices, see vol 37, no 2, of the *Journal of the Philosophy of Education*. I will not try to resolve this debate here; for my present purpose what matters is that there are internal goods involved in educational practices that ought to be distinguished from external goods.
10. Like, for example, perseverance, critical and independent thinking, and curiosity.

References

Brown, W. M. 1990. Practices and prudence. *Journal of the Philosophy of Sport* **17**:71–84.

Chatterjee, A. 2006. The promise and predicament of cosmetic neurology. *Journal of Medical Ethics* 32:110–13.

Farah, M. J., J. Illes, R. Cook-Degan, H. Gardner, E. Kandel, P. King, E. Parens, B. Sahakian, and P. R. Wolpe. 2004. Neurocognitive enhancement: what can we do and what should we do? *Nature Reviews: Neuroscience* **5**:421–5.

Gazzaniga, M. S. 2005. *The ethical brain.* New York: Dana Press.

Green, S. P. 2004. Cheating. *Law and Philosophy* **23**:137–85.

Juengst, E. T. 1998. What does *enhancement* mean? In: *Enhancing human traits*, ed. E. Parens, 29–47. Washington, DC: Georgetown University Press.

Loland, S. 1999. Justice and game advantage in sporting games. *Ethical Theory and Moral Practice* **2**:165–83.

Loland, S. 2005. The varieties of cheating—comments on ethical analyses in sport. *Sport in Society* **8**:11–26.

MacIntyre, A. 1985. *After virtue*, 2nd edition. London: Duckworth.

MacIntyre, A., and J. Dunne. 2002. Alasdair MacIntyre on education: In dialogue with Joseph Dunne. *Journal of Philosophy of Education* **36**:1–19.

McNamee, M. 1995. Sporting practices, institutions and virtues: a critique and a restatement. *Journal of the Philosophy of Sport* **22**:61–82.

Mehlman, M. J. 2004. Cognition-enhancing drugs. *Milbank Quarterly* **82**:483–506.

President's Council on Bioethics. 2003. *Beyond therapy*. New York: Dana Press.

Rose, S. 2005. *The future of the brain*. Oxford: Oxford University Press.

Sabini, J., and J. Monterosso. 2005. Judgements of the fairness of using performance enhancing drugs. *Ethics and Behavior* **15**:81–94.

Sandel, M.J. 2004. The case against perfection. *Atlantic Monthly* **3**:51–62.

Savulescu, J., B. Foddy, and M. Clayton. 2004. Why we should allow performance enhancing drugs in sport. *British Journal of Sports Medicine* **33**:666–70.

Spriggs, M. 2005. Hypoxic air machines: performance enhancement through effective training—or cheating? *Journal of Medical Ethics* **31**:112–13.

Tamburrini, C. 2005. Commentary. *Journal of Medical Ethics* **31**:114.

Tännsjö, T. 2005. Commentary. *Journal of Medical Ethics* **31**:113.

Turner, D., and B. Sahakian. 2006. The cognition enhanced classroom. In: *Better humans?*, ed. P. Miller and J. Wilsdon, 79–85. London: Demos.

Walzer, M. 1983. *Spheres of justice*. New York: Basic Books.

Whitehouse, P. J., E. Juengst, M. Mehlman M, and T. H. Murray. 1997. Enhancing cognition in the intellectually intact. *Hastings Center Report* **27**:14–22.

Chapter 14

Psychiatric nosology and cognitive enhancement

Dan J. Stein

14.1 Introduction

There has been renewed interest in psychiatric classification in recent years, with the development of the fifth edition of the *Diagnostic and Statistical Manual of Mental Disorders* (DSM-5) (American Psychiatric Association 2013), the 11th edition of the International Classification of Diseases (ICD-11) (Reed and Ayuso-Mateos 2011), and the Research Domain Criteria (RDoC) framework (Insel et al. 2010). The different approaches instantiated in DSM-5, ICD-11, and RDoC reflect somewhat different conceptual and empirical emphases. To some extent, discussions regarding the development of these classifications and the subsequent responses of critics mirror the central debates in contemporary psychiatry.

Debates about cognitive enhancement also mirror these key contemporary debates. I have elsewhere put forwards a conceptual framework which proposes that different positions in philosophy of science, language, medicine, and psychiatry accord with different perspectives on cognitive enhancement (Stein 2012). Contemporary psychiatry must straddle a line between a scientism that regards psychiatry as theory-free and value-neutral, and a skepticism that sees psychiatry as a pseudo-science that merely reflects sociopolitical dynamics (Stein 1991, 1998). An integrative perspective draws on aspects of these more classical and critical positions, to provide its own solutions (Stein 2008a).

In this chapter, I will consider issues relevant to cognitive enhancement from the perspective of debates in psychiatric classification, reviewing positions taken during the development of DSM-5, ICD-11, and the RDoC framework. I am concerned in particular with how a clinician should respond when a patient with subthreshold symptoms complains of distress or impairment and requests pharmacotherapeutic enhancement–treatment. Fuzzy nosological borders between normality and pathology undermine strict contrasts between enhancement and treatment, and so deserve careful consideration. I will again propose an integrative approach which emphasizes that clinicians must be both scientists with a knowledge of physiology and humanists who understand the experience of illness.

14.2 DSM-5

A great deal of effort was invested in the revision of DSM-IV. Thus, for example, a series of research conferences, funded by the National Institute of Mental Health in the United

States, brought together clinicians and researchers from around the world, prior to the establishment of DSM-5 working groups (Regier et al. 2009). At this early stage of the revision process, expectations were high that DSM-5 would be quite different from DSM-IV given advances in psychiatry research. In particular, DSM-5 leaders emphasized the importance of including neuroscientific findings, incorporating dimensional measures of symptoms, and doing away with the clinical significance criterion (i.e., the diagnostic criterion that describes the presence of distress or impairment).

As the process unfolded, however, it became increasingly clear that DSM-5 would not exemplify an entirely new approach to psychiatric nosology, but would rather comprise a refinement of DSM-IV (First 2010; Kendler and First 2010). Biomarkers for psychiatric disorders did not have sufficient sensitivity and specificity to be incorporated into the diagnostic criteria (Kupfer and Regier 2011). Dimensional measures are included in DSM-5, but they are tucked away in an appendix, and it is unclear whether busy clinicians will find the time to use them. The clinical significance criterion remains key; as in DSM-IV it is the ultimate, somewhat tautological, arbiter of the distinction between normality and pathology: those with a psychiatric disorder must demonstrate distress or impairment.

It is notable that a proposed diagnosis of prodromal psychotic disorder or attenuated psychosis syndrome was not accepted for inclusion in DSM-5 (Shrivastava et al. 2011). Proponents of the diagnosis argued that those individuals who met diagnostic criteria were at significant risk for developing schizophrenia, and deserved early treatment with antipsychotics (Woods et al. 2010). Critics of the diagnosis argued that there was insufficient evidence to justify pharmacotherapeutic prophylaxis, and that having such a diagnosis would lead to harmful prescription of antipsychotics, which are already widely overprescribed for a range of off-label indications, with growing use in children and adolescents, as well as in geriatric patients, where they may lead to substantial harmful effects (Yung et al. 2010). Over time it became clear that DSM-5 leaders were concerned about being criticized for overmedicalization, and that thresholds for the introduction of new diagnoses in the nomenclature were high.

On the other hand, proposals for inclusion of mild neurocognitive disorder, derived from work on mild cognitive impairment (MCI), were accepted for DSM-5 (Sachdev et al. 2014). Somewhat similar considerations apply to both MCI and prodromal psychosis. On the one hand, such diagnoses may lead to timely diagnosis and robust intervention; on the other hand, they may lead to overdiagnosis and overprescription. In the area of MCI, however, there have been substantial advances, including the development of radioligands for assessing the presence of amyloid plaques, and completion of a number of randomized controlled trials of pharmacotherapy for individuals with MCI (Sachs-Ericsson and Blazer 2014). That said, not everyone with radiological evidence of amyloid plaques has cognitive impairment; here too sensitivity and specificity are suboptimal. Society seems to have fewer qualms about medicalizing cognitive impairment in older adults than other forms of social deviance in younger adults.

One view is that there are currently insufficient data on which to base a revolution in psychiatric nosology. Such a revolution will emerge with future advances in basic and clinical neuroscience (Hyman 2007). Another view, however, suggests that we may be expecting too much from our nosology (Nesse and Stein 2012). In the rest of medicine, for example, there is no expectation that each syndrome has a unique pathophysiology. Cardiac failure is a useful clinical entity, for example, even though it may reflect a wide range of disturbances in cardiac physiology. Similarly, in the rest of medicine, even when physiological and laboratory measures are widely used, it is understood that such measures do not translate straightforwardly into cut-points for diagnosis of disorder. There is considerable debate, for example, about what cholesterol levels warrant initiation of treatment with statins, with diagnostic guidelines revised over time, along with the advice to assess a range of other variables (e.g., evidence of carotid plaques).

Another important consideration is the issue of clinical utility. While dimensional approaches to the evaluation of psychopathology may carry a great deal of scientific validity, they may simply be too cumbersome in the field (First 2005). It is notable, for example, that after much debate about replacing DSM-IV categories of personality disorders with dimensional approaches to the measurement of personality traits, the decision was taken to stay with the DSM-IV entities (Skodol et al. 2013). From a conceptual viewpoint, any dimension can be translated into a categorical cut-score, so that dimensional and categorical approaches are complementary (Kessler 2002). Nevertheless, from a clinical perspective, there continues to be a need for simple, user-friendly constructs.

What are the implications of DSM-5 in helping the clinician decide how to respond to a patient with mild or minimal symptoms, who is asking for psychopharmacological intervention? The continued reliance on categorical diagnoses means that the emphasis in DSM-5 is on treating recognized disorders, rather than on enhancing normal functioning. At the same time, the continued reliance on the clinical significance criterion for separating psychopathology from normality means that clinical judgment remains key in decision-making about diagnosis. Indeed, DSM-5 allows clinicians to diagnose "other specified" or "unspecified" disorders when patients have symptoms characteristic of a diagnostic class of disorders (e.g., anxiety disorders) and meet the clinical significance criteria, but do not meet the full criteria of any disorder in that class (e.g., because they have subthreshold symptoms). Furthermore, various DSM definitions of mental disorder have emphasized the fuzzy boundaries of this construct (Stein et al. 2010a).

Diagnosis does not necessarily imply treatment; sometimes the best intervention for a particular diagnosis is no treatment (Frances and Clarkin 1981). Furthermore, where an "other specified" or "unspecified" disorder is diagnosed, there may be a very limited evidence base on which to draw in deciding about interventions. However, often times, a particular diagnosis does lead to treatments based on the relevant evidence base. One clinician may regard social anxiety as a normative and potentially useful phenomenon and have a very high threshold for diagnosing social anxiety disorder; another may emphasize that patients with social anxiety are often impaired in ways that deserve early recognition and robust intervention (Campbell-Sills and Stein 2005). This is not to say that drawing a border between normal and abnormal social anxiety is entirely arbitrary; a range of clinical considerations, such as the pervasiveness of symptoms across contexts, may inform decision-making about diagnosis and about treatment (First and Wakefield 2013; Stein 2008a).

The failure of prodromal disorders to be included in DSM-5 might be taken to mean an ongoing emphasis on the more severe manifestations of psychopathology. While mild neurocognitive disorder was introduced into DSM-5, the threshold for diagnosis of this condition is not low. Still, the introduction of this disorder does raise the question for clinicians of whether some individuals with MCI require diagnosis and intervention. In the case of MCI, there is fortunately a growing body of data that can help clinicians in choosing appropriate interventions. At the same time, the limitations of these data should be emphasized. We currently lack sensitive and specific biomarkers of MCI and treatment effect sizes are low. Clinical judgment therefore remains key in determining the presence of disorder, and in comprehensively weighing the risks versus benefits of intervention.

14.3 **ICD-11**

The process of revising the International Classification of Disease has also been a considered one (Reed and Ayuso-Mateos 2011). Although the World Health Organization (which produces ICD-11) has considerably fewer financial resources than the American Psychiatric Association (which produces successive editions of the DSM), ICD-11 working groups have met, literature reviews have been undertaken, and field trials are under way. Given that the ICD-11 is mandated for use in many countries around the world, one can expect that this classification will be an important and influential development for global psychiatry.

ICD-11 leaders have emphasized the importance of clinical utility, global applicability, and the use of the manual in primary care settings by nonspecialized clinicians (Reed and Ayuso-Mateos 2011). These emphases mean that although ICD-11 and DSM-5 leaders have emphasized the importance of aligning the two classifications, ICD-11 will likely differ in a number of important respects from DSM-5. In particular, ICD-11 will focus on diagnostic guidelines rather than on operational criteria. Indeed, from an ICD-11 perspective, DSM-5 clinical criteria are often characterized by pseudo-scientific specificity.

ICD-11 will likely address the issue of clinical significance in several of its diagnostic guidelines. ICD-10 and ICD-11 have emphasized the importance of measuring impairment, in particular, making the argument that symptomatology and impairment measures are conceptually distinct. Thus, a patient with significant levels of symptomatology may not show a great deal of impairment, while conversely a patient with lower levels of symptomatology may nevertheless be quite disabled. It is incumbent on clinicians not to conflate, but rather to evaluate both symptomatology and impairment.

The distinction between symptomology and impairment is easiest to make in the case of serious mental disorders such as schizophrenia. In the case of common mental disorders, such as depression, anxiety, and substance use, there seems to be a necessary conflation of symptom severity and functional impairment. Given that depressive and anxiety symptoms and substance use are normative and widely distributed in the general population, a crucial way in which clinicians determine the clinical threshold for such symptoms is by assessing their associated distress and impairment.

ICD-11, like DSM-5, will not include the category of prodromal psychotic disorder or attenuated psychosis syndrome. Nevertheless, preliminary indications are that it will include new diagnostic categories that extend the boundaries of mental illness to include phenomena that some might argue are normative. In particular, it seems that ICD-11 will include a diagnosis of compulsive sexual behavior (Grant et al. 2014). While this diagnosis was proposed by some for DSM-5, it was not included in the final version of the manual (Kafka 2014). There is, however, growing evidence that excessive sexual activity may be accompanied by distress and impairment, and from a global health perspective it is important to recognize this phenomenon as a public health problem that is associated with sequelae such as sexually transmitted disease.

What are the implications of ICD-11 for diagnosis of subthreshold symptoms and for cognitive enhancement? When considering the ICD-11 in clinical practice, it is important to bear in mind the broader context of WHO initiatives to develop primary care treatment guidelines for common mental disorders, and to encourage task-sharing of interventions (i.e., use of nonspecialized personnel such as community health workers) that can be scaled up broadly throughout the world (Saxena and Setoya 2014). While it is theoretically possible that clinicians can use ICD-11 to diagnose and treat patients with subthreshold symptomatology, in effect the envisaged use of the manual focuses rather on reducing the treatment gap by addressing underdiagnosis and undertreatment of those with serious mental disorders or more severe common mental disorders.

14.4 **RDoC**

The Research Domain Criteria framework is an initiative of the National Institute of Mental Health in the United States (Insel et al. 2010). RDoC focuses on broad domains (e.g., negative affect, positive affect, cognition), which cut across a range of different mental disorders. These broad domains are, in turn, defined using a set of units of analysis that include components of circuits (genes, molecules, cells), circuits, and circuit outputs (behavior, physiological responses, and verbal reports or clinician-completed instruments), and that provide a bio-behavioral signature for each individual. RDoC principles include an emphasis on a translational approach, the use of dimensional measures, and an equal focus on behavior and neurocircuitry.

RDoC promises to be enormously influential because grants submitted for funding to the US National Institutes of Health are encouraged to incorporate the RDoC framework into their thinking. At the same time, RDoC leaders have had to emphasize that their vision is a long-term one, and that its impact on clinical practice remains to be seen. At the time of publication of DSM-5, Thomas Insel, Director of the National Institute of Mental Health and an advocate for RDoC, published a blog in which he emphasized the need for a new approach to nosology (Insel 2013). Within a few weeks, he and Jeffrey Lieberman, President of the American Psychiatric Association, issued a joint statement, in which they emphasized the need for both RDoCD and DSM-5, and described these as important complementary initiatives (Insel and Lieberman 2013).

The development of the RDoC framework seems particularly relevant, however, to issues regarding subthreshold symptomatology and cognitive enhancement. First, the RDoC framework emphasizes that normal and abnormal behaviors fall onto a continuum. This means that there is no clear cut-point separating them, and this in turn means that there is no clear cut-point delineating enhancement from treatment interventions. For example, treatment of subthreshold depression may be viewed as an enhancement by some, but as a valuable therapeutic intervention by others. Second, the RDoC framework emphasizes the enormous variation in physiological mechanisms that underlie behaviors. This means that each patient requires an individualized approach, and this in turn again blurs the boundaries between enhancement and treatment. For example, an individual who is at high genetic risk for depression perhaps warrants treatment with antidepressants; this would be viewed as an enhancement by some, but as a treatment by others.

I have elsewhere argued that the RDoC approach, while important, demands caution (Stein 2014a). First, a clear goal of medical and psychiatric classification is clinical utility, which is only partly related to underlying pathophysiology. In medicine, diagnosis of a syndrome, such as cardiac failure, may provide little information about precise etiology, but nevertheless helps guide treatment. In psychiatry, many entities are syndromic. While syndromes may have multiple causes, blurry boundaries, and absent biomarkers, they are also clinically useful (Nesse and Stein 2012).

It may be counter-argued that much of medicine focuses on specific etiologically based entities (e.g., viral pneumonia). Psychiatry too has a few specific diseases, such as psychosis due to medical conditions such as neurosyphillis. But these exceptions prove the rule; many diagnoses in medicine and psychiatry reflect the fact that patients present with variegated symptoms that are the result of multiple mechanisms. Some cases of hypertension, headache, and depression are due to single gene variants or other circumscribed pathophysiologies, but the majority of cases reflect multiple influences.

Second, given that multiple mechanisms play a role in producing psychiatric signs and symptoms, there are both pros and cons in foregrounding any particular diagnostic validator, such as "circuit-based behavioral dimensions." Science has progressed from Hippocrates's account of the "humors," to theories of the neurocircuitry basis of positive and negative valence, but it is possible that a century from now current circuitry concepts will be considered rudimentary. On the other hand, the construct of depression, which is based on several other validators, may continue to resonate with centuries of clinical descriptions.

DSM-5 distinguishes between anxiety and obsessive-compulsive related disorders partly on the basis of the putatively different neurocircuitry that underpins these conditions (Stein et al. 2014). But there are also strong arguments for lumping these disorders on the basis of considerations such as response to serotonin reuptake inhibitors and cognitive-behavioral treatments (Stein et al. 2010b). We need to accept that diagnostic systems cannot "carve nature at her joints"; rather, facts and values need to be continually reassessed in order to optimize classifications (Stein 2008b).

Third, we should be cautious in expecting that diagnostic criteria or thresholds will ultimately be based on behavioral dimensions or biological markers, given the multiple mechanisms underlying psychiatric complaints and the many considerations relevant to treatment decisions. Simple assessments, such as blood pressure measurement or mental status examination in medicine and psychiatry, can provide important information. Still, such information is partial. In medicine and psychiatry, deciding on whether and how to intervene necessarily requires a complex assessment of a range of factors, including the function of symptoms, their social context, and the risks versus benefits of treatments.

A public health perspective draws attention to one set of factors sometimes neglected by critics of the nomenclature (Stein et al. 2013). Psychiatric classifications focus on individual disorders, where underlying "endophenotypes" may be particularly relevant. However, it may also be important to address "exophenotypes": societal phenomena, such as socioeconomic inequality or interpersonal violence, that crucially contribute to the burden of disease. Furthermore, decisions about thresholds for psychiatric intervention need to include not only facts about underlying neurobiological mechanisms, but also considerations such as the cost-effectiveness of particular interventions.

Based on these considerations, it is doubtful that RDoC will soon replace categorical diagnoses. Even in the area of cognitive enhancement, when for example the clinician is faced with a patient requesting treatment for subthreshold or mild symptoms, this framework does not provide a personalized medicine approach that avoids the need for careful clinical judgment. Rather, for the foreseeable future, when faced with individuals with subthreshold symptoms asking for enhancement interventions, such clinical judgment will remain key. In particular clinical judgment will be needed to weigh up the multiple relevant considerations at hand such as how severe are the symptoms? How impairing are they? What is the evidence that intervention may help? What is the evidence that intervention may harm?

14.5 **Conclusion**

Much attention has been paid to revisions of psychiatric classification systems. Nevertheless, there remains significant dissatisfaction with various proposed approaches. From a neuroscience perspective, diagnostic criteria have failed to incorporate neurobiological data, and a focus on "circuit-based behavioral dimensions" will improve diagnosis (Cuthbert and Insel 2013). From a more critical perspective, diagnosis runs the risk of merely medicalizing "problems in living" since psychiatric disorders do not represent valid disease entities (Stein 1991). It is perhaps not surprising that DSM-5, ICD-11, and RDoC do not lead to a simple response to questions about cognitive enhancement, and that the clinician faced with a patient asking for intervention for subthreshold symptoms must exercise his or her best clinical judgment.

These specific debates echo larger conceptual debates about classification and medicine. Many emphasize notions of disease, arguing that clinicians must be scientists who understand physiology. Others emphasize the experience of illness, stressing that clinicians must be humanists who understand the patient's explanatory model of symptoms, and empathize with his or her suffering (Kleinman 1991). An integrative medicine and psychiatry arguably recognizes each of these as important aspects of being a good diagnostician and researcher (Stein 1993). While the boundaries of mental disorder may be fuzzy (Stein 2013, Stein 2014b), clinicians should aim to optimize decision-making about diagnosis and treatment. Thus a good clinician must weigh up multiple considerations, whether relying on DSM-5 criteria or ICD-11 guidelines for clinical disorders, or whether employing RDoC constructs to assess behavior and biology.

Such considerations may need to cover a range of facts and values, employing both medical and moral metaphors (Stein 2008a). On the one hand, epidemiological data may emphasize that subthreshold symptoms are associated with significant impairment (Kessler et al. 2003). On the other hand, treatment data may emphasize that effect sizes are lower in those with less severe symptoms, perhaps in part reflecting high placebo response rates in those with mild illness (Stein and Mayberg 2005). On the one hand, patients may value various aspects of pharmacotherapeutic treatment, including its convenience, its consistency with a neurobiological explanatory model, and its potential for restoring full health. On the other hand, patients may value other treatment modalities, such as psychotherapy, which are more consistent with explanatory models that encourage self-efficacy. The clinician needs to decide whether symptoms serve a useful defensive function and whether there is evidence of underlying harmful dysfunction, and needs to weigh up the risks versus benefits of medicalization and of various kinds of intervention.

While the focus of the discussion here has been on subthreshold symptomatology, and on the fuzzy border between treatment and enhancement in such cases, similar considerations also apply to thinking through whether particular thoughts, feelings, and behaviors are more indicative of a mental disorder, or of some other form of deviance (Stein 2015). Some disorders may be conceptualized as more typical (for example, in pneumonia and in post-traumatic stress disorder, the condition is precipitated by an external agent, the individual does not seem to bear responsibility, the sick role seems warranted), while others seem more atypical (for example, in substance use disorders, the individual seems to bear significant responsibility for both initiating the behavior and discontinuing it). While some requests for medical intervention, say surgery to look like a movie star, are clearly more like schmoctering than doctoring (Parens 1998), the debate about whether to medicalize an atypical condition such as a substance use disorder is a more complex one (Hall et al. 2003), in which the relevant facts and values must be carefully weighed.

In addition, while the analysis here has focused on disorder, analogous considerations would indicate that no clear-cut essentialist definition of optimal health is possible, and that the boundaries of constructs such as "well-being" must be decided judiciously (Stein 2012). On the one hand, humans do appear to have some design flaws, and there seem to be few a priori reasons for not using pharmaceuticals and psychotropics to improve well-being. On the other hand, given how well humans have been designed by evolution, the difficulties associated with developing new medications, and the high placebo response in those with only mild impairment, it is hard to generate data that interventions to improve physical or psychological welling are in fact efficacious and cost-efficient (Stein 2012). In a particular individual interventions may lead to real improvements, but raising false expectations about the existence and value of interventions that maximize well-being may also constitute bad doctoring. An individualist approach to clinician decision-making is needed, one which is neither overly optimistic nor unduly pessimistic about biotechnological advances (Stein 2012).

Ongoing work on nosological systems is needed to optimize clinical utility and diagnostic validity. DSM-5, ICD-11, and RDoC should be supported to the extent that each framework leads to research that allows such progress. However, it is important not to have unrealistically high expectations of diagnostic systems. DSM-5, ICD-11, and RDoC provide only partial guidance to the clinician faced with a patient who is asking for cognitive enhancement. RDoC arguably comes closest to providing guidance for personalized medicine. Still, we need to be cautious of medical strawmen, such as the physician who relies solely on laboratory tests to determine diagnoses, or the public health practitioner who eradicates pathogens by simple interventions such as hand-washing. No matter how many dollars we pour into behavioral neuroscience, we may have to accept that there are few diagnostic biomarkers for delineating psychiatric disorders from normality, and few mosquito nets to combat these conditions (Stein 2009).

Indeed, given the complexity of medicine, DSM-5 and ICD-11 provide a number of approaches worth emulating. Thus a physician faced with a patient with headache should be able, after a careful history and examination, to diagnose a particular headache syndrome (indeed, headache classification takes a DSM-like approach (Olesen 2008)). Then, based on neuroscience knowledge, as well as a range of other considerations, including an examination of the relevant clinical science and an understanding of the patient's views, and a weighing of the pertinent facts and values, one or another intervention, whether medical or not, may be a reasonable choice. Similarly a physician faced with a complex public health problem, such as substance abuse, knows that the causes are complex, that a range of responses are needed, and that, as in much of psychiatry, there is no mosquito net.

For the foreseeable future, an integrative approach to psychiatric diagnosis and treatment ought to incorporate DSM, ICD, RDoC, and a broad range of other constructs. In particular, when faced with a patient who asks for psychopharmacotherapeutic intervention for subthreshold or minimal psychiatric symptoms, the clinician will need to take into account a wide range of relevant considerations, in an attempt to exercise his or her best clinical judgment. DSM-5 and ICD-11 will predominate in clinical psychiatry for the foreseeable future. The RDoC framework, however, draws on a long tradition of research on the psychobiology of mental disorders (Van Praag et al. 1990), and is likely to continue to inform future investigations. Future work on different nosological approaches to psychiatric symptoms and disorders should be accompanied by ongoing research on their ethical, legal, and social implications (ELSI).

14.6 Acknowledgments

Professor Stein is supported by the South African Medical Research Council.

References

American Psychiatric Association. 2013. *Diagnostic and Statistical Manual of Mental Disorders Fifth Edition*. Arlington, VA.

Campbell-Sills, L., and M. B. Stein. 2005. Justifying the diagnostic status of social phobia: a reply to Wakefield, Horwitz, and Schmitz. *Canadian Journal of Psychiatry* **50**(6): 320–323; discussion 324–326.

Cuthbert, B. N., and T. R. Insel. 2013. Toward the future of psychiatric diagnosis: The seven pillars of RDoC. *BMC Medicine* **11**:126.

First, M. B. 2005. Clinical utility: a prerequisite for the adoption of a dimensional approach in DSM. *Journal of Abnormal Psychology* **114**(4):560–4.

First, M. B. 2010. Paradigm shifts and the development of the diagnostic and statistical manual of mental disorders: past experiences and future aspirations. *Canadian Journal of Psychiatry* **55**(11):692–700.

First, M. B., and J. C. Wakefield. 2013. Diagnostic criteria as dysfunction indicators: bridging the chasm between the definition of mental disorder and diagnostic criteria for specific disorders. *Canadian Journal of Psychiatry* **58**(12):663–9.

Frances, A., and J. F. Clarkin. 1981. No treatment as the prescription of choice. *Archives of General Psychiatry* **38**(5):542–5.

Grant, J. E., M. Atmaca, N. A. Fineberg, L. F. Fontenelle, H. Matsunaga, Y. C. Janardhan Reddy, H. B. Simpson, P. H. Thomsen, O. A. van den Heuvel, D. Veale, D. W. Woods, and D. J. Stein. 2014. Impulse control disorders and "behavioural addictions" in the ICD-11. *World Psychiatry* **13**(2):125–7.

Hall, W., L. Carter, and K. I. Morley. 2003. Addiction, neuroscience and ethics. *Addiction* **98**(7):867–70.

Hyman, S. E. 2007. Can neuroscience be integrated into the DSM-V? *Nature Reviews: Neuroscience* **8**(9):725–32.

Insel, T. 2013. Director's blog: transforming diagnosis. Available at http://www.nimh.nih.gov/about/director/2013/transforming-diagnosis.shtml.

Insel, T., B. Cuthbert, M. Garvey, R. Heinssen, D. S. Pine, K. Quinn, C. Sanislow, and P. Wang. 2010. Research domain criteria (RDoC): toward a new classification framework for research on mental disorders. *American Journal of Psychiatry* **167**(7):748–51.

Insel, T., and J. Lieberman. 2013. DSM-5 and RDoC: Shared interests. Available at http://www.nimh.nih.gov/news/science-news/2013/dsm-5-and-rdoc-shared-interests.shtml.

Kafka, M. P. 2014. What happened to hypersexual disorder? *Archives of Sexual Behavior* **43**(7):1259–61.

Kendler, K. S., and M. B. First. 2010. Alternative futures for the DSM revision process: iteration v. paradigm shift. *British Journal of Psychiatry* **197**(4):263–5.

Kessler, R. C. 2002. The categorical versus dimensional assessment controversy in the sociology of mental illness. *Journal of Health and Social Behavior* **43**(2):171–88.

Kessler, R. C., K. R. Merikangas, P. Berglund, W. W. Eaton, D. S. Koretz and E. E. Walters. 2003. Mild disorders should not be eliminated from the DSM-V. *Archives of General Psychiatry* **60**(11):1117–22.

Kleinman, A. 1991. *Rethinking psychiatry: From cultural category to personal experience.* New York: Free Press.

Kupfer, D. J., and D. A. Regier. 2011. Neuroscience, clinical evidence, and the future of psychiatric classification in DSM-5. *American Journal of Psychiatry* **168**(7):672–4.

Nesse, R. M., and D. J. Stein. 2012. Towards a genuinely medical model for psychiatric nosology. *BMC Medicine* **10**:5.

Olesen, J. 2008. The International Classification of Headache Disorders. *Headache* **48**(5):691–3.

Parens, E. 1998. Is better always good? In *Enhancing human traits: ethical and social implications*, ed. E. Parens. Washington, D.C.: George Washington University Press.

Reed, G. M., and J. L. Ayuso-Mateos. 2011. Towards a more clinically useful International World Health Organisation classification of mental disorders. *Revista de Psiquiatría y Salud Mental* 4(3):113–16.

Regier, D. A., W. E. Narrow, E. A. Kuhl, and D. J. Kupfer. 2009. The conceptual development of DSM-V. *American Journal of Psychiatry* **166**(6):645–50.

Sachdev, P. S., D. Blacker, D. G. Blazer, M. Ganguli, D. V. Jeste, J. S. Paulsen, and R. C. Petersen. 2014. Classifying neurocognitive disorders: the DSM-5 approach. *Nature Reviews: Neurology* **10**(11):634–42.

Sachs-Ericsson, N., and D. G. Blazer. 2014. The new DSM-5 diagnosis of mild neurocognitive disorder and its relation to research in mild cognitive impairment. *Aging and Mental Health* **19**(1):1–11.

Saxena, S., and Y. Setoya. 2014. World Health Organization's Comprehensive Mental Health Action Plan 2013–2020. *Psychiatry and Clinical Neurosciences* **68**(8):585–586.

Shrivastava, A., P. D. McGorry, M. Tsuang, S. W. Woods, B. A. Cornblatt, C. Corcoran, and W. Carpenter. 2011. "Attenuated psychotic symptoms syndrome" as a risk syndrome of psychosis, diagnosis in DSM-V: The debate. *Indian Journal of Psychiatry* **53**(1):57–65.

Skodol, A. E., L. C. Morey, D. S. Bender, and J. M. Oldham. 2013. The ironic fate of the personality disorders in DSM-5. *Personality Disorders* **4**(4):342–9.

Stein, D. J. 1991. Philosophy and the DSM-III. *Comprehensive Psychiatry* **32**(5):404–15.

Stein, D. J. 1993. Cross-cultural psychiatry and the DSM-IV. *Comprehensive Psychiatry* **34**(5):322–9.

Stein, D. J. 1998. Philosophy of psychopharmacology. *Perspectives in Biology and Medicine* **41**(2):200–11.

Stein, D. J. 2008a. *Philosophy of psychopharmacology*. Cambridge, UK: Cambridge University Press.

Stein, D. J. 2008b. Is disorder X in category or spectrum Y? General considerations and application to the relationship between obsessive-compulsive disorder and anxiety disorders. *Depression and Anxiety* **25**(4):330–5.

Stein, D. J. 2009. Is there a "mosquito net" for anxiety and mood disorders? *Current Psychiatry Reports* **11**(4):264–5.

Stein, D. J. 2012. Psychopharmacological enhancement: a conceptual framework. *Philosophy, Ethics, and Humanities in Medicine* 7:5.

Stein, D. J. 2013. What is a mental disorder? A perspective from cognitive-affective science. *Canadian Journal of Psychiatry* **58**(12):656–62.

Stein, D. J. 2014a. An integrative approach to psychiatric diagnosis and research. *World Psychiatry* **13**(1):51–3.

Stein, D. J. 2014b. Psychopharmacology and natural kinds: A conceptual framework. In: *Classifying psychopathology: Mental kinds and natural kinds*, ed. H. Kincaid and J. A. Sullivan. Cambridge, MA: MIT Press.

Stein, D. J., M. A. Craske, M. J. Friedman, and K. A. Phillips. 2014. Anxiety disorders, obsessive-compulsive and related disorders, trauma- and stressor-related disorders, and dissociative disorders in DSM-5. *American Journal of Psychiatry* **171**(6):611–13.

Stein, D. J. 2015. Disorder versus deviance: Where to draw the boundary? *Philosophy, Psychiatry and Psychology* **21**:262–5.

Stein, D. J., N. A. Fineberg, O. J. Bienvenu, D. Denys, C. Lochner, G. Nestadt, J. F. Leckman, S. L. Rauch, and K. A. Phillips. 2010b. Should OCD be classified as an anxiety disorder in DSM-V? *Depression and Anxiety* **27**(6):495–506.

Stein, D. J., C. Lund, and R. M. Nesse. 2013. Classification systems in psychiatry: diagnosis and global mental health in the era of DSM-5 and ICD-11. *Current Opinions in Psychiatry* **26**(5):493–7.

Stein, D. J., and H. Mayberg. 2005. Placebo: the best pill of all. *CNS Spectrums* **10**(6):440–2.

Stein, D. J., K. A. Phillips, D. Bolton, K. W. Fulford, J. Z. Sadler, and K. S. Kendler. 2010a. What is a mental/psychiatric disorder? From DSM-IV to DSM-V. *Psychological Medicine* **40**(11):1759–65.

Van Praag, H. M., G. M. Asnis, R. S. Kahn, S. L. Brown, M. Korn, J. M. Friedman and S. Wetzler. 1990. Nosological tunnel vision in biological psychiatry. A plea for a functional psychopathology. *Annals of the New York Academy of Science* **600**:501–10.

Woods, S. W., B. C. Walsh, J. R. Saksa, and T. H. McGlashan. 2010. The case for including attenuated psychotic symptoms syndrome in DSM-5 as a psychosis risk syndrome. *Schizophrenia Research* **123**(2–3):199–207.

Yung, A. R., B. Nelson, A. D. Thompson and S. J. Wood 2010. Should a "risk syndrome for psychosis" be included in the DSMV? *Schizophrenia Research* **120**(1–3):7–15.

Chapter 15

Will cognitive enhancement lead to more well-being? The case of people with disabilities

Heather Bradshaw-Martin

15.1 Introduction

In this chapter it will be argued that cognitive enhancements need to be assessed on a case-by-case basis using the morphological identity framework. The main issue is that cognitive enhancement is merely one example of the wider class of morphological changes humans can experience. As such, frameworks for assessing the impact on well-being of morphological changes are also relevant for cognitive enhancements. One such framework arises from work with people with disabilities who have experienced multiple morphological changes. This framework is based on the concept of morphological identity. This framework helps us better understand the moral value of cognitive enhancement technologies because it allows us to relate their use directly to the effects on the well-being of individual people, and the operation of societies more widely, as we might do for other life choices. The chapter will begin by defining terms relevant to morphological identity.

15.2 Cognitive enhancement is morphological change

A cognitive enhancement is an enhancement or raising up of cognitive abilities. Cognitive abilities are those relating to the "mental action or process of acquiring knowledge through thought, experience and the senses" (Pearsall 1999, viii). Morphology and morphological change are terms extended and developed from the definitions in Sandberg (2001). Here the morphology of an organism, conscious being, machine, or other entity will be used to mean the layout of the matter which makes it up. Any change to its material constitution will affect its morphology.

Many humans have a morphology which connects a body and a brain via a complex communication system of electrical pulses through nerves and chemical signals in fluid paths. Sandberg (2001) notes: "There is no dividing line between the body and our mentality, both are part of ourselves. *Morphological* freedom is the right to modify oneself" (emphasis added).

Therefore, changes to our mentality, to the mental actions or processes of acquiring knowledge, to our sensory equipment and processing, or in our patterns of thought and other morphological records of experience are changes to our morphology.

Morphological terminology can be helpful in thinking through the ethical implications of enhancements. For example, the terms enhancement and disability both include ethical judgements through their etymology and general usage. In two papers Anita Silvers has called for more ethically neutral terminology in the enhancement debate (Silvers 2003, 2008). Jackie Leach Scully has also called for an equivalent to the term gender for wide bodily variations. She has pointed out that for abled and disabled people we lack the equivalent of gender to male and female (Scully 2008, 34).[1] Morphology can serve adequately in this role, allowing us to separate the body type, or morphology, from ethically significant qualities such as desirability or voluntariness.

Morphology seems preferable to the term embodiment for two reasons. First, morphology makes clear that the internal arrangement of molecules within bodies matters as well as their surface and external appearance, and, second, because morphology does not distinguish between the body and the brain or mind whereas embodiment emphasizes the body and says less of the brain or mind.

Etymologically morphology is already rich in associations. Enhancement has reached us from Latin through Old French and meant initially "to raise up" (Harris 2007, 210). Morphology, in contrast, is more recently traceable back to an Arabic term "morphil," which came to us through Spanish. Ancient Greek also had "morphos," referring to structure or form (contrast with "amorphous"), and "logos," meaning word or speech. From logos we have the suffixes –logy and –ology in English. The former means saying or speaking about and the second formally names a science or department of study. One might thus understand morphology as being a science of or a speaking about form. So we have "morphology" in biology meaning "the branch of biology that deals with the form of living organisms and their parts and the relationships between their structures" (Oxford English Dictionary 2012). We also have Morpheus in literature. He is the god of sleep, and of dreaming in which forms and structures change. There is "morph" in computing, to shape shift, and the old "morphodite," used to refer to technologies such as ships and carts that had structures which crossed traditional type-boundaries, as well as to humans whose sex and sexuality was unclear.

Human morphologies can be changed by design or accident, as well as through the effects of everyday living. These changes include changes to the layout of the brain, the strength and connections of neurons for example, such as those produced by traditional learning, new sensory experience, drugs, physical interventions, or radiation and mutation. Many enhancements would be examples of deliberate, designed, voluntary morphological change.

A morphological change shall be defined here as:

> A change in the functioning of some subsystem of, or the addition or removal of a morphological capacity to, an individual organism; an ordered pair, change and trait, not necessarily desirable for the individual or society. (Bradshaw-Martin 2012, 206)

This is adapted from Bostrom and Sandberg's definition of an enhancement as are the definitions of enhancement and disability which I discuss in section 15.3 (Bostrom and Sandberg 2009, 378). Cognitive enhancements are morphological changes and I shall refer to them as such henceforth.

Both morphology and morphological change are more neutral than the terms disability or enhancement. They encourage us to ask questions such as whether the change was chosen, voluntary, and made by a consenting adult, or whether a given morphology is desirable for society, merely for the individual, or neither.

15.3 Experience of disability: its relevance

Simply put, people with experience of disability are more likely than others to have experience of morphological change. This might be as a result of the onset of an illness or disability. But it might also be as a result of a medical intervention. Further, these groups are also often seen as potential early adopters of new biotechnologies. Recent research on the definition of enhancement explored fourteen people's experiences of morphological change (Bradshaw-Martin 2012). Key findings suggested that whether individuals viewed a morphological change as an enhancement, therapy, or disability depended on their own previous life experiences and morphology rather than on what was common in their societies. The impact of their morphological changes on their well-being was more intricate.

Notably, in disability scholarship epistemological precedence is often given to first-person narratives (Bradshaw-Martin 2012, 42; Grönvik 2007; Scully 2008, 108). One reason for this which stands out in the literature is that those who have unusual morphologies learn early that others are poor at imagining what contributes to their well-being (Bradshaw-Martin 2012, 42). Such empathic and imagination limitations can also be expected to affect our estimations of the effects on our well-being of voluntary morphological changes; indeed, this was so in two cases in the research where the rest of society might expect the morphological change in question to have had positive impacts on well-being. In these cases participants with different impairments who had each gained significant improvements in their sensory capacities reported that there were many aspects of living with their new morphologies which they had not expected, despite those morphologies being ones which were very common in the general population. Both described negative experiences such as pain and disgust associated with sensory experiences utilizing the wider frequency range their morphological changes enabled them to perceive.

A participant who chose to have a cochlear implant, says, "It's too much sometimes … years on, it can still be quite painful sometimes to hear a noise that is [from the new frequencies] … er, I would have to say I would not choose to go to a concert where there was a flute being played. I just could not wish that upon myself" (Bradshaw-Martin 2012, 284). Another participant whose distorted vision was corrected by surgery which changed, among other things, the range of colors she could see, vividly described the effects some of those new colours had on her:

> "Colour," she says, "the dark reds, the browns were very much enhanced, and what was very much toned down were the more synthetic colours which I suppose were fairly new anyway, like turquoise, bright orange etc. When I had this op [*sic*] it sorted out colours for me but that was a shock. In fact the first thing that happened in the ward [when the bandages came off], the first thing I saw was some bottles with turquoise tops on and they were … so horrible … bright plastic turquoise. I'd *never* seen that colour …" … "Must have been just toned down. Yeah … well, it wasn't nasty. I was

> used to it. Whatever it was it was all right. Turquoise was nasty. And then very soon afterwards I went into Marks and Spencer and there were: oranges! A kind of orange, I thought they were plastic, I did *not* believe they were real, I hadn't seen ... I've got used to that, now just about, now, but I still look at it and think 'How disgusting!' " She laughs disgustedly. "I've still got [all the rich colours] but they are now a little bit offset by these *ghastly* other colours that are a little bit much." (Bradshaw-Martin 2012, 277)

This effect of the unexpected might be more marked for morphologies which were rare or nonexistent in the general population.

The emphasis on the relationship between a person's own previous morphological experience and a new morphology led me during this research to contrast morphological changes relative to identity and those relative to the wider environment.

Enhancements have been considered to cause a discontinuity in a person's identities (Degrazia 2005; Schechtman 1996). Explorations of disability from social and psychological perspectives have explored similar effects (Cole 2004). In contrast much work in disability studies has been concerned with political identity and bodily difference (Oliver 2004; 2009).

Theoretical approaches to understanding effects of morphological change on sense of self and collective identity have come from at least three traditions. In analytic philosophy the question of sameness or when a thing is the same thing has been transferred to persons as, for example, in the work of Parfit (1987). In social theory and sociology ideas of the formation of the self through interaction with others have grounded multiple interpretations of identity including narrative and self-creation approaches (Lawler 2008; Ricoeur 1994). These latter theories generally try to find constants or identifications and are challenged by morphological discontinuity. Psychology provides understandings of how we cope with development and change through the life cycle as, for example, in the work of Erikson (1995).

In this research the focus was on continued successful existence through change. Participants varied greatly in age and reproductive experience as well as experience of morphological change. Their understanding of the impact of morphological change on their lives was often articulated in terms of only their own previous morphological experience—initially referred to as their identity—and not in comparison to other morphologies common in their societies. This sense of self-reference led to an initial distinction in classifications of morphological changes:

We can define an identity relative disability (or impairment) as the following:

> A decrease in the functioning of some subsystem below its usual state as experienced by that individual or the removal of a capacity familiar to that individual: an ordered pair, decrease, and trait not necessarily desirable for the individual or society.

Compare an environment relative disability/impairment:

> A decrease in the functioning of some subsystem below its usual state as experienced by that individual or the removal of a capacity familiar to that individual: an ordered pair, decrease, and trait not necessarily desirable for the individual or society but in which the power of the individual is reduced relative to their physical or social environment.

Similarly enhancements might be identity relative:

> An improvement in the functioning of some subsystem *beyond its usual state as experienced by that individual* or the addition of a capacity *new to that individual*: an ordered pair, intervention, and trait not necessarily desirable for the individual or society.

Or environment relative:

> An improvement in the functioning of some subsystem beyond its normal healthy state or the addition of a new capacity: an ordered pair, intervention, and trait not necessarily desirable for the individual or society but which increases the power of the individual relative to their physical or social environment.

And, for completeness, therapy would be defined as:

> An intervention to restore a previously experienced morphology needed to fulfil an existing identity.

Applying Erikson's theories to the analysis of the research led to the development of a more sophisticated concept capturing the ideas of loss and regaining of equilibrium among morphology, social environment, and physical environment of which participants spoke. This concept of "morphological identity" is described further in Bradshaw-Martin (2012, 73–106) and briefly here in section 15.6, Morphological identity.

Given this complexity of factors affecting whether a given morphological change might be an enhancement I shall prefer the more neutral term morphological change as we are considering the relationship of morphological changes to well-being to be an open question.

15.4 **Morphological change and well-being**

Those experiencing a change to their morphology will be referred to here as morphees. The circumstances in which a morphological change comes about are also likely to affect whether it will lead to more well-being. Morphological changes which are perceived by the morphee as being imposed on them by others have different effects on well-being to those of morphological changes resulting from chance events or those which are perceived to be freely chosen. There are nevertheless some morphological changes which Western society today considers acceptable to impose on members. For example, education. Education is a morphological change because, on a naturalistic, materialistic understanding of the brain and learning, education changes the physical structure and connections between neurons, thus changing the being's morphology. But we also impose severe sanctions for other forms of imposed morphological change such as deliberate injury of others.

There are also questions of what I shall call desirability. One might assume that if a person chooses a morphological change it is because they desire something about the life which they believe the new morphology will enable them to live. Morphees might thus be expected, if they are choosing the change, to anticipate increased well-being. This may come from exercise of their new abilities, from achievement of a goal through their increased abilities, or from increases in others' well-being as a result of the application of their new abilities. For example, from sheer joy of thinking, from completing crossword puzzles or solving technical problems, and from creating technologies which reduce others' pain. A potential morphee—that is, one who is considering choosing such a change for themselves—can be expected to desire, that is to have some motivation connecting back somehow to their well-being, the change. They may be mistaken about the morphological change bringing them the life they desired, and they may be mistaken about whether that life will bring them increased well-being. But if they choose a morphological change we can deduce that they desire something they expect to follow from it.

If they do not choose the morphological change though, it may still result in an improvement in well-being. For example, a man who became tetraplegic after an accident in early middle age talked about the positive effects on his spirituality. "What I've found with my disability is it slowed me down hugely and in slowing down, which I wouldn't have most likely done, unless it'ud been enforced upon me ... it gives one time to reflect and think and you see the quality and the wonder of the life around you all the time" (Bradshaw-Martin 2012, 230).

Morphological changes can be desired by one person for another. For example, one might wish that a child or loved one had a different morphology. One might wish this because one envisages, perhaps mistakenly, that they would have higher well-being with the different morphology. Perhaps one wants them to be fitter so that they gain the same satisfaction from physical exercise that oneself does. Maybe one may wish that their morphology were different because it would be in some way more convenient for oneself. Perhaps one wishes that one's elderly parents' hearing was better as it is hard work having to repeat and shout everything one needs to communicate to them. Perhaps one wishes occasionally that one's partner or children were more alike physically so that clothes, bikes, toys, and sports equipment could be shared rather than duplicated.

We can begin to see from these examples and lines of thought how complicated the relationship between morphological changes and well-being is. Well-being also has a complex relationship with ethical theory.

15.5 **Well-being and ethics**

Some ethics scholars have argued that ethical theories should be independent of theories of well-being (Savulescu 2006). Others, such as the Canadian philosopher Larry Sumner, argue that the two are intimately connected. Sumner's book *Welfare, Happiness and Ethics*, constitutes a defence of the concept of "welfarism" defined as the moral theory in which "individual welfare ... is the only thing with final or ultimate ethical value, the only state of affairs which we have a moral reason to promote for its own sake" (Sumner 2003, viii).

One's position on this question may depend on whether one favors objective or subjective theories of well-being. Within a naturalistic metaphysical framework well-being can be thought of as an in-built internal indicator of how well an organism is doing, crudely: how good its chances of survival and reproduction are, whether it is thriving.[2] Moreover the nature of this indicator is such that low survival and reproduction chances are so unpleasant for the system, or organism, that the system is motivated to act to improve its well-being, thus correlating external circumstances with internal states.

Theories of well-being tend to focus on either the *external circumstance* or the *internal state* of the organism. Objective theories of human well-being seek to describe the external circumstances which will best provide for human thriving. If a human has all the things on the list, then the human has high well-being whatever their internal reading may be. There lie behind this two assumptions. First, that beings of similar morphology will thrive in similar circumstances. Second, that humans are sufficiently morphologically similar for the first assumption to hold. Objective theories of well-being have been popular in bioethics, even with those writing in favor of enhancement (Bradshaw-Martin 2012, 58–71, 111; Harris 2007; Kahane and Savulescu 2009).[3]

In contrast, subjective, or "welfarist" theories of human well-being emphasize instead the priority of the internal reading, accepting that a human with all the things on the objective list may still genuinely be suffering poor well-being. Subjective theories, however, must contend with the notorious apparent unreliability of human well-being reports (Albrecht and Devlieger 1999). Subjective approaches are more able to cope with wide morphological range and morphological diversity within a society. Sumner describes a subjectivist concept of well-being in a life as "how well it is going for the individual whose life it is" (Sumner 2003, 20). Because of its potential for application where there is morphological diversity this subjective theory of well-being will be used here in a Sumnerian welfarist ethical theory.

15.6 Morphological identity

Other writers have noted a connection between the morphology of individual humans and their interactions with their social and physical environments in their own personal experience and in the experience of others (Amundson 1992; Scully 2008). The psychologist Erik Erikson, whose work on ego identity was seminal, wrote the following thought-provoking, and now somewhat controversial, words in 1968:

> in every technology and in every historical period there are types of individuals who ("properly" brought up) can combine the dominant techniques with their identity development, and *become* what they *do*. Independently of minor superiorities or inferiorities, they can settle on that *cultural consolidation* which secures them what joint verification and what transitory salvation lies in doing things together and in doing them right—a rightness proven by the bountiful response of "nature" ... the goods produced, the money made, or the technological problems solved. (Erikson 1994, 32).

During my own research with people of varied morphologies and ages I noticed relationships between morphology, life path, and well-being in what people told me about their own well-being following morphological changes. Well-being varied depending on whether the change fitted with their previous experience and way of life. This led me to the concept of morphological identity which I currently understand as:

> The technologically mediated tripartite relationship between the morphology of an organism and the social and physical milieu in which they live. This relationship influences, but does not constitute, the well-being of the organism.

Erikson also connects morphology, or his "types of individuals," with what he refers to as "doing things together" to elicit the "bountiful response of nature." Identity's nested depth shows here as we start to see how individuals at one level cooperate to form another single identity at a higher level.

If a morphological change enables an individual to return to a morphological identity which they previously had, then the effect on well-being is likely to be positive. Morphological identities include all aspects of our lives including relationships, work and contribution to society, religion and spirituality, housing, sport and other hobbies, and so on, which enable us to thrive as organisms.

But if the morphological change disrupts an established morphological identity then the effect on well-being is more likely to be negative. A process of adaptation and identity seeking begins. In our teenage years we undergo a similar search for an identity. It takes us a number of years. It is the same with later changes to morphological identity. Many descriptions and reports suggest an adjustment period of 4 or 5 years at least. During this period well-being can be low, sometimes very low.[4] If morphology continues to change during this period, or if the social or physical environment changes, then no new morphological identity may be formed and despair may set in.

We may find that aspects of life, which, before the morphological change, we thought we would value after it, do not turn out to be satisfying or even possible afterwards. This can leave gaps in the new morphological identity and result in lowered well-being.

15.7 The case of people with disabilities

Evidence from people with disabilities about different forms of morphological change suggests morphological identity is important (Bradshaw-Martin 2012). For example, Astrid, a middle aged lady with a long employment career and who is also a successful sportswoman found her morphological identity as a disabled person disrupted when she developed a chronic lung illness. She considered herself to have been both disabled and healthy before her lung illness: "I was a physically fit disabled person, and now I'm not ... And life as a disabled person isn't lazy but it's just become for me, over the last two years, quite a bit harder" (Bradshaw-Martin 2012, 383).

In contrast Louisa, a Deaf academic, had a cochlear implant in her late teens as a result of a complex history of pressures and freedoms within her family. When the implant operation was finally done Louisa was already a member of the Deaf community with a settled morphological identity there. After a number of years wrestling with the ambiguous identity the implant gave her she chose not to use the cochlear implant any more. She can't take it out, but she no longer places the exterior parts on her scalp to allow radio transmissions into her head. She says of this choice, "It might have made communicating with hearing people more difficult but it was a sense of relief and socially I felt more at home with myself" (Bradshaw-Martin 2012, 385).

Louisa's morphological identity as Deaf—her work and husband are from this social environment—proved stronger than the social ties with her birth family which drew her toward the hearing world. In contrast, Neil, another early adulthood implantee, found a morphological identity which included a new job in which he must rely on his new hearing abilities and a partner who was hearing. But he emphasized the disruption of the change period during which many of his preconceptions about how the world worked, both scientifically and socially, were destroyed.

Doreen, a grandmother who has been adjusting to her immobility for some years, gives a telling description of how a new identity forms. She says:

> "Well I'm no different to how I was before."

But her husband, Don, isn't having this:

> "Well, you weren't happy at first were you?"
> "No," says Doreen slowly, "Um ..."
> Don interrupts again:
> "You're a lot happier now."
> "I'm happier now ..." says Doreen slowly, "because I ... I get really frustrated because I used to walk ... I used to do a lot of walking. I used to ... we used to go dancing."
> "I would rather be walking," she says eventually, "because before all this happened, um, before we had the children, we used to be dancers. We used to compete. Modern. Latin. Old time. We used to go all around the country didn't we?"
> She turns to Don.
> "We used to, yeah," he says.
>
> (Bradshaw-Martin 2012, 386–7)

The point here is not about the basic mobility of walking. It is about a morphological identity as a dancer which she shared with her life partner. But she does dance again now, using her wheelchair. And she has had time to adjust, to build a new identity, find the things she can do around the house, find ways of still being able to care for her grandchildren, and so on. She has begun to form a new identity.

And she plays down the effect on her well-being of the morphology by itself, emphasizing instead the fuller morphological identity and thus drawing out the positive well-being:

> "We're just a normal couple," says Doreen.
> "We have a laugh, we've got a family. We're happy with the family."
>
> (Bradshaw-Martin 2012, 387)

Another contrast case is provided by Tami and Elisa. They have both lost careers as painters, Tami through damage to her eyes and painting hand, and Elisa as her unusual eyesight was medically modified to make it more species-normal and her more independent. Tami is not able to let go of her identification as an artist, despite its conflict with her inability to use her painting hand and to see. "It was a terribly shattering thing to be told you're going to lose your ... specially somebody like me, where my eyes, well, everybody's eyes mean everything, but in my career it meant a lot" (Bradshaw-Martin 2012, 387).

Whereas Elisa says of her previous art and craft based career: "I used to paint," she says, "I was always a painter. And a very interesting thing is that when I see paintings that I did before, I don't recognise them." "I truly cannot relate to them at all.... I don't see like that any more" (Bradshaw-Martin 2012, 388).

But Elisa has been able to create a new professional identity in a line of work not dependent on vision (as she knows she could lose hers again any moment), and to build her new identity around her family and children.

As Neil did, some people may choose morphological changes on the basis of a desire to develop some part of their lives. In Neil's case it was largely his work life. His new morphological identity is dependent on his cochlear implant and his ability to interpret speech. Other parts of the new morphological identity followed from this. But were his hearing to be threatened again he may have to give up the job he now loves. He is dependent on the technology which enabled his morphological change. Similarly, someone taking drugs to improve their decision-making or memory so that they can compete at a profession they love is dependent on the technology. Were it to cease to be available or effective, they would have to give up the morphological identity depending on it.

This dependence can itself affect well-being in negative ways. A number of participants in my research spoke of living "under the sword of Damocles" (Elisa's phrase). They were insecure about the reliability of the technologies or biological changes which enabled their new and valued morphological identities. They expected catastrophic morphological change to befall them at any moment.

The deceptive nature of estimations of the value of a new morphological identity was also a recurrent theme. Elisa valued her new independence in the physical world but she mourned both the expected and unexpected losses of aspects of her identity she had treasured. In particular the uniqueness of her art and the pride she had taken in her international travel. She learned that traveling was much, much easier for others than she had ever expected. In this she experienced something of a movement of her values as an activity which she had valued and taken pleasure in exercising because she believed herself to be exceptional at it became, when she realized it was not worth this kind of value, no longer a positive part of her identity. Neil valued his ability to hear and understand spoken language but found no pleasure in the new frequency ranges in music he could hear because they were simply painful. And, to some extent, they spoilt his previous enjoyment in music.

We are particularly susceptible to these kinds of mistakes when the aspect of morphology we alter is connected with intelligence—activities which are enjoyable to brains trained and shaped in certain ways can become far less satisfying to brains capable of more complex tasks. Just as an infrequent cyclist can take great satisfaction from a 30-mile ride while a professional would consider this too short to be challenging, so our intellectual and other activities can become less meaningful as our brains change.

Take our existing experience of education and maturing. Reading comics and children's books can become less exciting as we mature and develop much greater abilities to take in complex information. What once seemed fascinating and complex can become mundane, predictable, and boring. And we cannot always see this coming—we do not realize which things that once had meaning and gave us pleasure will cease to do so when we have different morphologies.

15.8 Implications for the moral value of cognitive enhancement technologies

Evidence has been given in section 15.5 which supports the use of a subjective welfarist interpretation of well-being and ethics in circumstances where there is variation of morphology. Sumner's welfarist ethical theory says we have moral reason to promote individual welfare. So if morphological changes which improve cognition—the acquisition of knowledge through thought, experience, and the senses—improve individual welfare, then these kinds of morphological changes, and thus the technologies which generate them, would have moral value.

Both the evidence given in section 15.7 from people with experience of disability and Sumner's theoretical framework emphasize that we cannot generalize about the effects of these kinds of technologies when evaluating them. We ought to examine the application of the technologies to each case individually. The evidence from people with experience of disability shows us that specifically what we ought to examine is how the application of the technology will affect each user's morphological identity. We ought to do this with their participation and through their own subjective assessments of what will improve their well-being.

The moral value of an application of a cognitive enhancement technology is, therefore, relative to the morphological identity of the morphee. No wider conclusions can be drawn about the moral value of cognitive enhancement in principle, nor about classes of application of these technologies. The effect on well-being will be dependent on the details of the morphee's existing life and experience, and the precise milieu—social and physical—in which they expect to exist for the future. In this, cognitive enhancement is no different to any other morphological change.[5] This is the implication of the subjective and individual aspects of what we learn from those with experience of disability, and is also captured in Sumner's theory.

In practice what happens might be spelt out as follows:

> From the morphology M1, with x abilities, the morphee experiences a level of well-being, WB1. They expect that a morphology M2 giving them x + 1 abilities will give them a higher level of well-being, WB2. They have this expectation because within the morphological identity MI1, M2 type morphologies are valued; that is, morphologies allowing x + 1 abilities can be observed in others, or are considered by the social group sharing that identity, to correlate with higher status or well-being. So having more of those abilities, and the associated morphology, appears to lead to higher well-being within MI1.

We might conversely, expect that the reverse would also hold true. If one underwent a morphological change which reduced one's ability count to x − 1, one's well-being would permanently reduce. The case of people with disabilities shows us that this does not hold. When we seek an explanation we find that this is because the morphological identity also shifts, giving a new value framework. Comparisons of well-being before and after a morphological change are often not comparing like with like. Similarly, evidence from people who have experienced an increase in their abilities suggests that morphological identity often changes then too.

So M1 may lead to WB1 with a transfer function (people's values) dependent on MI1, but after the morphological change M2 the morphee may no longer share MI1, but instead be forming a new morphological identity MI2 in which a different set of values determines how their morphology contributes to their well-being. WB2 is therefore not necessarily greater than WB1—the relative levels will depend on the morphological identities. These have both an individual and a social aspect.

The practical implication is that potential morphees need to build a detailed picture of their existing morphological identity when evaluating a morphological change, including a cognitive enhancement. This should include mapping out personal relationships including listing shared values and activities, listing the physical environments available to them and those not available to them and why, and examining their employment, hobbies, educational, healthcare, and other interactions with social institutions. They should also list the technologies they currently use to enable their existing morphological identity. A second map then needs making, including the effects of the predicted morphological change on each of the elements from the first map. The key to the process is recognizing that the new morphology will give rise to a different morphological identity—and not merely to a different positioning in the original morphological identity as is usually imagined. Some elements of the first MI will be lost along with the gains.

The influence of the MI on well-being will vary for different MIs. The transfer function between morphology and well-being, if you like, will be different for different morphological identities, partly according to what is valued in a given morphological identity. This is unlikely to be easy—or even possible—for a potential morphee to appreciate without considerable experience of living in the culture of the new MI. This is what makes therapeutic applications of morphological change technology easier to evaluate—the morphee in such a case knows very well what to expect in the target MI because they have previously lived it. But for someone attempting a morphology entirely new to them, there is as yet no way for them to simulate the new lifestyle.

When we move from the importance of using individual's subjective assessments of well-being to focusing on helping them predict future effects on well-being we need to consider again the physical and especially the social aspects of their morphological identity. This is complicated by the obvious, but often overlooked, way in which the well-being of one human depends on their relationships and connections with others. The well-being of one human can depend on the well-being of other humans they care about. Earlier it was argued that a morphee can expect well-being to increase after a morphological change in three ways: firstly from the sheer joy of exercising the new abilities; secondly, from putting them to use to achieve a goal—winning a competition, gaining a better paying job, getting promoted to a position with more status; and thirdly, from achieving specific sorts of goals—those in which the new abilities are used to improve the well-being of a human other than the morphee themselves. So, when examining morphological identity it is important to consider relationships and social networks.

In the case of outwardly visible morphological changes the effects of exclusion or inclusion into activity groups—sports teams, for example—may be reasonably easy to point out. But it may be much more difficult to imagine how changes to cognitive abilities will affect which people one enjoys spending time with … or not.

Considering the well-being of others also leads to the idea that one's own well-being might increase if others underwent a morphological change of some sort. Immediately we can see that these kinds of circumstance reduce a morphee's ability to choose freely—their social milieu may put pressures on them. And from here the interconnected intricacy of morphological identity is easy to grasp—social environments may affect the range of morphological identities which are available. To some extent societies have the ability to influence this. As more technologies which can be applied to change morphologies become available, policy-makers may find themselves having to grapple with the levers they have to hand to affect the morphological identities available to their members. They may need to do this in order to reduce social pressures and frictions which arise as an indirect result of people's choosing morphological changes.

For example, in traditional Amish society collective singing was valued, but individual performance and the use of musical instruments was not. The morphological identity of virtuoso violinist, for example, was not available in this society—it was forbidden by collective decision. Anyone born into this society and wishing to pursue this morphological identity would have to pursue it outside the society—as they have the choice to do (Kraybill 1989). In other societies there might be increased pleasure for many in having a few more virtuoso violinists. We could, similarly, imagine that a society as a whole might benefit from having more people with very high cognitive skills if those skills are directly aimed at improving the well-being of other members of the society, as well as improving the well-being of the morphees. (Clearly, on the subjective welfarist framework it is not ethical for the morphological change to reduce the well-being of the morphees in order to benefit others!)

So, in the morphological identity framework, not only is the moral value of the application of cognitive enhancement technologies relative to the morphological identity of individual morphees, but we can also see why certain morphologies—and associated morphological changes—may be considered morally valuable in some societies but not in others. The moral value of applying cognitive enhancement technologies is also relative to society.

It is necessary to have some framework capable of encompassing the social aspects of morphology before the costs and risks of morphological change can be placed in context. The discussion here has been focused entirely on avoiding the major risk highlighted by the research with people with experience of disability: the unexpected loss of the original MI without sufficient planning or understanding of the need to form a new MI around the new morphology. The cost of failing to mitigate this risk is a waste of money and other resource on providing morphological changes which do not lead to increased well-being, the obvious suffering and lost well-being of the disappointed morphees, and the disruption to their productivity as members of the society which occurs while they seek a feasible morphological identity—a task taking on the order of 5 years. But there are other potential costs.

For example, people with experience of disability who had gained abilities new to them often discussed their fear of losing their new abilities. This might be from deliberate action of others—the theft of the exterior part of a cochlear implant, or the power supply of an artificial heart valve (Bradshaw-Martin 2012; Moore 2008). Or from the riskiness and novelty of the morphological change—surgery could fail, the underlying condition could reassert itself. The fear in these cases is of the new MI becoming suddenly impractical—Elisa's "living under the sword of Damocles." Some participants in the research were unwilling to even attempt morphological change for fear of such double disruption. In the case of, say, the use of pills for cognitive enhancement, a reliable and affordable supply of the drugs would be necessary to mitigate these kinds of uncertainties. Individuals might have a strong motivation to secure these. Societies relying on their members having abilities which in turn depend on these kinds of aids would also need to secure and institutionalize their supply.

As discussed elsewhere in this volume it is entirely possible that a particular technology will not result in the intended morphological change for a particular person. Modafinil may not improve cognition. For some technologies, the placebo effect may be the only effect. It is also possible that the intended morphological change may result, but that it would not lead to the intended morphological identity. And, as we have discussed at length, the connection between morphological identity and well-being may also not be as expected. Each of these risks can break the complex chain of results needed for a morphological change to result in a positive change to an individual's well-being. By making this chain clearer the morphological identity framework does assist us to evaluate the moral value of these complex new morphological change technologies.

15.9 Conclusion

This chapter began with an argument based on definitions of morphology, cognition, and morphological change that showed cognitive enhancements are morphological changes. Next, work with people with experience of disability was used to show how the effects of morphological change on well-being are often counterintuitive and seem to be relative in some way to the particular past experiences of each individual. This counterintuitiveness was discussed in the context of chosen changes to morphology, such as cognitive enhancement. The concept of morphological identity was proposed to explain the counterintuitive disconnects between morphological change and well-being. A compatible theory of ethics based on subjective welfarism was introduced and combined with the concept of morphological identity to form the morphological identity framework. This was then applied to the data from people with experience of disability and to the proposed case of cognitive enhancement to demonstrate how the framework could be used to help us assess the moral value of applications of morphological change technologies. The conclusions are that cognitive enhancement must be evaluated relative to each individual's morphological identity and that the moral value of these technologies is also relative to the particular society in which they are to be applied.

Notes

1. Scully (2008, 34) writes, "Disability is a manifestation or subcategory of something more encompassing, as women are a subcategory of the wider concept of gender.... English and other European languages, as far as I know, lack a concise way of saying 'the general category of characteristics, one manifestation of which is what we mean when we talk about impairment.' "
2. Papineau (2009) explains that in metaphysics naturalism rejects all supernatural causes—gods, magic, spirits—in favor of the causal order of nature. But Schwandt (2007, 205) tells us that in philosophy of science naturalism may mean that all forms of science should follow the model and methods of the natural (or physical) sciences.
3. Griffin (1986) attempted to find a position between these in his book-length study of well-being. His "fully informed desire" account is closely related to concepts which have become important in economics and decision theory. It is not clear to me that this account escapes the objections against objective accounts for morphologically diverse societies. This is because of the difficulties of being "fully informed" about life with a different morphology.
4. For a collection of case studies see, for example, Cole (2004) and Hunt (1966).
5. Most morphological changes will affect how a human acquires knowledge because most morphological changes will affect the senses, how we experience the world, and thus our thought patterns.

References

Albrecht, G. L., and P. J. Devlieger. 1999. The disability paradox: high quality of life against all odds. *Social Science and Medicine* **48**:977–88.

Amundson, R. 1992. Disability, handicap and the environment. *Journal of Social Philosophy* **23**:104–19.

Bostrom, N., and A. Sandberg. 2009. The wisdom of nature: An evolutionary heuristic for human enhancement. In *Human enhancement*, ed. J. Savulescu and N. Bostrom, 375–416. Oxford: Oxford University Press.

Bradshaw-Martin, H. G. 2012. Defining enhancement, disability and therapy: How technology affects identity and the ethical implications of this. PhD thesis, University of Bristol.

Cole, J. 2004. *Still lives: Narratives of spinal cord injury.* Cambridge, MA: MIT Press.

Degrazia, D. 2005. Enhancement technologies and human identity. *Journal of Medicine and Philosophy* **30**:261–83.

Erikson, E. H. 1994/1968. *Identity: Youth and crisis.* New York: Norton.

Erikson, E. H. 1995. *Childhood and society.* London: Vintage.

Griffin, J. 1986. *Well-being: Its meaning, measurement, and moral importance.* Oxford: Clarendon Press.

Grönvik, L. 2007. The fuzzy buzz word: conceptualizations of disability in disability research classics. *Sociology of Health and Illness* **29**:750–66.

Harris, J. 2007. *Enhancing evolution: The ethical case for making better people.* Princeton, NJ: Princeton University Press.

Hunt, P. 1966. *Stigma: The experience of disability.* London: Geoffrey Chapman.

Kahane, G., and J. Savulescu. 2009. The welfarist account of disability. In *Disability and disadvantage*, edited by K. Brownlee and A. Cureton, 14–53. Oxford: Oxford University Press.

Kraybill, D. B. 1989. *The riddle of Amish culture.* Baltimore: Johns Hopkins University Press.

Lawler, S. 2008. *Identity: Sociological perspectives.* Cambridge, UK: Polity.

Moore, P. 2008. *Enhancing me: The hope and the hype of human enhancement.* Chichester, UK: Wiley.

Oliver, M. 2004. The social model in action: if I had a hammer. In *Implementing the social model of disability*, ed. C. Barnes and G. Mercer. Leeds: Disability Press.

Oliver, M. 2009. *Understanding disability: From theory to practice.* Houndsmills UK: Palgrave MacMillan.

Oxford English Dictionary. 2012. "morphology, n." Oxford: Oxford University Press Available at: http://www.oed.com/view/Entry/122369?redirectedFrom=morphology&amp.

Papineau, D. 2009. Naturalism. In *The Stanford encyclopedia of philosophy*, Spring 2009 ed., edited by E. N. Zalta. Available at http://plato.stanford.edu/archives/spr2009/entries/naturalism/.

Parfit, D. 1987. *Reasons and persons.* Oxford: Clarendon Press.

Pearsall, J. ed. 1999. *Concise Oxford English dictionary*, 10th ed. Oxford: Oxford University Press.

Ricoeur, P. 1994. *Oneself as another.* Chicago: University of Chicago Press.

Sandberg, A. 2001. *Morphological freedom—Why we not just want it, but need it.* Available at: http://www.nada.kth.se/~asa/Texts/MorphologicalFreedom.htm.

Savulescu, J. 2006. Justice, fairness and enhancement. *Annals of the New York Academy of Science* 1093:321–38.

Schechtman, M. 1996. *The constitution of selves.* Ithaca, NY: Cornell University Press.

Schwandt, T. A. 2007. *The Sage dictionary of qualitative inquiry.* Los Angeles: Sage.

Scully, J. L. 2008. *Disability bioethics: Moral bodies, moral differences.* Lanham, MD: Rowman and Littlefield.

Silvers, A. 2003. On the possibility and desirability of constructing a neutral conception of disability. *Theoretical Medicine* **24**:471–87.

Silvers, A. 2008. The right not to be normal as the essence of freedom. *Journal of Evolution and Technology* **18**:79–85.

Sumner, L. 2003. *Welfare, happiness and ethics.* Oxford: Clarendon Press.

Chapter 16

The legal aspects of cognitive enhancement

Imogen Goold

16.1 Introduction

This chapter will examine a range of legal issues that have arisen as a result of the emergence of putative cognitive enhancement technologies. It focuses on pharmaceuticals such as modafinil which, some studies suggest, have the ability to redress some of the impacts of tiredness and fatigue, and to improve attention and focus in those who are well-rested.[1] It also applies to new technologies that make claims that they will improve other cognitive abilities, such as transcranial direct current stimulation (tDCS) devices, which are marketed as a means of improving a person's capacity to concentrate for long periods of time, as well as aiding in memory, learning, and facial recognition.

The implications of cognitive enhancement are myriad and expanding, and a chapter of this kind cannot hope to provide an in-depth treatment of all the relevant legal issues. Its aim is therefore to provide an overview of the areas of challenge, and present some of the essential legal principles that may be relevant in the context of new cognitive enhancement technologies. In doing so, it is hoped that this chapter will spark further discussion and exploration of the legal issues surrounding cognitive enhancement, many of which currently remain under-researched. While a burgeoning literature has developed around the ethical issues related to cognitive enhancement, there has been little similar discussion in the legal literature.

This chapter covers a broad range of areas of law. It examines how product liability rules might apply to the sale and supply of enhancement products, particularly given the growing non-medical use of such products, including devices. Related to this are questions about sale of goods and fitness for purpose, where drugs or devices are marketed with claims about what they may achieve for users. This necessarily involves some discussion of the law of contract.

Another key area in which cognitive enhancement technologies pose legal challenges is tort law, most particularly negligence, where enhancement may have an impact on the standard of care expected of persons. If enhancement enables us to improve our capacities, and if that enhancement becomes normalized or widespread (whether safe or not), this might have implications for how we define "reasonable care" and the "reasonable person." What if we improve our reaction times? Or our capacity to maintain our attention? And if this occurs, should the law apply different standards to the enhanced and the unenhanced? This leads on to questions of whether we might even oblige people to enhance themselves, which is explored under the heading of negligence but also picked up again in the context of the employment law.

The chapter also briefly examines the criminal law implications of enhancement, focusing on the mental element of crimes and issues of consent. Questions of what it means to be able to form an intention and the relevance of enhancement to this is explored, as are some of the implications of possible future enhancement for our understanding of capacity.

Finally, some attention is paid to human rights law, as increasingly scholars are asking questions about how enhancement technologies may have an impact on how we understand and how we protect those rights. This discussion touches on issues of privacy, particularly the emerging idea of mental or psychological privacy and how we should (if we should) protect this.

Before delving into the specific issues and questions, some points about how the law can and should respond to new technologies should be made. In the face of new developments with both considerable promise for improving human capacities (and perhaps the human condition), but also attendant risks, it might be tempting to develop new, targeted laws to manage these developments. Perhaps this might be necessary, but it is important that we think carefully about whether new or different regimes are really needed, or indeed whether any change is needed at all. David Friedman (2001–2, 71) argues that technological chapter affects law in at least three ways:

(1) by altering the cost of violating and enforcing existing legal rules;
(2) by altering the underlying facts that justify legal rules; and
(3) by changing the underlying facts implicitly assumed by the law, making existing legal concepts and categories obsolete, or even meaningless.

Cognitive enhancement technologies might have all three effects, but particularly the latter two. The potential impact of modafinil use on the standard of care in negligence is a good example of such possible impacts. The same might become true of intentionality in the criminal law, and we might see implications in the area of employment as well.

He makes the very salient point that broadly defined laws will rarely, if ever, need to be adapted to new technologies—they can simply encompass them. But, as he goes on to say, these principles can at times be too broad "to apply with predictable results at a reasonable cost" (Friedman 2001–2, 85). The gains in terms of flexibility and future-proofing must be balanced against the costs of lack of certainty and not providing citizens with clear guidance as to what is legal.

When we think about how our current laws apply to cognitive enhancement, we should bear Friedman's points in mind in a range of ways. It may be that sufficiently broad rules or principles from which we can extrapolate are already in place, and so we might be able to apply these to address the new issues. Human rights principles might be one means to do this. We should also be mindful of not promulgating new laws (whether by legislation or case law decisions) that operate at a high level of generality while specifically focused on enhancement technology. To do so might lead us to make laws that cannot be appropriately applied with sufficient specificity, serving only to create a new rule of law problem.

In determining where there is a need for legal change, we might ask ourselves: just how have cognitive enhancers created legal challenges and what is their exact form? We might consider which norms or facts they may have affected, and the degree to which such changes mean that existing legal approaches or categories are in fact in need of updating or jettisoning. Here, again, the example of negligence is a good one. If everyone is eventually enhanced, that might well mean that the reasonable person standard has fundamentally changed. But if only some are enhanced, and by choice, this might not be true. The answers to these questions will inform us as to whether we need new laws, need to change existing laws (to avoid obsolescence or to account for changes in under-pinning facts or norms), or need to amend laws that now rest on outdated or irrelevant justifications.

A further element of how the law should respond to cognitive enhancement is to resist creating new, technology-specific or challenge-specific legislation unless completely necessary. It may be that this turns out to be preferable, or at least unavoidable, but doing so runs the risk that we will create a regime that lacks flexibility to change as the technology develops or becomes rapidly out of date (see, e.g., Bennett Moses 2007, 415). Lyria Bennett Moses (2011), for example, has pointed out that there are significant problems with *sui generis* regulatory frameworks tailored to specific technologies. The scheme may not successfully capture all the possible impacts of enhancement, meaning that appropriate controls and protections may not be provided. We can avoid this temptation by leaving the courts to respond reactively to enhancement issues by interpreting relevant laws to apply as needed.

We can also avoid the need for new laws by ensuring that before creating new schemes or rules we identify where enhancement technologies present challenges, what they are, and how they might be addressed within existing legal frameworks. This chapter aims to identify potential areas of challenge and briefly outline those challenges.

16.2 **Product liability**

Product liability law and some other measures protect consumers against products likely to do harm. Under the common law, negligence in producing or supplying a product may attract liability if the manufacturer can foresee that "the absence of reasonable care in the preparation or putting up of the products will result in an injury to the consumer's life or property" (*Donoghue v. Stevenson*, AC 562 [1932]). Therefore, someone who produces an enhancer negligently may be liable for the damage it causes due to that negligence. This is one area where the law as it stands might apply fairly well to the supply of enhancers, as the broadly drawn requirements of negligence can be extrapolated to cover enhancing drugs or products. However, the common law only protects against products that are negligently made. For products that might simply be inherently dangerous or risky, a legislative scheme applies.

The Consumer Protection Act 1987 covers any goods, including those which are components of other products. This encompasses things we ingest and devices we use (Consumer Protection Act 1987, s 1(2)(c)). Therefore, cognitive-enhancing drugs like modafinil would fall within its remit, as would devices like tDCS. Where a defect in the product causes damage, multiple parties may be liable: (a) the producer, (b) a person who puts his name or mark or trademark on it, or (c) any person who has imported the product from outside the European Union (EU) (Consumer Protection Act 1987, s 2(2)).

A productive is defective if it is not as safe "as persons generally are entitled to expect" (Consumer Protection Act 1987, s 3(1)). Where the public is considered to be aware of the potential harms the product may cause, it is not defective. For example, a condom that splits is not defective under the legislation as it is widely known that this may occur. The risk of splitting is therefore part of what the public may expect from a condom, so one that splits is not defective (*Richardson v. LRC Products Ltd*, 59 BMLR 185 (QB) [2000]). The same is true of drugs with well-known side effects, such as the contraceptive pill (*XYZ v Schering Health Care Ltd*, EWHC 1420 (QB) [2000]). A producer can ensure a product is not defective by including warnings about potential side effects or risks (*Worsley v Tambrands*, PIQR 95 (QB) [2000]).

One of the strengths of the legislation is that it protects consumers against products that may cause them harm for reasons other than negligence—it is not fault-based, but harm-oriented. That is, where a consumer suffers harm, he or she has a potential claim, regardless of how much care was taken in producing the product. If it is defective because it was unexpectedly unsafe, liability will probably attach. For cognitive enhancers that cause unknown side effects, such protections will be important if they are marketed to consumers. The legislation would capture drugs or devices sold to consumers on the basis that they will produce enhancing effects where in reality they damage the user in ways that the user would not be expected to know might occur.

While the scheme is one of strict liability (i.e., there is no need to prove negligence or fault), producers can protect themselves by providing warnings. Therefore, the producer of an enhancer could inform users of possible side effects and avoid liability, providing such warnings were sufficient. These warnings allow users to make informed choices and prevent producers from being placed under unreasonable liability burdens. Product liability law, then, is likely to apply to cognitive enhancers by giving producers an incentive to provide warnings and information, and finding them liable when vulnerable consumers are unwittingly harmed.

It might seem that this places too great a burden on producers in the case of risks or side effects that no one could have foreseen. If this were the case, producers would be likely to act very cautiously in bringing products to market, particularly drugs like cognitive enhancers. This might have a chilling effect on development. The legislation takes this into account via the "development risks defense." Under the defense, the producer will not be liable for defects that they could not have been expected to discover given the state of the scientific and technical knowledge at the time (Consumer Protection Act 1987, s 4(1)(e)). The knowledge requirement is an objective one. This means that it will be the level of knowledge that the producer could be assumed to possess, provided the knowledge was accessible at the time, when the product was put into circulation. It does not depend upon what the producer actually believed or knew (C 300/95 *Commission v UK* All ER 481 (EC) [1997]). The defense only applies if the risk is not known about, not if it is impossible to eliminate it (this would need to be covered by a warning) (*A v National Blood Authority*, 3 All ER 289 (QB) [2001]).

The defense covers producers against liability for risks of which they simply could not be aware. The burden of these risks therefore falls on the consumer but the producer is expected to be aware of current scientific and other relevant knowledge. For makers of cognitive-enhancing substances and devices, this pushes them to be responsible in their production of these substances and puts the onus on them to inform themselves of possible risks, or otherwise bear the legal consequences of any harm they cause. The defense allows them to create new products without fear of liability for risks that they could not be expected to know about. This is important for innovation in the field of new cognitive-enhancing products.

But there are good reasons why simply dealing with enhancers via standard product liability might be insufficient. Product liability laws respond after defective products have caused harm. They act as a deterrent only to the extent that producers know that they will be liable if the product can be shown to cause harm. If this incentive is ineffective, whether because they simply do not respond to it, or because they include the costs of compensating those who suffer harms into their pricing, then product liability laws will not prevent unsafe enhancers from entering the market. If a case is brought for harm, this may create an added deterrent but that is no guarantee.

At present, most putatively enhancing drugs are already medicinal products, and hence have been vetted for safety. Before coming to market, all medicines supplied in the EU must undergo clinical trials to demonstrate their safety and efficacy as required by both EU and domestic legislation (Directive 2001/83/EC as amended by Directives 2002/98/EC, 2003/63/EC, 2004/24/EC, and 2004/27/EC; Medicines for Human Use (Clinical Trials) Regulations 2004 (SI 2004 no 1031), implementing Directive 2001/20/EC relating to good clinical practice in the conduct of clinical trials on medicinal products for human use—the European Clinical Trials Directive). Similar structures are in place to regulate medical devices. However, many of the medicinal substances used for cognitive enhancement are used off-label and so prescribed for a purpose for which they have not been licensed. In the UK, off-label prescribing is legal for those who are licensed prescribers, that is doctors (or nurses or pharmacists working under a clinical management plan conducted by a doctor) (Department of Health 2006). As long as the prescriber is competent and operating within the professional codes and ethics of their relevant statutory body, the decision to prescribe for an off-label purpose will be lawful (Medicines Act 1968, s 9(1)).

The fact that medications prescribed in this way have been assessed for safety and efficacy may not be sufficient to identify potential harms when they are used in unapproved ways. These protections may not prevent harms if users take different dosages, or use them when they do not have the condition they were developed to treat (such as healthy individuals using modafinil, which is primarily used to treat sleep disorders such as narcolepsy).

Product liability laws will step in only once those harms have occurred. So will regulatory measures that control prescribing if the doctor's decision to do so was not competent (and might also be legally negligent). If the risks of taking enhancers are substantial, then we may need to do more to regulate them before they reach the market. This could be done by expanding the regulatory mechanisms already in place for medicines, by modifying product liability regimes, or by some other oversight mechanisms.

More problematic are devices that are now used for putatively enhancing purposes and that have not been assessed under the medical devices regulatory framework. One example is the marketing of tDCS devices for nonmedical purposes, like *Foc.us*, which is aimed at the computer gamer market (Foc.us 2014). Such devices are not subject to safety checks because they are nonmedicinal. This leaves them solely within the ambit of products liability laws. Given this, calls have recently been made to include enhancing devices within the much more stringent regimes covering medical devices. In EU these would be covered by the Medical Devices Directive, which at present covers only devices used for diagnostic or therapeutic purposes (Medical Devices Directive 93/42/EEC).

This would bring such devices within schemes that check the safety and efficacy of medical devices before they can be used. As with the regulation of medicines, this ensures that checks are made before such drugs become available and a much more stringent regulatory approach is used to prevent harms from occurring. There are of course some problems with extending the scheme to nonmedical devices. These are explored by Maslen et al. (2014, 2015a), who argue that they can be surmounted. These concerns are important for the law to consider, both locally and at the EU level. A similar approach might be considered for drugs that are created and sold for enhancing purposes that fall outside the regulation of medications (which is a similar stringent scheme).

On balance, it seems that the current structure for managing products that might cause harm can encompass cognitive-enhancing drugs and devices. The implications of this coverage is that producers are protected against unknowable risks that their products might carry. The consumer will bear these. Given the concerns identified in this section, there is clearly work that needs to be done to improve our current regulatory approach to ensure that users are not left unprotected from harm. We ought also to think in terms of devising regulatory approaches that prevent harms rather than merely compensate consumers after the fact.

16.3 **Contract law**

Contract law deals with voluntary obligations made between parties. In relation to cognitive enhancement, the most relevant questions will revolve around the sale of devices and substances that are advertised as having an enhancing effect. The law implies terms into contracts for sale to protect buyers. Under the Sale of Goods Act 1979, where there is a contract for the sale of goods by description, there is an implied term that the goods will correspond with the description (Sale of Goods Act 1979, s 13(1)). Where those goods are sold in the course of a business, there is also an implied term that they will be of satisfactory quality. This requires that they meet the standard that a reasonable person would regard as satisfactory, taking account of any description of the goods, the price (if relevant), and all the other relevant circumstances (Sale of Goods Act 1979, s 14(2), (2A)). Quality includes its fitness for all the purposes for which goods of the kind sold are commonly supplied, including safety (Sale of Goods Act 1979, s 14(2B)). The act, as well as the law of contract generally, provides a range of remedies if these implied terms are breached, including a right to rescind the contract of sale (see, further, Sale of Goods Act 1979, s 243(A)).

Such rules can be easily applied to cognitive enhancers that are sold commercially. The technology involved does not alter the fundamental aspects of contract law—the supply of an item that has a claimed effect. While there is as yet little or no case law on how the Sale of Goods Act will apply to enhancing drugs and devices, the sections are likely to apply to persons who sell enhancers on the basis of claims about what they can do, potentially rendering them liable for breach of contract if the putative enhancers fail to do as claimed. This means that in the future we may well see cases where purchasers bring claims when a purported enhancer does not do as promised. Where the purchased product causes actual harm, this will fall under product liability or common law negligence. Therefore, purchasers will be protected against both harms and snake-oil sellers.

16.4 **Negligence**

Tort law encompasses numerous contexts in which one person may raise the assertion that another person should be held responsible for harm that they have caused (or have been involved in causing). The most likely area of law in which the use of enhancers may have legal implications is the law of negligence. Negligence is "the omission to do something which a reasonable man, guided upon those considerations which ordinarily regulate the conduct of human affairs, would do, or doing something which a prudent and reasonable man would not do" (*Blyth v Birmingham Waterworks Company*, 11 Ex Ch 781 (Ct Ex) (1856)).

The use of enhancers poses a number of challenges of different kinds. First, it may affect how we understand the "reasonable person" standard. This may mean that the standard is raised, but still acts as a threshold standard. Or it may mean that we move away from the idea of a threshold toward the notion of "responsibility tracking capacity," in which those who have greater capacities have greater responsibility (Vincent 2011). This would involve holding enhanced people to higher standards of care. Second, and relatedly, if the reasonable person standard shifts, this may effectively create *obligations* on people to self-enhance. These issues are now examined in turn.

At present, the standard of care in negligence in both tort and the criminal law is a threshold standard. This means persons must take at least the care a reasonable person would take, based on normal human capacities. If someone has extra capacities, this is not relevant in the same way that being of less than average intelligence or having lower impulse control is no defense (unless they also cross a relevant threshold, as in the case of mental capacity). For example, the standard of care expected of road users is that of a reasonably competent and experienced driver. A learner driver is held to this standard, despite their lack of experience. As Lord Denning put it in *Nettleship v Weston*, 2 QB 691, 699 (CA) [1971], the civil law "eliminates the personal equation and is independent of the idiosyncrasies of the particular person whose conduct is in question."

Vincent (2011, 515) observes that the capacity to enhance might affect this approach:

> once a person becomes cognitively enhanced, may they then be legitimately expected to observe a higher standard of care than their non-enhanced counterparts, and should their breaches of such higher standards attract regulatory, civil and criminal sanctions?

Given that there are good reasons for maintaining the threshold approach, this seems unlikely. As Lord Denning points out, the law eliminates the person, and it does so because rule of law demands clarity and certainty. An infinitely variable standard of care based on individual characteristics would create considerable legal uncertainty and move the focus away from what is expected of persons in protecting others. At present, at best it can be said that a higher standard is imposed on some *classes* of person who are expected to have higher skill or ability levels (which is what defines the class), such as medical practitioners. If in some way enhancement becomes an expected aspect of such a profession (e.g., if all members will be enhanced), then this might imply a higher standard of care for that class.

A move away from a threshold standard would also necessarily work in both directions, demanding that we drop the standard of care for those with lower capacities. At present, there is good reason why this is not the approach in the United Kingdom and other common law jurisdictions. For example, it was said in *McHale v Watson*, 115 CLR 199, 213 (HCA) (1966) by Kitto J.:

> the standard of care being objective, it is no answer for … an adult to say that the harm he caused was due to his being abnormally slow-witted, quick-tempered, absent minded or inexperienced.

A major reason for this approach is that those harmed by a person with lower capacities would, through no fault of their own, be left without a remedy. This would seriously undermine the claimant-focused elements of negligence.

It is important to note that, absent a special duty of care under which a person has assumed responsibility, the law does not oblige us to act in the interests of others, only to refrain from acting in ways that will harm them. As Lord Goff noted in *Smith v. Littlewoods* [1987] AC 241, 271 (HL), "the common law does not impose liability for pure omissions," and the mere fact that a person could foresee that a failure to enhance would harm others is not enough for liability to attach (see also *Stovin v. Wise* [1966] AC 923, 943–944 (HL)). Therefore, a pure failure to enhance is unlikely to lead to liability. That said, there might be contexts in which someone assumes responsibility in such a way that an obligation to enhance does arise. In these contexts there will still be barriers to liability being imposed, as is explored later in this chapter.

16.5 **Criminal law**

The use of enhancers—or even simply their availability—might affect expectations about behavior that falls within the remit of the criminal law. It might cause shifts in what we expect of drivers in terms of attention paid, or the level of impulse control we can expect of people who commit crimes. However, such shifts will only occur if the use of enhancers becomes so widespread that everyone is enhanced. If this is not the case, and it is not likely to be the case for some time, then we come back to the question of whether people ought to be expected to enhance to avoid criminal responsibility. This faces many of the same concerns already raised in the context of negligence.

To consider how the law might approach these issue, we can extrapolate from the case law on road accidents caused by defendants' failure to take needed medication, such as those who suffer from epilepsy or hypoglycemia. What is telling about those cases is that almost invariably, the wrong was the decision to drive or continue to drive, not the failure to medicate. Goold and Maslen (2014a, 82) have explained why in cases such as *R v. Remi Akinyeme* [2007] EWCA Crim 3290, in which the defendant had an epileptic fit while driving (after not taking his medication) that led him to hit and kill a cyclist:

> The decision of whether to self-medicate is one for the individual, whereas the decision to drive is one in which the court takes an interest. That decision not to self-medicate formed part of the circumstances in which the wrong occurred; it was not the wrong in itself.

This would suggest that the courts will focus on the allegedly criminal act, rather than the possibility of changing one's capacity.

Vincent's ideas about responsibility tracking capacity present many of the same concerns when we think about the implications for the criminal law of the availability of enhancers. As in negligence, only certain classes of people will be held to higher standards, and there the standard is that applied to the class on the basis of shared characteristics.

Chandler (2013) has drawn attention to another important way in which a parallel enhancer use might raise issues in the legal context—in relation to punishment options. What if an offender could avoid punishment by taking an enhancer? Chandler explores how the availability of drugs could influence punishment of criminal activity. Courts have for some time offered various treatments to prisoners as a condition of release: addiction treatment programs (Greely 2008), antilibidinal drug treatment ("chemical castration") (*Deacon v. Canada (Attorney General)* 2005 FC 1489, aff'd 2006 FCA 265; Stinneford 2008), and surgical castration (Russell 1996–7; Vanderzyl 1994–5), for example, have all proved influential in sentencing. With the availability of drugs such as cyproterone acetate, which has been used to reduce sex drive in sex offenders (Fisher et al. 2011), the potential for a medical "fix" for criminality, at potentially lower costs than imprisonment, may becoming increasingly relevant.

There are numerous ways enhancers might raise issues here, depending on what drugs and devices are developed. Perhaps they might be used simply to bring a defendant up to the capacity required to stand trial and understand proceedings. But what of an enhancer that improves empathy? Or enhances memories? It is not difficult to imagine how these might be used punitively on an otherwise unrepentant criminal. We might even see their use advocated as part of reintegrative shaming approaches to better bring home to the perpetrator the harm their actions have caused. But would this be right? Would this be an appropriate way for the law to respond to crime, and if not, why not?

16.5.1 Mental element of crimes

What the accused was thinking at the time of committing the crime is almost always relevant to determining which crime might have been committed and whether the accused is guilty of that crime. For many crimes, "intention" must be shown in its ordinary meaning as "having the purpose or aim" of doing what the person actually did (*R v. Moloney* [1985] AC 905 (HL)). This is the case for the offenses of murder, rape, and battery. Sometimes, recklessness will be sufficient to meet the intention criteria, depending on the offense. Someone is "reckless" if he or she foresees that particular consequences may occur and proceeds with the given conduct, not caring whether those consequences occur (*R v. Cunningham* [1957] 2 QB 396 (CA)). People are held to the standard of a sober, ordinary person. A defendant who was voluntarily intoxicated at the time of an offense, and therefore failed to foresee a risk they would have foreseen if they had been sober, will be treated by the court as having foreseen the risk (*R v. Bennett* [1995] Crim LR 877 (CA)). Whether this works in the other direction, with people being held less or more liable depending on whether they have been enhanced is difficult to know. It is probably a question of capacity.

Only those with the capacity to form the relevant intention will be convicted of an intention-based offense. The defendant will lack this capacity if he or she did not understand what they were doing. Voluntary intoxication will not, however, excuse a defendant even if in fact it means they did not fully appreciate the nature of their actions or did not have the ability to form the required intention. Indeed, it is no defense even if the intoxication was involuntary (*R v. Kingston* [1995] 2 AC 355 (HL)). As we saw earlier, the law takes a threshold approach to capacity in both the civil and criminal laws. It is unlikely, then, that any increase in capacity will affect criminal responsibility.

In some cases, the defendant will have some capacity, but their responsibility will have been diminished in some way. For example, a person who is suffering from "abnormality of mental functioning" is considered to have diminished responsibility, and so will not be guilty of murder (but may be convicted of other offenses). This abnormality must stem from "a recognized medical condition" (Coroners and Justice Act 2009, s 52(1)(a)). Diminished responsibility does not apply where the cognitive abnormality is due to the influence of drugs. The same is likely to be the case if an enhancer actually reduces capacity, such as by leading the defendant to overfocus on one task at the expense of another (such as listening to music while driving). If that reduction was the result of ingesting drugs voluntarily, it is no defense, regardless of the nature of those drugs.

The flipside of intention is that it is a defense to some offenses that the victim consented. This is what absolves surgeons from what would otherwise be criminal batteries when they cut into patients on the operating table. Consent can only be given by someone who is mentally competent, and all adults are presumed to be competent (Mental Capacity Act 2005, s 1(2); *Re MB (An Adult: Medical Treatment)* [1997] 2 FCR 541, 553 (CA)), unless they are demonstrated to lack capacity. This will be the case when, at the material time, he or she is unable to make a particular decision for himself because of an impairment of or a disturbance in the functioning of the mind or brain (Mental Capacity Act 2005, s 2(1)). This may be due to an inability to understand relevant information, to retain it, to use or weigh it in order to make the decision, or simply to communicate the decision once made (Mental Capacity Act 2005, s 3(1)). Here, we might think there is a role for enhancers to enable people to give competent consent when they might otherwise be incapable of doing so. Imagine an enhancer that enables someone to take in and understand information sufficient for competent consent when they might otherwise not be able to do so—such an enhancement might be useful, but given concerns about bodily integrity and the like, it is hard to imagine their use would ever be made compulsory for self-regarding decision-making.

16.6 **Human rights law**

Human rights law might have a role to play in how we respond to cognitive enhancements. Pariotti and Ruggiu (2012) argue that the flexibility of human rights law, particularly in Europe, may provide an effective avenue for governance. In their view, soft law instruments of both of the Council of Europe and the EU such as the Oviedo Convention 1997 (entered into force on December 1, 1999 CETS no. 164) and its protocols are good ways of managing the impact of scientific and technological developments (see, e.g. the European Social Charter (Strasbourg, May 3, 1996 CETS no. 163); the European Agreement on the Exchange of Therapeutic Substances of Human Origin (Paris, December 12, 1958, (entered into force, January 1, 1959 CETS no. 026). These, in combination with the European Convention on Human Rights (ECHR) (Strasbourg, November 4, 1950, entered into force, September 3, 1953), are flexible instruments that guide the case law of the European courts. Their interpretation may help in how we approach the challenges of cognitive enhancement.

Human rights have relevance to the questions of discrimination considered elsewhere and so might provide a useful lens through which to consider whether enhancement should be obligatory, and how far the state should require us to change our bodies. The ECHR protects each person's right to respect for private and family life, and this includes protection against interference with an individual's bodily integrity (*Glass v. UK* [2004] 1 FCR 553 (ECHR); *YF v. Turkey* (2004) 39 EHRR 34, [33] (ECHR)). This includes freedom from interference with a person's psychological integrity (as well as their physical integrity) (*YF v. Turkey* (2004) 39 EHRR 34, [33] (ECHR)).

This suggests that proposals to directly or indirectly oblige people to enhance themselves might be challenged under human rights principles. Someone who comes under pressure to do so in the workplace (or elsewhere) might argue that a law that allows such pressure is in breach of a country's commitment to human rights. As Hagger and Hagger-Johnson (2011, 152–3) point out, Article 8(1) of the ECHR protects dignity, and hence if personality traits are considered a core aspect of identity, which may be altered by cognitive enhancement intentionally or otherwise, this could possibly be seen as an infringement of dignity (see, e.g., *Pretty v. United Kingdom* [2002] ECHR 427, [65] (ECHR)).

One response to such a challenge might come from the restrictions placed on the protection of that right. Infringements of privacy or family life can be made where it is necessary in a democratic society in the interests of national security, public safety, or the economic well-being of the country, for the prevention of disorder or crime, for the protection of health or morals, or for the protection of the rights and freedoms of others (Human Rights Act 1998, Sch 1, Art 8(2)). In cases where enhancement is obliged on the ground of the protection of others, this exception might support such an obligation, but whether this is the case will depend heavily on the particular facts, the degree of risk associated with the enhancement, and the impact of the failure to enhance on others. In the absence of relevant case law, it remains to be seen how such a situation would play out. What we can say is that when considering how to regulate enhancers, these fundamental background rights need to be considered when assessing which obligations will be both acceptable and lawful.

In the future, we might also see the emergence of new human rights in response to our changed capacities, or at least our ability to change those capacities that may come with the emergence of enhancement technologies. Bublitz and Merkel (2014) point out that to date the law has paid little attention to protecting interference with our mental lives. They argue that the bright line between bodily interferences and mental interferences is dimming in the age of neuroscience, and that it is time for the law to consider how such interferences might be managed. They note that at present, legal systems distinguish between bodily and mental injuries, and that this is right—they should not be elided.

In their view, human rights law can provide a foundation for protecting against such interferences. They note that that European Union has recently introduced a right to respect for "mental integrity" in Article 3 of the Charter of Fundamental Rights. The European Court of Human Rights includes mental integrity under the scope of Article 8 (privacy), but it has not played an important role (yet). In their view, problems such as the possibility of unwanted neuroenhancements (such as parents enhancing children against their will, or coerced enhancement in the employment context), could be addressed via a human right to self-determination (or cognitive liberty). This, they suggest, could be deployed as an underpinning assumption within the law. It would then be necessary for the law to ensure people are protected from severe interferences in their mental liberty by the state and third parties, and to maintain their freedom to self-determine their own inner realm (Bublitz 2013, 233–64; Bublitz and Merkel 2014).

From this brief overview, it is evident that human rights principles already have a role to play in how we address issues raised by cognitive enhancement. This role is likely to be increasingly important as our capacity to enhance improves.

16.7 Children

The key issue in the context of children is that it is the parents who make most decisions about what foods, treatments, and medications a child will receive. Most children whose parents want them to take enhancers will have little choice in the matter, either in reality or sometimes even legally. To a degree, some older children can refuse: following the case of *Gillick v. West Norfolk & Wisbeck Area Health Authority* [1986] AC 112 (HL) if a child under the age of 18 can show that they have sufficient maturity to make the decision in question then they will be deemed able to provide their own effective consent to treatment. This can, however, be overridden where there is a treatment need: in *Re R* [1991] 4 All ER 177, [24H] (CA), Lord Donaldson made it clear that a child's refusal of consent could be overridden by the provision of another valid consent (for example, that of their parents), a view reiterated in *Re W (A Minor) (Medical Treatment: Court's Jurisdiction)* [1993] FLR 64, 88 (Fam).

It is unlikely that a court would enforce an enhancement on a child over the parent's refusal. It is most likely that the parent will simply give the child enhancers in the privacy of their own home. In some cases, a child might protest and action might be taken, but this is unlikely. The issue, then, is whether parents should be permitted to give their children enhancers, in cases of both a willing and an unwilling child.

Hagger and Hagger-Johnson (2011, 148) comment that "so long as there is a lack of evidence of longer term substantial harm, ostensibly enhancing interventions are likely to be regarded as legally permissible." They make the point that for the most part parents act with their child's welfare in mind, but:

> clearly, parents do sometimes harm their children.... Salvemini discusses the risks of children taking growth hormone and notes that 42 per cent of prescriptions are off-label in the US, which indicates that parents are willing to take these risks. She describes parents' aesthetic preferences for taller children ... They are prepared to give these drugs to their children even though they carry significant risks such as kidney failure.... Given that parents are willing to enhance their offspring in this way, despite the possibility of dangerous side-effects, it follows that many would be willing to cognitively enhance.

Parental pressure on children to succeed is prevalent in many countries, including much of Europe and the United States. This pressure may come close to duress, as Singh and Kelleher (2010) comment. Hagger and Hagger Johnson (2011) argue in favor of promoting participatory decision-making for younger children when considering whether enhancers should be given, while using doctors as gatekeepers on access to enhancing drugs as far as possible. They support calls for a code of good practice, arguing correctly that a ban is unlikely to succeed.

16.8 Employment and discrimination law

The employment context is one of the main areas in which cognitive enhancement is likely to become an issue for the law. The use of enhancers in the workplace raises issues of safe working practices as well as pressures (and even obligations) to enhance. It also raises issues about the selection of employees and discrimination.

16.8.1 Selection of employees

One concern that might emerge, and which might require a legal response, is discrimination in favor of those who have chosen to enhance themselves. It might also be the case that in the face of such discrimination, people feel pressured to self-enhance in order to make themselves competitive in the job market. Discrimination on many grounds in the selection of employees is prohibited under both United Kingdom law and EU law, including human rights protections. Article 14 of the ECHR prohibits discrimination on the grounds of sex, race, color, language, religion, political or other opinion, national or social origin, association with a national minority, property, birth, or other status.

In the United Kingdom, discrimination in the selection of employees protections prevent employers from choosing one person over another on the basis of a "protected characteristic," namely age, disability, gender reassignment, race, religion or belief, sex, or sexual orientation (Equality Act 2010, s 4 and s 13(1)). However, a preferential decision on the basis of age may be permitted if doing so is a proportionate means of achieving a legitimate aim (Equality Act 2010, s 13(2)). Employers must also not discriminate against (or victimize) job applicants in the terms on which they offer employment (Equality Act 2010, s 39(1)), nor in the way an employee is afforded access (or not) to opportunities for promotion, transfer, or training or for receiving any other benefit, facility, or service, or by subjecting the employee to other detriments (Equality Act 2010, s 39(2)).

None of these protected characteristics naturally encompasses enhancement. This suggests that a decision based on whether a person has enhanced characteristics will not *prima facie* be discriminatory under the current law. It is more likely that it will simply be regarded as part of selecting the person best placed to perform the job. The reason for their being in this position is not relevant.

On its face, this position seems reasonable. However, we may need to revisit this exclusion if we are concerned about the impact of forms of enhancement that are potentially harmful. This will be particularly the case if people experience pressure to enhance by taking potentially unsafe drugs or by using unsafe techniques. Those who defend the exclusion might reply that employees already take risks in training to undertake a job, such as the financial risks of investing in higher education, and so accept any risks associated with their training, such as learning to fly, or being exposed to much higher levels of potentially infectious agents during medical training. On this argument, would being pressured into taking modafinil while studying be sufficiently different to warrant special treatment? Probably not, and we must also be mindful in such cases of the risks raised by Bennett Moses earlier in this chapter in tailoring a legislative approach to a new technology.

But we might then wonder whether in fact the law *should* do this. In many other ways, people do things to gain a workplace advantage and we do not legally prevent them. Imagine the person who works very long hours, or studies all weekend taking extra courses. These entail risks—they may exhaust themselves; they may waste their resources;

they may put their marriages at risk, or neglect their children. The law rightly does not consider these issues its business. Perhaps, then, the main focus should remain on keeping the workplace safe once the person is in employment. Currently, in the United Kingdom, there are a range of workplace safety laws that might be applicable to cognitive enhancement, which are considered in the next section.

It is possible that the law may simply not be able to manage all of the concerns that arise here. What of people who have taken enhancers in the past, and reaped the benefits, but no longer use enhancers? Their past use has given them a present competitive advantage—a good example would be someone who used modafinil during university studies to achieve better grades and obtain a better job after they stopped taking it. It is unlikely that such advantage could be detected when this person applies for a job or that the law could (or even should) provide an effective response. Where enhancement is ongoing, as this is done privately via taking medication, it is unlikely to be detected, even if it should be.

Perhaps the better legal response would be to address the pressures to enhance that arise from competition for employment. It might put in place similar restrictions to those in place to address discrimination on other grounds (such as a woman's plans to start a family). Or it might not regard enhancement as a ground for discrimination but as a good thing, or a safe, acceptable thing a person might choose to do to improve his or her employment prospects. Indeed, if enhancement better fits a person for a job, and is not harming to them, it could well be a reasonable basis on which to select in favor of them. At this stage, we can really only speculate. The key point seems to be to avoid supporting forces that pressure people to do things that are self-harming in order to improve their employment prospects.

16.8.2 Requiring employees to self-enhance

How should the law address the situation that may potentially arise if employees are expected to cognitively enhance themselves as a condition of their employment?. A crucial question in the ethical literature on cognitive enhancement is whether a person might be *obliged* to enhance him or herself. If the answer is yes, in some contexts this might raise the question of whether the law should oblige people to enhance themselves.

Suggestions that there might be pressure on people to cognitively enhance themselves in the workplace are not fanciful. These sort of calls are increasing. For example, in a recent report from the Joint Academies' workshop on "Human Enhancement and the Future of Work," it was suggested that in some "high-responsibility occupations," such as medicine and aviation, where the lives of others are at stake, "enhancement could be seen as a moral obligation, or even demanded by the public." (Academy of Medical Sciences et. al. 2012, 38).

Such calls have come not only from outside professions, but from within them. For example, the editors of the *Mayo Journal of Clinical Proceedings* (Rose and Curry 2010, 302), both medical doctors, have asked:

> What if a legal stimulant that is shown to be safe could be used to improve medical care during periods of fatigue, regardless of the number of hours worked? Would not the more ethical

> choice be to promote the reduction of errors … With publication of original research … use of prescribed drugs to promote wakefulness in the absence of comorbidities, such as narcolepsy and sleep apnoea, will undoubtedly move from the military into the public arena.

In their view, the use of drugs to promote wakefulness would be the ethically correct course of action in medicine. Warren et al., surgeons writing in the *Journal of Surgical Research*, have supported such a stance and suggested that surgeons having to take enhancers may come to be required practice. They write (2009, 167–72):

> The prospect of fatigued surgeons taking a … drug … to allow them to operate for longer, and possibly to a higher standard, is perhaps not as far-fetched as some may suggest. [Modafinil] has already been trialled in emergency physicians, when performing non-medical-related tasks at the end of a nightshift.

One objection in the workplace context is that employees should not be expected to risk their health for others. Indeed, this is the general position of English law that there is no duty of easy rescue (see, e.g., *Home Office v. Dorset Yacht Co* [1970] AC 1004, 1060 (HL)). But as Warren et al. (2009, 170–1) point out, the concept of surgeons risking their health to benefit patients is not an alien one. They cite operating on patients with blood-borne transmissible diseases as an example of where the risk to the surgeon is felt justifiable to improve the patient's chances of recovery. Having noted that there are "useful and warranted forms of coercion" forcing surgeons to undertake practices such as hand washing and sterility prior to and during survey, they ask:

> What will our employers feel about a drug that makes us less prone to error, able to work longer hours, or to operate more efficiently? … Will a day arise where they can recommend or even insist on surgeons being artificially enhanced? This may seem fanciful, but recent work has suggested that a mixture of napping and caffeine attenuates fatigue in interns and thus should be adopted by hospital administration. Why not other types of stimulant?

Such practices do not, however, make legal obligations. But if enhancers became sufficiently safe, it may be that their use becomes a standard of care for a responsible medical practitioner.

That some regulators are already beginning to consider using stimulants like modafinil as a fatigue-management strategy suggests that lawyers need to begin thinking about whether such practices will be legal and how they should be addressed, if at all. One example is Queensland Health, a state medical regulatory body of one of the states of Australia, which in its Fatigue Risk Management System Resource Pack (2009) compared the use of caffeine with "other psychoactive drugs e.g. modafinil" as a means of managing the risks associated with fatigue in medical professionals. The report considers caffeine to be "supported in its use" because it is "more readily available and less expensive" than drugs like modafinil. There was no suggestion that modafinil would be problematic to use for this purpose, so it seemed implicitly to be open to that use if the cost and availability issues were to be resolved. Commenting on this, Maslen et al. (2015b) have suggested that if these problematic aspects (as well as the risk of side effects and safety concerns) no longer existed, then "it is perfectly conceivable that a future report may recommend that such drugs be used". It should be noted that in its 2012 Resource Pack, Queensland Health removed the comparison, but the point still stands and it is possible that future reviews may recommend the use of these drugs in the workplace. As Ravelingien and Sandberg (2008, e.9 p.3) suggest, "[i]t is conceivable that the availability of safe and effective wake enhancers will create or fortify a responsibility to ensure that fatigue no longer affects performance, particularly within professional contexts."

Indeed, they point out that there is already one employment arena in which the taking of modafinil can be legally compelled—the military (in the United States, at least). They note that the US Uniform Code of Military Justice requires soldiers to accept medical interventions that make them fit for duty, which might include taking cognitive enhancers. There are already reports of the use of drugs such as modafinil and Ritalin within the US Air Force, where pilots may be being indirectly coerced into enhancing (Farah 2012, 584 citing Borin 2003; Sample 2004; Spears 2003).

Is it likely that the law will oblige people to enhance outside the military? The answer may depend in part on the context. For example, Chandler (2013) has convincingly argued that the civil law might indirectly oblige surgeons to cognitively enhance, if the prevalence of their use redefines what is regarded as reasonable care practice. She comments that if many in the workplace who are responsible for the safety of others begin to use enhancers to redress cognitive deficits, this might become reasonable practice and thereby render those who fail to do so negligent. Her point, which is a strong one, is that this may well happen indirectly (Chandler 2013, 256):

> Cognitive deficits may also raise the risk of liability if they cause a physician to make errors that would not be made by the reasonably prudent practitioner in the field, or to fail to keep up with developments in the field to an extent that is considered to fall below the reasonable standard of care in the profession. In such cases, the courts would simply find there had been a failure to maintain the standard, without necessarily commenting on cognitive deficits or therapeutic methods to alleviate them.

Chandler's argument is convincing in many ways, but we should also keep in mind that at least at present, many barriers to obliging people to take enhancers, even indirectly, remain. Their safety remains uncertain, as does their efficacy. It is unlikely given this that the law will require them to be taken, even if their use becomes prevalent. Goold and Maslen (2014a) detail such concerns, and outline the public policy arguments that speak against obligating surgeons and other professionals to enhance. In their view, creating a culture of enhancement (specifically enhancements that redress the impacts of fatigue through the use of modafinil) will have problematic knock-on effects. These include increasing the load on surgeons, who are already under pressure, and also creating a situation in which surgeons accrue long-term sleep debts, which carry known health risks. Such surgeons are unlikely to be given time off to catch up sleep, they note, and hence will be expected to work as usual following a period of using an enhancer. This simply saves up fatigue for later, creating dangerous knock-on effects. They also point to the problems inherent in proving the requisite legal standard, that failing to enhance can be said to legally have caused a harm (that is, it will be difficult to actually bring a successful claim) (Goold and Maslen 2014b).

It is true that there are contexts in which people are effectively required to do things to their bodies as part of their job. The medical profession is the prime example, as medical practitioners and students must be vaccinated if they are to practice medicine (or, as students, demonstrate that they are fit to practice medicine) (General Medical Council 2009, para. 38). While vaccination protects the medic, it is also required as evidence of the capacity to protect patients and colleagues, and themselves, from infectious diseases. Practicing medical professionals are expected to be vaccinated against common serious communicable diseases (unless otherwise contraindicated), including receiving TB, measles, mumps, rubella, and varicella (General Medical Council 2013, para. 29).

But these requirements need to be balanced against the necessity to ensure that the workplace is safe for employees. The 1974 Health and Safety at Work etc Act makes it the duty of every employer to ensure, so far as is reasonably practicable, the health, safety, and welfare at work of all his or her employees (Health and Safety at Work etc Act 1974, s 2(1)). This creates an obligation on employers to ensure a safe working environment (Health and Safety at Work etc Act 1974, s 2(1)(a)), adequate information and training of staff to ensure health and safety (Health and Safety at Work etc Act 1974, s 2(1)(c)), and adequate welfare provisions for staff at work (Health and Safety at Work etc Act 1974, s 2(1)(e)). These protections are bolstered by the provision of a nondelegable duty under the common law to maintain a safe system of work (*Wilsons and Clyde Coal Ltd v. English* [1937] UKHL 2 (HL)).

If it emerges that enhancers carry risks, then such protections will apply to any employers who require their use. This does not mean that employees might not be expected to take such enhancers as part of their employment. Clearly, there are risky jobs that people take on. Vaccinations are a good example—they are not risk-free. But it does mean that there is a legal framework within which to manage expectations to take on such risks when they are placed on employees, and to ensure that any risks or side effects to enhancing are taken into account. This suggests that there is no need for new laws; those we have can encompass enhancement technologies, but it remains to be seen exactly whether that will be the case. At this stage, perhaps we might tentatively say that it might be considered unsafe to expect employees to take enhancers or use enhancing devices when the risks of doing so remain uncertain.

16.9 **Obliging people to enhance?**

An obligation to enhance might also arise outside the employment context. One suggested arena is the courtroom. For example, Sandberg et al. (2011, 281) claim:

> As our range and knowledge of cognitive enhancers increases, some enhancers should be provided, as we now provide coffee, if they are safe and do not bias judgement. Substances like modafinil may well meet this criterion. Given the stakes, we have a moral imperative to investigate ways to utilize these new technologies to improve cognitive performance in the courtroom.

In their view, jurors may be morally obliged to self-enhance to improve memory and retain the complex and voluminous information they must manage in the course of a trial. If such enhancements become sufficiently safe, then perhaps the law could require the use of enhancers as part of one's civic duty to sit on a jury. But such a move would be subject to the same concerns about bodily integrity and privacy raised throughout this chapter.

Of course, it is not only jurors who might benefit from enhancement during a trial. If we find that some enhancing drugs can affect memory, and might be capable of improving eyewitnesses' capacity to remember the events of a crime, then promoting the use of cognitive enhancement could be of value in ensuring the accuracy of prosecutions. Vedder and Klaming (2012) suggest that this may serve to further the common good. However, these arguments presuppose that we can easily decide whether cognitive enhancements will improve the accuracy of testimony when this is in fact very doubtful. Even if they can be, we must ask would it be reasonable to require witnesses to take such enhancers, even if it might make the difference to a conviction? And what if those enhancers sharpen their memories in a way that makes it harder for them to psychologically cope with the offense that they witnessed, such as a violent crime? Would it be acceptable for the law to impose such an adverse psychological impact on a witness? These questions remain unanswered, and it remains to be seen if they will in fact arise.

16.10 **Some final regulatory issues**

There are some general points to be made about how to regulate enhancers. Bostrom and Sandberg (2009) note that the current regulatory approval system for pharmaceuticals is focused on medications to treat health conditions, not on enhancing treatments. One consequence is that "drug companies can find it difficult to get regulatory approval for a pharmaceutical whose sole use is to improve cognitive functioning in the healthy population" (Bostrom and Sandberg 2009, 331). If the law wishes to support the development and production of enhancers, this will need to be addressed.

There is also the question of how the law should manage the risks associated with enhancements. At present, in most cases the law leaves us with the liberty to take risks in our lives as we choose. It does not prevent us riding horses or bungee-jumping or whatever other dangerous activity we might wish to partake in. Only in limited cases does it do so, and even then mostly by regulating the provision of the activity rather than our choice to engage in it. The use of banned drugs like cocaine and heroin, euthanasia, and some forms of consensual physical harm are the exceptions, rather than the rule (Misuse of Drugs Act 1971, s 7; *R v. Brown* [1994] 1 AC 212 (HL); Suicide Act 1961, s 2). One potential way to manage these risks is the inclusion of enhancement drugs and devices within regulatory schemes designed to ensure the safety of medical devices and drugs as noted earlier. Others, suggested in the literature (predominantly the philosophy literature rather than by legal commentators), are licensing schemes for users of cognitive-enhancing drugs, which would be provided only if the user could demonstrate sufficient understanding of the risks of the drug (Bostrom and Sandberg 2009, 332); and provision of access to enhancers only after the person had attended a course (for which he or she would have to pay) which provided information about the risks and benefits of the enhancers (Dubljević 2012). Such measures are akin to the way in which the law regulates drivers, and so the system could be modeled on the one already in place. However, there are major obstacles to these sorts of proposals, namely, international drug control treaties that prohibit the use of most stimulant drugs for other than medical and scientific purposes (see, further, Dubljević 2013). Even absent such obstacles, it is unlikely such steps would be taken given the cost of the infrastructure required, and the largely self-regarding nature of the risks of enhancement. More likely, the law would regulate via the prescription system, where medical practitioners would provide enhancers only after obtaining informed consent from the user,[2] or in the manner cigarettes are sold, with age limits and warnings on the packet.

One problem is that enhancing drugs are widely available on the internet. In many cases, the user obtains the drugs without a prescription and uses them without medical guidance. In the UK, the sale or supply of prescription drugs by a person who is not a licensed prescriber is an offense (Medicines Act 1968, s 7(2)). Only once a drug has been approved for general sale will supply by unlicensed persons be lawful (Human Medicines Regulations 2012, r. 62). Clearly, there are powers to punish such sale and supply locally, but when drugs that have not been approved for general sale are sold and supplied via the internet, it is much more difficult to control this form of supply. This is how many people avoid the controls the prescription system puts in place. This creates complex regulatory challenges. As Mossialos et al. (2010) note, the absence of a common approach to public health across the EU would make effective regulation of internet supply difficult. This leaves single jurisdictions to manage the issues via local laws, with the EU funding projects to create and disseminate information that is used in legislative processes to promote public health.

Regulatory approaches to enhancement will also be complicated by the difficulties associated with determining what actually constitutes enhancement. As Mehlman (2004) points out, it will be difficult if not impossible to draw a line between therapeutic and nontherapeutic uses of drugs. This is because the concept of enhancement is necessarily dependent on our idea of normal human capacities that are putatively improved. As most of these capacities exist on a spectrum, it will not always be clear when treatment is merely to bring a person up to a level regarded as normal (more likely to be seen as treatment of a deficit), and when it raises them above a level such that they are not treated but enhanced. IQ is a good example used by Mehlman. Would it be a treatment that brought IQ up to the population average, or would it be enhancement? Would it matter what the person's original IQ was? And how would we distinguish cases where two people with different original IQs, one below average, one above, were both raised to the same level? This brings us back to Bennett Moses' point about technology-specific regulatory schemes. If we use enhancement as the focal point of our regulation, we may hit hurdles or create unnecessary problems.

16.11 **Conclusion**

This chapter has perhaps raised more questions than it answers. As yet the future impact of cognitive-enhancing drugs on the law is difficult to ascertain. What can be said is that difficult, complex issues may arise. In some ways, cognitive enhancements challenge accepted aspects of the law's approach, as in the case of negligence and the mental element of crimes. In others, they may simply become another aspect of an area that is already subject to regulatory instruments that can potentially encompass the impacts of cognitive enhancers, such as workplace safety regulations.

There is a need for further research into the legal implications of cognitive enhancement from many perspectives and in many areas of law. We must, on the one hand, be mindful to avoid creating new schemes of regulation focused solely on cognitive enhancement, as these may become outdated as the science improves. We must also ensure that any legal developments are grounded firmly in what is technologically possible or likely to be possible. Laws based on science fiction, or ill-founded speculation (or fear) in the absence of evidence, may create as many problems as they solve. We must also ensure that any new laws respect the wider principles to which the law is committed, such as the right to bodily integrity, freedom from coercion, and protections for privacy. The newness of the science is not a reason to jettison these principles as we move forward to encompass new technologies that may potentially alter our human capacities.

Notes

1. There is also evidence to the contrary, which is examined elsewhere in this collection.
2. Although it should be noted that not all jurisdictions have a doctrine of informed consent, including the United Kingdom.

References

A v. National Blood Authority [2001] 3 All ER 289 (QB).

Academy of Medical Sciences, the Royal Society, British Academy, the Royal Academy of Engineering. 2012. *Human enhancement and the future of work.* [pdf] Place: Academy of Medical Sciences. Available at: http://www.acmedsci.ac.uk/p47prid102.html#downloads. Accessed July 22, 2014.

Bennett Moses, L. 2007. The legal landscape following technological change: Paths to adaptation. *Bulletin of Science, Technology and Society* **27**(5):408–16.

Bennett Moses, L. 2011. Sui generis rules. In: *The growing gap between emerging technologies and legal-ethical oversight: The pacing problem*, ed. G. E. Marchant, B. R. Allenby, and J. R. Heckert, 77–94. The International Library of Ethics, Law and Technology 7. Dordrecht: Springer.

Blyth v. Birmingham Waterworks Company (1856) 11 Ex Ch 781 (Ct Ex).

Bostrom, N., and A. Sandberg. 2009. Cognitive enhancement: Methods, ethics, regulatory challenges. *Science and Engineering Ethics* **15**:311–41.

Bublitz, J.-C. 2013. My mind is mine!? Cognitive liberty as a legal concept. In: *Cognitive enhancement: An interdisciplinary perspective*, ed. E. Hildt and A. G. Frank, 233–64. Dordrecht: Springer.

Bublitz, J-C., and R. Merkel. 2014. Crimes against minds: On mental manipulations, harms and a human right to mental self-determination. *Criminal Law and Philosophy* **8**:51–77.

Chandler, J. 2013. Autonomy and the unintended legal consequences of emerging neurotherapies. *Neuroethics* **6**(2):249–63.

C 300/95 ***Commission v UK*** [1997], All ER 481 (EC).

Consumer Protection Act 1987 (c. **43**). London: The Stationery Office.

Coroners and Justice Act 2009 (c. **25**). London: The Stationery Office.

Council Directive 93/42/EEC of 14 June 1993 concerning medical devices. *OJL* 169, 12/07/1993:1–43.

Council Directive 2001/83/EC of the European Parliament and of the Council of 6 November 2001 on the community code relating to medicinal products for human use. *OJL* **311**, 28/11/2004:67–128.

Council of Europe. Convention for the protection of human rights and dignity of the human being with regard to the application of biology and medicine (Convention on Human Rights and Biomedicine or the Oviedo Convention) (CETS no. 164), (Oviedo, 4 April 1997) (entered into force on December 1, 1999).

Council of Europe. The European Agreement on the Exchange of Therapeutic Substances of Human Origin (**CETS no.** 026) (Paris, December 12, 1958) (entered into force, January 1, 1959).

Council of Europe. European Convention on Human Rights (ECHR) (Strasbourg, November 4, 1950) (entered into force, September 3, 1953).

Council of Europe. The European Social Charter (CETS no. 163) (Strasbourg, May 3, 1996).

Deacon v. Canada *(Attorney General)* 2005 FC 1489, aff'd 2006 FCA 265.

Department of Health. 2006. *Improving patients' access to medicines: A guide to implementing nurse and pharmacist independent prescribing within the NHS in England.* London: HMSO.

Donoghue v. Stevenson [1932] AC 562.

Dubljević, V. 2012. Principles of justice as the basis for public policy on psychopharmacological cognitive enhancement. *Law, Innovation and Technology* **4**(1):67–83.

Dubljević, V. 2013. Prohibition or coffee shops: Regulation of amphetamine and methylphenidate for enhancement use by healthy adults. *American Journal of Bioethics* **13**(7):23–33.

Equality Act 2010 (c. 15). London: The Stationery Office.

Farah, M. 2012. Neuroethics: The ethical, legal, and societal impact of neuroscience. *Annual Review of Psychology* **63**:571–91.

Fisher, A.D., E. Bandini, H. Casale, and M. Maggi. 2011. Paraphilic disorders: Diagnosis and treatment. In: *Hormonal therapy for male sexual dysfunction*, ed. M. Maggi, Chapter 8. Chichester: Wiley.

Foc.us. (2014). *Foc.us.* [Online] Available at: http://www.foc.us. Accessed July 22, 2014.

Friedman, D. 2001–2. Does technology require new law? *Harvard Journal of Law and Public Policy* **25**:71–85.

General Medical Council. 2009. *Medical students: Professional values and fitness to practise.* [pdf] London: General Medical Council. Available at <http://www.gmc-uk.org/education/undergraduate/professional_behaviour.asp. Accessed July 22, 2014.

General Medical Council. 2013. *Good medical practice: Domain 2: Safety and quality.* [pdf] London: General Medical Council. Available at: <http://www.gmc-uk.org/guidance/good_medical_practice.asp. AccessedJuly 22, 2014.

Gillick v. West Norfolk & Wisbeck Area Health Authority [1986] AC 112 (HL).

Glass v. *UK* [2004] 1 FCR 553 (ECHR).

Goold, I., and H. Maslen. 2014a. Must the surgeon take the pill? Negligence duty in the context of cognitive enhancement. *Modern Law Review* 77(1):60–86.

Goold, I., and H. Maslen. 2014b. Responsibility enhancement and the law. In: *Handbook of neuroethics*, ed. J. Clausen and N. Levy, 1363–80. Dordrecht: Springer.

Greely, H. 2008. Neuroscience and criminal justice: not responsibility but treatment. *Kansas Law Review* **56**:1103–38.

Hagger, L., and G. Hagger Johnson. 2011. Super kids: Regulating the use of cognitive and psychological enhancement in children. *Law, Innovation and Technology* **3**(1):137–66.

Health and Safety at Work etc Act 1974 (c. **37**). London: The Stationery Office.

Home Office v. Dorset Yacht Co [1970] AC 1004, 1060 (HL).

Human Medicines Regulations 2012 (no. 1916). London: The Stationery Office.

Human Rights Act 1998 (c. **42**). London: The Stationery Office.

McHale v. Watson (1966) **115** CLR **199**, 213 (HCA).

Maslen, H., T. Douglas, R. C. Kadosh, N. Levy, and J. Savulescu. 2014. *Mind machines: The regulation of cognitive enhancement devices*. Oxford Martin Policy Paper. Oxford: University of Oxford. Available at: http://www.oxfordmartin.ox.ac.uk/downloads/briefings/Mind_Machines.pdf. Accessed July 22, 2014.

Maslen, H., T. Douglas, R. C. Kadosh, N. Levy, and J. Savulescu. 2015a. Do-it-yourself brain stimulation: a regulatory model. *Journal of Medical Ethics* **41**(5):413–14.

Maslen, H., F. Santoni de Sio, and N. Faber. 2015b. With cognitive enhancement comes great responsibility?. In *Responsible innovation 2: Concepts, approaches and applications*, ed. B.-J. Koops, I. Oosterlaken, H. Romijn, T. Swierstra, and J. van den Hoven, Chapter 7. Amsterdam: Springer.

Medicines Act 1968 (c. 67) London: Department of Health.

Medicines for Human Use (Clinical Trials) Regulations 2004 (SI 2004 no. 1031). London: The Stationery Office.

Mehlman, M. 2004. Cognition-enhancing drugs. *Milbank Quarterly* **82**(3):483–506.

Mental Capacity Act 2005 (c. 9). London: The Stationery Office.

Mossialos, E., G. Permanand, R. Baeten, and T. K. Hervey. 2010. *Health systems governance in Europe: The role of EU law and policy.* 2010. Cambridge: Cambridge University Press.

Nettleship v. Weston [1971] 2 QB 691, 699 (CA).

Pariotti, E., and D. Ruggiu. 2012. Governing nanotechnologies in Europe: Human rights, soft law, and corporate social responsibility. In: *Little by Little: Expansions of Nanoscience and Emerging Technologies*, ed. H. Van Lente, et al. Heidelberg: IOS Press/AKA-Verlag.

Pretty v. United Kingdom [2002] ECHR 427 (ECHR).

Queensland Health. 2009. *Fatigue Risk Management System Resource Pack.* [pdf] Brisbane: Queensland Health. Available at: http://enhancingresponsibility.com/wp-content/uploads/2014/01/Queensland-Health-Fatigue-Risk-Management-System-resource-pack-2009.pdf. Accessed October 23, 2015.

R v. Bennett [1995] Crim LR 877 (CA).

R v. Brown [1994] 1 AC 212 (HL).

R v. Cunningham [1957] 2 QB 396 (CA).

R v. Kingston [1995] 2 AC 355 (HL).

R v. Moloney [1985] AC 905 (HL).

R v. Remi Akinyeme [2007] EWCA Crim 3290.

Re MB (An Adult: Medical Treatment) [1997] 2 FCR 541 (CA).

Re W (A Minor) (Medical Treatment: Court's Jurisdiction) [1993] FLR 64 (Fam).

Ravelingien, A., and A. Sandberg. 2008. Sleep better than medicine? Ethical issues related to "wake enhancement". *Journal of Medical Ethics* **34**(9) e9.

Richardson v. LRC Products Ltd [2000] 59 BMLR 185 (QB).

Rose, S., and T. Curry. 2010. Fatigue countermeasures, and performance enhancement in resident physicians—reply. *Mayo Clinic Proceedings* **85**(3):301–2.

Russell, S. 1996–7. Castration of repeat sexual offenders: an international comparative analysis. *Houston Journal of International Law* **19**:425–59.

Sale of Goods Act 1979 (c. 54). London: The Stationery Office.

Sample, I. 2004. Wired awake. *The Guardian*, 29 July. Available at: http://www.theguardian.com/education/2004/jul/29/research.highereducation. AccessedOctober 29, 2015.

Sandberg, A., W. Sinnott-Armstrong, and J. Savulescu. 2011. Cognitive enhancement in courts. In: *Oxford Handbook of Neuroethics*, ed. J. Illes and B. J. Sahakian, Chapter 17. Oxford: Oxford University Press.

Singh, I., and K. Kelleher. 2010. Neuroenhancement in young people: Proposal for research, policy, and clinical management. *AJOB Neuroscience* **1**(1):3–16.

Smith v. Littlewoods [1987] AC 241, 271 (HL).

Spears, T. 2003. New drug may help soldiers stay awake: Doctors unsure of long-term effect. *The Ottawa Citizen*, 11 October. Available at: https://modapharma.com/article/soldiers.html. Accessed October 29, 2015.

Stinneford, J. F. 2008. Incapacitation through maiming: chemical castration, the eighth amendment, and the denial of human dignity. *University of St. Thomas Law Journal* 3:559–99.

Stovin v. Wise [1996] AC 923 (HL).

Vanderzyl, K. A. 1994–5. Castration as an alternative to incarceration: an impotent approach to the punishment of sex offenders. *Northern Illinois University Law Review* **15**:107–40.

Vedder, A., and L. Klaming. 2012. Human enhancement for the common good: Using neurotechnologies to improve eyewitness memory. *Tilburg Law School Legal Studies Research Paper Series*, no. **14**/2012:22–33.

Vincent, N. 2011. The challenges posed to private law by emerging cognitive enhancement technologies. In: *The law of the future and the future of law*, ed. S. Muller, S. Zouridis, M. Frishman, and L. Kistemaker, Chapter 6.1. Oslo: Torkel Opsahl Academic EPublisher.

Warren, O. J., D. R. Leff, T. Athanasiou, C. Kennard, and A. Darzi. 2009. The neurocognitive enhancement of surgeons: an ethical perspective. *Journal of Surgical Research* **152**(1):167–72.

Wilsons and Clyde Coal Ltd v. English [1937] UKHL 2 (HL).

Worsley v. Tambrands [2000] PIQR 95 (QB).

XYZ v. Schering Health Care Ltd [2002] EWHC 1420 (QB).

YF v. Turkey (2004) 39 EHRR 34 (ECHR).

Chapter 17

Students and "smart drugs": Empirical research can shed light on enhancement enthusiasm

Brad Partridge

Recent enthusiasm about pharmacological cognitive enhancement has often focused on the use of prescription stimulants as a study aid by US university students. Despite the illegality (in most countries) of nonprescription use of stimulants and the potential for addiction, it is often claimed that healthy students without any diagnosed disorder are increasingly using drugs such as methylphenidate and modafinil to increase their alertness, concentration, or memory, in the belief that they will improve their performance during examinations or when studying (Greely et al. 2008).

Cognitive enhancement in this sense is really a form of nonmedical use of prescription drugs, in that the drug is not being used for its indicated purpose and is not being used "off-label" to treat some other medical condition (Boot et al. 2011). Other terms used to describe this practice include "neuroenhancement" and "cosmetic neurology" (Chatterjee 2004). The latter implies a discretionary, nontherapeutic motive to be better than wellness that distinguishes it from "treatment"—analogous to other bodily enhancement practices such as botox injections (to remove wrinkles), breast enlargement surgery, and the use of steroids by athletes.

Terms such as "smart drug" are somewhat loaded when referring to currently available pharmaceutical stimulants, because this tends to assume their efficacy for the purpose of improving cognition in healthy people. The assumed efficacy of so-called smart drugs and their use by US students has fueled hyped media portrayals of the practice, and led some to conclude that this is a trend increasing in universities across the globe (Partridge et al. 2011). The perception that the enhancement use of stimulants is common and increasing might give rise to one of two regulatory positions. Those who are alarmed by it may argue for measures to restrict or prohibit this practice, perhaps beyond the current prescription system. Alternatively, those who are excited by the prospect of enhancing cognitive capacities may argue for policies that may facilitate the practice. For example, a number of prominent bioethicists and neuroscientists have recommended a relaxation of laws that prohibit nonprescription use of stimulants and suggest that pharmaceutical companies ought to be permitted to market drugs to healthy people for cognitive enhancement (Greely et al. 2008).

17.1 Do smart drugs enhance cognition?

In healthy people, stimulants may help a person stay awake, but there is only weak evidence that they improve cognition, or prevent it from deteriorating. A recent review of studies exploring the enhancing effects of prescription stimulants found the evidence to be equivocal—there were some improvements on various cognitive tasks; however, findings are mixed with many studies showing no improvement and even impairment (Smith and Farah 2011). The evidence for the enhancing effects of prescription stimulants is even less convincing given that there may be a bias in the literature toward publishing positive findings, and there is considerable variation in study design, participant numbers, dosage, and tasks used to measure "cognition" (Smith and Farah 2011). It is not clear how much application many of the studies reviewed have to real-world situations, or whether improvements are the result of a placebo effect or possibly improvements in mood rather than cognition per se. A systematic review by Repantis and colleagues found that in healthy people, methylphenidate may improve memory, but there was no consistent evidence for other cognitive enhancements (Repantis et al. 2010). Similarly underwhelming results were found for modafinil:

> Modafinil ... was found to improve attention for well-rested individuals ... but repeated doses were unable to prevent deterioration of cognitive performance ... possibly inducing overconfidence in a person's own cognitive performance. (Repantis et al. 2010)

Of concern are the possible side effects of using stimulants for nonmedical purposes. Amphetamines certainly have the potential for addiction and other side effects (e.g., heart/blood pressures problems, anxiety) (Bell et al. 2012), and there is also evidence that modafinil has addictive potential (Mohamed 2012).

17.2 Student use of smart drugs

A number of influential papers have described an apparent increase in the use of drugs for cognitive enhancement, particularly by students and academics (Greely et al. 2008; Maher 2008; Sahakian and Morein-Zamir 2007). In the academic literature and media, one of the most commonly cited pieces of evidence for widespread cognitive enhancement is a poll conducted by *Nature* in 2008 (Maher 2008). However, its findings can be misleading and its limitations are significant. The poll was an informal online survey of 1420 self-selected readers of *Nature*, with the headline finding that 20% of respondents had ever used a drug for cognitive enhancement, most commonly modafinil and methylphenidate. The survey was "triggered by a commentary by behavioural neuroscientists Barbara Sahakian and Sharon Morein-Zamir, of the University of Cambridge, UK, who had surveyed their colleagues on the use of drugs that purportedly enhance focus and attention." That commentary was provocatively titled "Professor's Little Helper" and claimed that not only was nonprescription use by the general public becoming more commonplace, but also

> In academia we know that a number of our colleagues in the United States and United Kingdom already use modafinil to counteract the effects of jetlag, to enhance productivity or mental energy, or to deal with demanding or important intellectual challenges. (Sahakian and Morein-Zamir 2007 , p. 1178)

Interestingly, when the 20% of respondents to the *Nature* poll who had engaged in cognitive enhancement were asked why they used the drug, among the top responses were to improve concentration, alleviate jetlag, and "[t]o actually see if there was any validity to the afore-mentioned article [by Sahakian and Morein-Zamir]" (Maher 2008). The *Nature* poll, although widely referred to, hardly represents an accurate or reliable piece of evidence about the true prevalence of cognitive enhancement among academics, students, or the general public.

Recent editorials have called for more empirical evidence about the use of drugs for cognitive enhancement in order to better inform policy discussions (Lucke 2012). The largest survey of the nonmedical use of prescription stimulants among US college students comprised over 10,000 students from 119 different US colleges (McCabe et al. 2005). That study found that lifetime prevalence was 6.9% and past-month prevalence was 2.1%, but did not ask why students used stimulants nonmedically—presumably it could include purposes such as neuroenhancement, recereation, weight loss, or self-treatment. Some studies have found higher prevalence rates, but they tend to have smaller samples or survey only a small number of institutions (Smith and Farah 2011). The results of US studies indicate that nonmedical users of prescription stimulants are rarely older than 25 years, use other illicit drugs at higher rates, tend to live in fraternities/sororities, and have lower grade point averages.

But we should not assume that data from US college students are globally transferrable. A fair appraisal of the country-specific cultural and regulatory factors that may influence the extent of cognitive enhancement and attitudes toward it will help to ensure that ethical discussions are well informed, and policy directions are responsible. Encouragingly, a number of empirical studies exploring attitudes and prevalence have been conducted outside the USA, meaning that a more global perspective is beginning to take shape. These studies tend to indicate a lower prevalence of nonmedical use of stimulants for cognitive enhancement compared with that in the US; however, there is little evidence from Asian countries about the use of pharmacological cognitive enhancers.

A survey of 1547 German students and pupils by Franke and colleagues found very low rates of prevalence (Franke et al. 2011). Lifetime prevalence of nonmedical prescription stimulant use for cognitive enhancement was only 1.29%, and past month prevalence was 0.06%. Another German survey of over 1000 university students by a separate group of researchers found similarly low levels of prescription stimulants use for cognitive enhancement (Forlini and Racine 2012). Only 2% had ever used methylphenidate for cognitive enhancement, 1.1% had used amphetamines, and 0.6% had used modafinil. A recent study of Italian students found higher rates of prevalence but the study had a number of limitations (Castaldi et al. 2012). While 16% had ever used a "cognitive enhancement medication," the study included only 77 participants from one university. Furthermore, it is not clear that the enhancing "medications" were prescription stimulants. Of the 12 (out of 77) who indicated they had ever engaged in cognitive enhancement, 8 claimed to acquire the drug from a pharmacy *without* a prescription (Castaldi et al. 2012). This would indicate that drugs other than methylphenidate, modafinil, and amphetamines were mostly being used.

Aside from students, there have been some estimations of nonmedical use of prescription stimulants among the US general population using data from the National Survey of Drug Use and Health. One study of over 68,000 members of the US public found fairly low rates of lifetime nonmedical Ritalin and Adderall use, and included use for any nonmedical reasons (Pilkinton and Cannatella 2012). Our research team at the University of Queensland recently gathered data about the prevalence of cognitive enhancement among the Australian general public through the Queensland Social Survey (Partridge et al. 2012b). We surveyed 1,265 adults from Queensland about their use of prescription drugs—not for a diagnosed disorder—but to improve concentration or alertness. Only 2.4% had ever done so, and a further 8% knew someone who had. Slightly higher rates of use (6%) were found among those aged 18–34 (Partridge et al. 2012b). In Australia, the prevalence of cognitive enhancement among university students is unknown and a recent qualitative study with Queensland students found that the perceived use of prescription stimulants was typically low (Partridge et al. 2012a).

Misperceptions of prevalence can, however, occur. McCabe surveyed of over 3500 students at a large US university (McCabe 2008). Prevalence of nonmedical stimulant use in this sample was slightly higher than their previous study—6% in the past year. They then asked participants to estimate the prevalence of nonmedical stimulant use by their peers. The average estimate was 20%, and those students who had actually engaged in nonmedical stimulant use estimated it to be even higher—34%.

17.3 The media and smart drugs

Researchers often express concerns about media reporting of neuroscience applications because misleading media reports can unrealistically raise (or lower) public expectations about their potential harms and benefits. Forlini and Racine (2009b) found that media reports about neuroenhancement often displayed enthusiasm about the potential effectiveness of drugs as well as giving the impression that the practice was common. A larger study of newspaper portrayals of neuroenhancement found that 95% of articles mentioned at least one benefit of using prescription stimulants for enhancement, although only 58% mentioned at least one side effect (Partridge et al. 2011). As many as 82% portrayed cognitive enhancement as common, increasing, or both. Interestingly, two-thirds of the newspaper articles referred to the academic literature to support these claims. However, the most commonly referred to papers from the literature were bioethics commentaries, which tended to be enthusiastic about the prospect of neuroenhancement and give the impression that the practice is widespread (Partridge et al. 2011). Rarely did newspaper articles cite empirical surveys of nonmedical prescription drug use when describing the prevalence of cognitive enhancement. Poor reporting about the effects and prevalence of cognitive enhancement may unwittingly normalize such drug use and it would be wise for neuroscientists and bioethicists to help debunk uninformed media reports that may encourage the nonmedical use of drugs for cognitive enhancement.

17.4 **Attitudes toward cognitive enhancement**

Aside from evidence of the potential benefits and risks of using purported cognitive enhancers, and how many people are using them, there have been a small number of studies exploring attitudes toward the practice. Although wanting to be smarter and more productive is fairly uncontroversial, the desire to pursue these goals by using currently available prescription stimulants is not. Only 7% of members of the Australian public thought that it is acceptable for healthy people to use prescription drugs to enhance their concentration or alertness (Partridge et al. 2012b). However, the same sample exhibited much higher rates of support for drug treatment of disorders such as attention deficit hyperactive disorder (ADHD) and depression (Partridge et al. 2012c), suggesting that lay people may make a moral distinction between enhancement and therapy. Interestingly, people who had used prescription drugs for cognitive enhancement (or knew someone who had) were more likely to think the practice is acceptable.

Student attitudes toward stimulant use for enhancement are mixed. In their sample of over 1500 German students and pupils (that had low rates of prevalence), Franke and colleagues found that 16% would not take cognitive enhancers in any circumstances (Franke et al. 2012). However, they did find that 80% would consider it if the drug was guaranteed to be safe and did not lead to addiction. This may indicate a substantial degree of interest in cognitive enhancement, but crucially, 95% of participants thought that currently available substances would lead to addiction.

Data on student attitudes from other parts of the world have come mostly from qualitative studies. A series of interviews with Canadian students, parents, and healthcare professionals found a considerable amount of ambivalence in attitudes, but stakeholder groups converged on a number of themes (Forlini and Racine 2009a, 2010, 2012). First, cognitive enhancement was often regarded as a matter of individual choice, although people also thought that there are tremendous social pressures to succeed and this may play a role in the decision to use drugs for enhancement. It seems that depending on the circumstances, we might expect to have conditions that facilitate people's willingness to engage in cognitive enhancement. Students also emphasized the importance of personal integrity and this clearly relates to beliefs about whether an "enhanced" performance is a genuine one.

Qualitative interviews with Australian students found that cognitive enhancement was thought to be uncommon, although a small number of participants perceived it to be very common among certain friendship groups (Partridge et al. 2012a). This may indicate the use of drugs for cognitive enhancement is not a uniform practice among students, but rather occurs in concentration—more comprehensive surveys with larger numbers of students will help to explore this. Most students doubted the effectiveness of prescription stimulants as a study aid, but nevertheless thought it was unfair for others to use. When asked why a student would engage in cognitive enhancement, two main motivations related to study emerged:

(1) To "get ahead"—a way for already high-performing students to keep succeeding; and

(2) To "keep up"—a method of coping and cramming for students who are struggling to meet the demands of their course.

17.5 What should we do about the use of stimulants for cognitive enhancement?

We should be aware of the potential risks to health from the nonmedical use of prescription drugs. Uncritical appraisals about the prevalence and risk–benefit profile of drugs for cognitive enhancement could give rise to unwarranted policy decisions about the practice. For example, facilitating the practice by removing laws that prohibit the use of stimulants without a prescription (Greely et al. 2008) assumes that cognitive enhancement is likely to beneficial to the user and society. But this would appear to ignore the public health imperative that underpins regulation of these drugs in the first place.

Conversely, calls for tighter regulations can also be unwarranted if not grounded in evidence. One speculated measure is for universities to drug test students prior to examinations, just as professional athletes are dope tested. However, aside from the logistical difficulties that might impede the implementation of such a policy, it also raises serious privacy concerns for those involved. A more pertinent criticism is that such a policy would assume that currently available "cognitive enhancers" do in fact improve exam performance, giving users an unfair advantage over nonusers, and that there are a large number of users that can be "caught" this way. And yet there is little evidence that any of these things are true. In fact, in some situations, such a policy might only increase the prevalence of cognitive enhancement by giving nonusers the impression that stimulants do improve performance and that many of their colleagues are using them.

We should also be cognizant of what a "cognitively enhanced society" that has a greater capacity to work might be like. In some industries, increased working hours have already resulted in unhealthy work practices that impinge upon other important responsibilities and goals such as family, interpersonal relationships, recreation, and physical well-being. Policy-makers should be aware that pharmacological cognitive enhancement may be attractive because of the ease that a "solution in a pill" seems to offer. Although there appear to be low rates of prevalence in a number of countries, it does not stand to reason that this will always be case. Given this, healthier ways of maintaining optimum cognitive functioning should be promoted such as adequate sleep, physical exercise, a balanced diet, reduced substance use (alcohol and other drugs), and good mental health (Lucke and Partridge 2013). There is considerable evidence that modification of these lifestyle factors can contribute to good cognitive functioning, and unlike nonmedical use of stimulants, they also benefit physical health.

Neuroscientists and bioethicists should be careful not to encourage potentially harmful use of prescription drugs through uncritical dialogue about their benefits and prevalence. Discussions of "best case scenarios" can be misleading when making real policy decisions—an effective drug taken by many healthy students to markedly improve cognitive capacities and exam performance without any risk of side effects bears little resemblance to currently available purported "cognitive enhancers" and ignores the history of many substances (e.g., amphetamines, cocaine) that were once touted as smart drugs (Bell et al. 2012).

17.6 Acknowledgments

Bradley Partridge is supported by a National Health and Medical Research Council (NHMRC) post-doctoral fellowship. This chapter is an amalgamation of two previously published articles: 'Deflating the Neuroenhancement Bubble' by Jayne C. Lucke, Stephanie Bell, and Brad Partridge, *AJOB Neuroscience* Volume 4, Issue 2, pp. 38–43, with permission of the publisher (Taylor & Francis Ltd, http://www.tandfonline.com) and 'Students and "Smart Drugs": Empirical Research Can Shed Light on Enhancement Enthusiams', by Bradley Partridge, *Asian Bioethics Review*, Volume 4, Issue 4, pp. 310–319, with permission of the publisher (The Hastings Center, http://www.thehastingscenter.org).

References

Bell, S. K., J. C. Lucke, and W. D. Hall. 2012. Lessons for enhancement from the history of cocaine and amphetamine use. *AJOB Neuroscience* **3**(2):24–9.

Boot, B. P., B. Partridge, and W. Hall. 2011. Better evidence for safety and efficacy is needed before neurologists prescribe drugs for neuroenhancement to healthy people. *Neurocase* 1–4.

Castaldi, S., U. Gelatti, G. Orizio, U. Hartung, A. M. Moreno-Londono, M. Nobile, and P. J. Schulz. 2012. Use of cognitive enhancement medication among northern Italian university students. *Journal of Addiction Medicine* **6**(2):112–27.

Chatterjee, A. 2004. Cosmetic neurology: The controversy over enhancing movement, mentation, and mood. *Neurology* **63**(6):968–74.

Forlini, C., and E. Racine. 2009a. Autonomy and coercion in academic "cognitive enhancement" using methylphenidate: Perspectives of key stakeholders. *Neuroethics* **2**(3):163–77.

Forlini, C., and E. Racine. 2009b. Disagreements with implications: Diverging discourses on the ethics of non-medical use of methylphenidate for performance enhancement. *BMC Medical Ethics* **10**:9.

Forlini, C., and E. Racine. 2010. Stakeholder perspectives and reactions to "academic" cognitive enhancement: Unsuspected meaning of ambivalence and analogies. *Public Understanding of Science* **21**: 606–625.

Forlini, C., and E. Racine. 2012. Added stakeholders, added value(s) to the cognitive enhancement debate: Are academic discourse and professional policies sidestepping values of stakeholders? *AJOB Primary Research* 3(1):33–47.

Franke, A. G., C. Bonertz, M. Christmann, S. Engeser, and K. Lieb. 2012. Attitudes toward cognitive enhancement in users and nonusers of stimulants for cognitive enhancement: A pilot study. *AJOB Primary Research* 3(1):48–57.

Franke, A. G., C. Bonertz, M. Christmann, M. Huss, A. Fellgiebel, E. Hildt, and K. Lieb. 2011. Non-medical use of prescription stimulants and illicit use of stimulants for cognitive enhancement in pupils and students in Germany. *Pharmacopsychiatry* **44**(2):60–6.

Greely, H., B. Sahakian, J. Harris, R. C. Kessler, M. Gazzaniga, P. Campbell, and M. J. Farah. 2008. Towards responsible use of cognitive-enhancing drugs by the healthy. *Nature* **456**(7223):702–5.

Lucke, J., and B. Partridge. 2013. Towards a smart population: A public health framework for cognitive enhancement. *Neuroethics* **6**:419–27.

Lucke, J. C. 2012. Empirical research on attitudes toward cognitive enhancement is essential to inform policy and practice guidelines. *AJOB Primary Research* **3**(1):58–60.

McCabe, S. E. 2008. Misperceptions of non-medical prescription drug use: a web survey of college students. *Addictive Behaviors* **33**(5):713–24.

McCabe, S. E., J. R. Knight, C. J. Teter, and H. Wechsler. 2005. Non-medical use of prescription stimulants among US college students: prevalence and correlates from a national survey. *Addiction* **100**(1):96–106.

Maher, B. 2008. Poll results: Look who's doping. *Nature* **452**(7188):674–5.

Mohamed, A. D. 2012. Modafinil has the potential for addiction. *AJOB Neuroscience* **3**(2):36–8.

Partridge, B., S. Bell, J. Lucke, and W. Hall. 2012a. Australian University students' attitudes towards the use of prescription stimulants as cognitive enhancers: perceived patterns of use, efficacy and safety. *Drug and Alcohol Review* **32**(3):295–302.

Partridge, B., J. Lucke, and W. Hall. 2012b A comparison of attitudes toward cognitive enhancement and legalized doping in sport in a community sample of Australian adults. *AJOB Primary Research* 3(4):81–6.

Partridge, B., J. Lucke, and W. Hall. 2012c. Public attitudes towards the acceptability of using drugs to treat depression and ADHD. *Australian and New Zealand Journal of Psychiatry* **46**(10):958–65.

Partridge, B. J., S. K. Bell, J. C. Lucke, S. Yeates, and W. D. Hall. 2011. Smart drugs "as common as coffee": Media hype about neuroenhancement. *PLoS ONE* **6**(11):e28416.

Pilkinton, M., and A. Cannatella. 2012. Nonmedical use of prescription stimulants: Age, race, gender, and educational attainment patterns. *Journal of Human Behavior in the Social Environment* **22**(4):409–20.

Repantis, D., P. Schlattmann, O. Laisney, and I. Heuser. 2010. Modafinil and methylphenidate for neuroenhancement in healthy individuals: A systematic review. *Pharmacological Research* **62**(3):187–206.

Sahakian, B., and S. Morein-Zamir. 2007. Professor's little helper. *Nature* **450**(7173):1157–9.

Smith, M. E., and M. J. Farah. 2011. Are prescription stimulants "smart pills"? The epidemiology and cognitive neuroscience of prescription stimulant use by normal healthy individuals. *Psychological Bulletin* **137**(5):717–41.

Chapter 18

Lessons for enhancement from the history of cocaine and amphetamine use

Stephanie Bell, Jayne Lucke, and Wayne Hall

An increased understanding of the neurobiological actions of neuropharmaceutical drugs has raised the possibility that drugs can be developed that will modify brain processes in ways that enhance memory, mood, and attention, in people who do not have an illness or disorder that impairs their mental functioning. This type of use has been termed "cognitive enhancement." In recent years, articles exploring and advocating for the use of pharmaceuticals for such purposes have been published (e.g., Cakic 2009; Sahakian and Morein-Zamir 2010). Articles in prestigious journals, such as *Nature*, have suggested that "cognitive enhancement" use of stimulant drugs is already common and increasing (Greely et al. 2008; Maher 2008). Many of these articles implicitly support the use of these drugs to improve concentration and productivity and alleviate the symptoms of sleep deprivation by suggesting ways in which such use may be allowed or facilitated.

These claims for the cognitive enhancement effects of methylphenidate and modafinil echo earlier claims for the nonmedical use of the older central nervous system (CNS)-stimulant drugs cocaine and amphetamines to improve concentration and attention in normal persons. This article outlines the history of such use. We argue that experience with these drugs provides strong grounds for caution when advocating or implicitly supporting the cognitive enhancement use of stimulant drugs such as methylphenidate. There is an absence of empirical information on the cognitive enhancement effects and the risks and benefits of using psychostimulant pharmaceuticals by healthy individuals for cognitive enhancement. The extent of their use for these purposes needs to be better studied and carefully monitored.

18.1 Cocaine

Cocaine was originally used by Peruvian Indians, who chewed the leaf of the coca plant to enhance endurance when laboring at high altitude (Haddad 1979). The coca leaves were believed to have enhancing properties that increased individuals' stamina (Cohen 1975). Observations of such effects among South American users led to the extraction and characterization of the alkaloid cocaine by Albert Nieman (Haddad 1979), and this was introduced into medical practice in the industrialized world at the end of the nineteenth century (Grinspoon and Bakalar 1981). In 1884 cocaine was first successfully used as a local anaesthetic in eye surgery by Koller (Greely et al. 2008; Karch 2006).

18.1.1 Lessons from history for enhancement

Its use soon expanded well beyond this indication (Grinspoon and Bakalar 1981), as cocaine came to be seen as the drug of choice to treat asthma, toothaches, and depression, and, in 1897, morphine addiction (Goldstein et al. 2009; Gootenberg 1999). The use of cocaine was enthusiastically promoted by the pharmaceutical company Parke-Davis, and by many physicians who emphasized its benefits and minimized risks arising from its use.

Cocaine became a medically approved drug for alleviating "overwork and exhaustion." Pharmaceutical companies claimed that it "could make the coward brave, the silent eloquent and render the sufferer insensitive to pain" (Goldstein et al. 2009, 7). Cocaine was portrayed as a cultural necessity in a rapidly developing modern society that was seen as increasingly hectic and fast-paced (Gootenberg 1999). It gave people energy, made them feel happy and satisfied, and made them better able to carry out the demanding tasks of everyday life (Grinspoon and Bakalar 1981). Users reported improved mental performance and the ability to work for extended periods of time with sustained high levels of attention (Spillane 2000). These effects are very similar to claims nowadays made about the cognitive enhancement effects of drugs such as methylphenidate. For example, student populations identify reducing fatigue and increasing cognition, interest, and memory as the primary motivators for nonmedical use of stimulant drugs prescribed for attention deficit hyperactivity disorder (ADHD) (DeSantis et al. 2008).

Given the broad range of medical and quasi-medical indications for cocaine, it is unsurprising that the medical community perceived it as a major scientific accomplishment that would greatly improve health and well-being (Spillane 2000). Parke-Davis asserted that "if these claims are substantiated [cocaine] will indeed be the most important therapeutic discovery of the age, the benefit of which to humanity will be incalculable" (Grinspoon and Bakalar 1981, 152). Cocaine was universally endorsed as a benefit to society by medical journals, the Pope, and prime ministers (Goldstein et al. 2009; Grinspoon and Bakalar 1981). Positive attitudes toward cocaine and its use in such a variety of situations cemented its routine place in medicine (Grinspoon and Bakalar 1981). Among the enthusiastic advocates of cocaine use was a young Sigmund Freud, who lyrically described the effects of cocaine:

> After a short time (10–20 minutes), he feels as though he had been raised to the full height of intellectual and bodily vigor, in a state of euphoria, which is distinguished from the euphoria after consumption of alcohol by the absence of any feeling of alteration. However astonishing this effect of ingestion of coca may be, the absence of signs that could distinguish the state from the normal euphoria of good health makes it even more likely that we will underestimate it … one can perform mental and physical work with great endurance, and the otherwise urgent needs of rest, food and sleep are thrust aside as it were. (Byck 1974, 114)

Cocaine use followed a typical drug use cycle. Increased prescribing and nonmedical use exposed large numbers of people to the drug often enough and for long enough for its adverse effects to became apparent.

Medical awareness gradually increased about the adverse effects of regular use, especially by injection. Hallucinations and delusions in heavy cocaine users were recognized (Gootenberg 1999), followed by addiction during the first decade of the twentieth century (Goldstein et al. 2009). These effects prompted a more critical perspective toward cocaine use in the medical profession, whose members became less inclined to prescribe it. Publicity about its adverse effects in medical journals and the popular press led to a decline in the initiation of new medical and nonmedical users and increased cessation among existing users (Musto 1992).

Societal concerns were enhanced by the popularity of recreational cocaine use among antisocial and socially marginalized groups. Traditionally, the cost of cocaine meant that use was largely the preserve of well-to-do members of society, but as its use spread beyond these groups, it became strongly associated with recreational users in the criminal milieu and the sex industry (Gootenberg 1999). The social profile of a cocaine addict changed from that of an otherwise respected member of the community who had become addicted as a result of medical treatment to a young male who was lacking in life direction and responsibility and who engaged in antisocial behavior and crime (Jonnes 1995). Regular users of cocaine were seen as "coke drunks," a changing perception that encouraged increasingly negative societal reactions to cocaine use and prompted policies that prohibited its nonmedical use (Gootenberg 1999).

Increased societal concern and medical awareness of harm prompted restrictive regulations that reduced access to cocaine and thereby contributed to the decline in its use. In the United States, the Harrison Narcotics Tax Act was passed in 1914 as a first step toward controlling and regulating a number of problematic drugs that included cocaine and heroin (Goldstein et al. 2009). By the 1920s the medical use of cocaine had declined and cocaine was primarily considered a drug of abuse among socially marginal groups (Greely et al. 2008; Haddad 1979). Cocaine's decline was particularly pronounced in the 1930s. As cocaine use dropped, the medical use of newer pharmaceutical stimulants, the amphetamines, became increasingly popular and came to play a role similar to that of cocaine (Goldstein et al. 2009).

18.2 **Amphetamines**

The amphetamines—dexamphetamine and methamphetamine—that were approved for medical use in the late 1930s by the American Medical Association followed a cycle similar to that of cocaine. Doctors began to prescribe them for an increasing range of medical indications, and they were eventually seen as sufficiently safe to be sold over the counter to be used for a wide variety of purposes without medical supervision (Fleming 1998).

Amphetamines were the first medication used to treat mood disorders (Rasmussen 2008). They came to be regarded as mood elevators, not just for those who were depressed but for persons who needed "a pick-me-up" (Rasmussen 2008). Their medical applications expanded to include relief of nasal congestion, the treatment of narcolepsy, and obesity, because they suppressed appetite (Iversen 2006).

Amphetamines became widely available and were promoted as safe substances that would increase energy and improve alertness in ordinary people (Rasmussen 2008). Like cocaine before them, amphetamines were seen as essential for maintaining optimal performance in an increasingly fast-paced modern life (Iversen 2006). Similarly, contemporary students who take modern stimulants (such as methylphenidate) for cognitive-enhancing purposes report the need to take these drugs to keep up with expectations and requirements of a university (DeSantis et al. 2008).

Amphetamines were widely used in the military forces during World War II. US troops received regular supplies of dexamphetamine to maintain alertness and performance in the face of sleep deprivation under battle conditions (Iversen 2006). Supply was not well controlled, and soldiers used these drugs at their own discretion (Rasmussen 2008) in "quasi-medical efforts to tune mind and body beyond normal human capabilities" (Rasmussen 2008, 3). An article published in *The Lancet* in 1936 reported increases in intelligence tests scores that were attributed to amphetamine use (Rasmussen 2008).

Prescriptions and sales of amphetamines continued to rise during the 1940s and 1950s as social acceptance of nonmedical amphetamine use increased. By the mid-1940s population use was equivalent to more than two tablets for every adult per annum in the United States. As consumption grew in the 1950s and 1960s, the adverse effects of regular amphetamine use also became evident. Servicemen had returned home addicted to these drugs after repeated use under wartime conditions. The recognition of the adverse health consequences including addiction did not occur until late in the 1960s, when use of amphetamine peaked. At this point 1 in 20 Americans had a prescription for the drug, and half as many again were estimated to be using amphetamines without a prescription (Rasmussen 2008).

As the negative effects of amphetamine use became clearer in the late 1960s—most notably, drug-induced psychoses and dependence—disapproval of their use increased. A decline in amphetamine prescriptions followed, initially voluntarily by prescribers, and in the following decade as a result of US government policies that restricted supply and discouraged their prescription.

Off-label use of amphetamines and similar substances by students remained popular in cramming for examinations. Early reports indicated that suppressing tiredness and enhancing performance was one of the most common reasons students gave for their use. As reported by a student who took Benzedrine prior to undertaking an exam, "My mind worked very rapidly and seemed to be able to consider one idea after another with great speed" (Grinspoon and Hedblom 1975, 88).

Staying awake to study and concentrate on work are also the two most common motivations for the cognitive enhancement use of prescription stimulants in the student population (DeSantis et al. 2009). Students have been quoted as stating, "It helps me cram if I stay up all night" (DeSantis et al. 2008).

The reasons for the decline in amphetamine resembled those for cocaine, namely, a combination of changing characteristics of users and increased recognition of the adverse effects of nonmedical use. Illegal use of drugs such as "speed" (amphetamines) continued after increased restrictions on use, and stimulant use came to be seen as another contributor to illegal drug culture and its associated social problems.

The history of cocaine and amphetamines draws attention to the recurrent cycle of initial acceptance and enthusiasm for a new psychoactive drug, followed by later rejection as adverse effects become apparent. Recent claims about the cognitive enhancement use of methylphenidate and modafinil show a worrying similarity to those that were made for cocaine and amphetamines. It is accordingly important to have information about the risks and benefits of these drugs (and any other drugs) that may be used for cognitive enhancement purposes before widespread use or advocacy for such use.

18.3 The neurobiology of stimulants

The neurobiology of stimulants such as cocaine and amphetamines helps to explain the trends in their use. These drugs increase the release of the neurotransmitter dopamine, which produces feelings of pleasure, enhanced attention, and endurance. The downside of these drugs is that tolerance rapidly develops with their regular use, so users need to increase their average dose to get the same dopamine release (Chamberlain et al. 2007; National Institute on Drug Abuse 2009). With increasing dose, adverse side effects occur such as increased blood pressure and heart rate, and decreased sleep and appetite. Repeated use can result in cardiovascular complications and sustained binge use can produce symptoms of profound fatigue and depression (National Institute on Drug Abuse 2009).

Many of these negative side effects from regular cocaine and amphetamine use do not reveal themselves immediately. When taken under a doctor's instructions, typically in small doses by oral means, these drugs are slowly absorbed and the dose can be controlled allowing safe use (National Institute on Drug Abuse 2009). Problems arise when the drug is self-administered in higher doses, especially by nonoral routes (e.g., smoking and injection). These routes hasten the release of dopamine, accelerate the development of tolerance, and increase the risks of addiction and other adverse health effects (National Institute on Drug Abuse 2009).

18.4 Newer stimulant drugs: methylphenidate

Methylphenidate (MPH) is a CNS stimulant that has amphetamine-like effects. Patented in the 1950s, MPH was prescribed for a variety of conditions, including depression, fatigue, and narcolepsy. Like cocaine and other amphetamines, optimism surrounded the therapeutic uses of this drug and its ability to enhance mental functioning. MPH was more attractive because of its higher level of safety and more limited side effects than cocaine and amphetamines. MPH has found its major use in the treatment of ADHD because it reduces symptoms of inattentiveness, restlessness, and impulsivity (Morton and Stockton 2000; Rasmussen 2008).

As the numbers of MHP prescriptions and hence availability have risen, so too has the opportunity for abuse (Morton and Stockton 2000). As with cocaine and amphetamines, route of administration and dose strongly influence the effects and side effects of MPH. When inhaled or taken intravenously the drug is absorbed quickly and can produce a high and thus has abuse potential (Vastag 2001; Volkow 2006). Reviews have indicated that methylphenidate when used in large doses can increase the risk of hallucinations and mania (as can other stimulants such as amphetamine) (FDA 2007).

The current expert view is that methylphenidate is a safe and effective drug when used to treat ADHD (Iversen 2006) by the oral route in prescribed doses (Lingford-Hughes et al. 2003). Globally there has been a steep rise in ADHD medication prescriptions, much of it for MPH, and increasingly for sustained-release formulations. Between 1993 and 2000, rates of use rose by 13.2% each year, and 16.8% each year from 2000 to 2003 globally (Scheffler et al. 2007).

18.4.1 Enhancement uses of methylphenidate

In the bioethics literature, methylphenidate is often mentioned as a drug that can be safely used for cognitive enhancement purposes, that is, used by healthy individuals who do not have a clinical diagnosis that would warrant a prescription to improve their cognitive performance (Hall 2004). Those bioethicists who are more positively disposed to enhancement uses of methylphenidate claim that such use is already widespread (e.g., Farah et al. 2004; Greely et al. 2008). They often appeal to anecdotal reports (Lucke et al. 2010), such as the following:

> In all honesty, I haven't written a paper without Ritalin since my junior year in high school. I even wrote my Harvard essay on it. It keeps you up when you're tired, and makes you much more aware of what you are doing. Although there are certain risks involved, I think it's worth it. (Iversen 2006, 68)

Those who are critical of these uses point to the lack of empirical data on the safety and effectiveness of stimulant pharmaceuticals when used for cognitive enhancement purposes. They are particularly concerned about the lack of information on the safety of long-term regular use (de Jongh et al. 2008; Outram 2010).

The evidence for the enhancement efficacy of these drugs is limited (Lucke et al. 2011). Elliott and colleagues (1997), for example, reported that methylphenidate improved performance on tasks already learned and improved spatial working memory but not attention. A review by the British Academy of Medical Sciences (British Medical Association 2007) identified a range of drugs, many of them pharmaceuticals, and concluded that there was limited evidence that they enhance cognition in healthy individuals. The limited short-term laboratory studies of enhancement effects report mixed results (Turner et al. 2003a,b). It is also unclear how well the acute effects in these laboratory settings translate into cognitive performance in real-world settings (Outram 2010) and whether any such effects persist with longer term use.

18.4.2 Lessons from history

The historical experience with cocaine and amphetamines should give pause to advocates of the enhancement use of methylphenidate and other drugs. As these experiences have shown, the use of new stimulant drugs often follows a cycle of uncritical enthusiasm that encourages their widespread use. It is only when use becomes common that the adverse effects of regular use can be identified.

Methylphenidate use for cognitive enhancement purposes could currently be considered to be at the beginning of such a typical drug cycle. There has been an increase in prescriptions of MPH in many countries for ADHD (Scheffler et al. 2007). Increased rates of prescription increase opportunities for nonmedical use, with US studies showing that most young people who use prescription stimulants without a prescription obtain their drugs from a friend with a prescription (McCabe et al. 2006). Thus, the number of students in a class with a medically prescribed stimulant predicts the prevalence of nonmedical stimulant use in that class (Poulin 2007). Around half of college students with an MPH prescription report that they have been approached to "divert" (sell, exchange, or give) their medication (McCabe et al. 2006).

Methylphenidate has less pronounced stimulant effects than cocaine and methamphetamine and so is unlikely to produce problems on the same scale of these older drugs. Nonetheless, the history of CNS stimulants highlights the need for more caution by bioethicists who advocate or implicitly condone their nonmedical use for enhancement purposes. This is especially the case when we lack an understanding of the effects of the long-term use of these drugs in the general population in larger than typical therapeutic doses.

18.5 Conclusions

The historical experiences with amphetamines and cocaine suggest that claims about the neuroenhancing potential of drugs such as methylphenidate and modafinil should be approached with more caution than has been the case to date. Cocaine and amphetamine were once considered to have enhancement capabilities and their use for this purpose was regarded in a wholly positive light. They were seen as safe and effective wonder drugs that increased alertness and mental capabilities, thereby allowing users to cope better with the increasing demands of modern life. It is important that we avoid repeating this experience with an uncritical advocacy of the enhancement use of newer therapeutic stimulants. We should advise people who insist on using these drugs off-label to be aware of their possible adverse effects when used regularly and in unregulated ways.

More specifically, we suggest that bioethicists should not unwittingly encourage the cognitive enhancement use of these drugs by inflating their prevalence of use and suggesting ways of facilitating such use (e.g., Greely et al. 2008; Maher 2008). It is arguably irresponsible to do so in the absence of research on their efficacy and safety when used for cognitive enhancement purposes. It is important that we monitor rates of use of prescription stimulants for cognitive enhancement. While we do not suggest that MPH will produce problems on the scale produced by cocaine and amphetamines, these earlier experiences highlight the possibility of adverse side effects if these drugs are regularly used on a broad scale by healthy young people.

18.6 Acknowledgments

Reprinted from "Lessons for enhancement from the history of cocaine and amphetamine use," Stephanie K. Bell, Jayne C. Lucke, and Wayne D. Hall, *AJOB Neuroscience*, 3 (2), pp. 24–29 © 2012, Taylor and Francis. Reprinted by permission of the publisher (Taylor & Francis Ltd, http://www.tandfonline.com).

This work was supported by an Australian National Health and Medical Research Council Fellowship (Grant 569 738) awarded to Professor Wayne Hall. We also thank Sarah Yeates for her assistance.

References

British Medical Association. 2007. *Boosting your brainpower: Ethical aspects of cognitive enhancements (discussion paper).* London, UK: BMA. Available at: http://www.bma.org.uk/ethics/health technology/CognitiveEnhancement2007.jsp. Accessed July 1, 2010.

Byck, R., ed. 1974. *Cocaine papers by Sigmund Freud.* New York: Stonehill Publishing.

Cakic, V. 2009. Smart drugs for cognitive enhancement: Ethical and pragmatic considerations in the era of cosmetic neurology. *Journal of Medical Ethics* **35**(10):611–15.

Chamberlain, S. R., T. W. Robbins, and B. J. Sahakian. 2007. The neurobiology of attention-deficit/hyperactivity disorder. *Biological Psychiatry* **61**(12):1317–19.

Cohen, S. 1975. Cocaine. *Journal of the American Medical Association* **231**(1):74–5.

DeSantis, A., S. M. Noar, and E. M. Webb. 2009. Nonmedical ADHD stimulant use in fraternities. *Journal of Studies on Alcohol and Drugs* **70**(6):952–4.

DeSantis, A. D., E. M. Webb, and S. M. Noar. 2008. Illicit use of prescription ADHD medications on a college campus: A multimethodological approach. *Journal of American College Health* **57**(3):315–24.

Elliott, R., B. J. Sahakian, K. Matthews, A. Bannerjea, J. Rimmer, and T. Robbins. 1997. Effects of methylphenidate on spatial working memory and planning in healthy young adults. *Psychopharmacology* **131**(2):196–206.

Farah, M. J., J. Illes, R. Cook-Deegan, H. Gardner, E. Kandel, P. King, E. Parens, B. Sahakian, and P. R. Wolpe. 2004. Neurocognitive enhancement: What can we do and what should we do? *Nature Reviews Neuroscience* **5**(5):421–5.

Fleming, P. M. 1998. Prescribing amphetamine to amphetamine users as a harm reduction measure. *International Journal of Drug Policy* **9**(5):339–44.

Food and Drug Administration (FDA). 2007. *FDA directs ADHD drug manufacturers to notify patients about cardiovascular adverse events and psychiatric adverse events.* Silver Spring, MD: U.S. Food and Drug Administration. Available at: http://www.fda.gov/NewsEvents/Newsroom/PressAnnouncements/2007/ucm108849.htm. Accessed November 2, 2010.

Goldstein, R. A., C. DesLauriers, and A. M. Burda. 2009. Cocaine: history, social implications, and toxicity—A review. *Disease-a-Month* **55**(1):6–38.

Gootenberg, P., ed. 1999. *Cocaine: Global histories.* London, UK: Routledge.

Greely, H., B. Sahakian, J. Harris, R. C. Kessler, M. Gazzaniga, P. Campbell, and M. J. Farah. 2008. Towards responsible use of cognitive-enhancing drugs by the healthy. *Nature* **456**(7223):702–5.

Grinspoon, L., and J. B. Bakalar. 1981. Coca and cocaine as medicines: An historical review. *Journal of Ethnopharmacology* **3**(2):149–59.

Grinspoon, L., and P. Hedblom. 1975. *The speed culture: Amphetamine use and abuse in America.* Cambridge, MA: Harvard University Press.

Haddad, L. M. 1979. 1978: Cocaine in perspective. *Journal of the American College of Emergency Physicians* **8**(9):374–6.

Hall, W. 2004. Feeling "better than well": Can our experiences with psychoactive drugs help us to meet the challenges of novel neuroenhancement methods? *EMBO Reports* **5**(12):1105–9.

Iversen, L. 2006. *Speed, ecstasy, Ritalin: The science of amphetamines.* Oxford: Oxford University Press.

de Jongh, R., I. Bolt, M. Schermer, and B. Olivier. 2008. Botox for the brain: Enhancement of cognition, mood and pro-social behavior and blunting of unwanted memories. *Neuroscience Biobehavioural Reviews* **32**(4):760–76.

Jonnes, J. 1995. The rise of the modern addict. *American Journal of Public Health* **85**(8 Pt 1):1157–62.

Karch, S. 2006. *A brief history of cocaine: From Inca monarchs to Cali cartels: 500 Years of cocaine dealing.* Boca Raton, FL: CRC Press.

Lingford-Hughes, A. R., S. J. C. Davies, S. McIver, T. M. Williams, M. R. Daglish, and D. J. Nutt. 2003. Addiction. *British Medical Bulletin* **65**:209–22.

Lucke, J. C., S. Bell, B. Partridge, and W. Hall. 2010. Weak evidence for large claims contribute to the phantom debate. *Biosocieties* **5**:482–3.

Lucke, J. C., S. Bell, B. Partridge, and W. D. Hall. 2011. Deflating the neuroenhancement bubble. *AJOB Neuroscience* **2**:38–43.

McCabe, S. E., C. J. Teter, and C. J. Boyd. 2006. Medical use, illicit use, and diversion of abusable prescription drugs. *Journal of American College Health* **54**(5):269–78.

Maher, B. 2008. Poll results: Look who's doping. *Nature* **452**(7188):674–5.

Morton, A. W., and G. G. Stockton. 2000. Methylphenidate abuse and psychiatric side effects. *The Primary Care Companion to the Journal of Clinical Psychiatry* **2**(5):159–64.

Musto, D. F. 1992. Cocaine's history, especially the American experience. *Ciba Foundation Symposium* **166**:7–14; discussion 14–19.

National Institute on Drug Abuse. 2009. DrugFacts: Stimulant ADHD medications: methylphenidate and amphetamines. Bethesda, MD: U.S. Department of Health and Human Services. Available at: http://www.drugabuse.gov/Infofacts/ADHD.html. Accessed May 24, 2011.

Outram, S. M. 2010. The use of methylphenidate among students: The future of enhancement? *Journal of Medical Ethics* **36**(4):198–202.

Poulin, C. 2007. From attention-deficit/hyperactivity disorder to medical stimulant use to the diversion of prescribed stimulants to non-medical stimulant use: Connecting the dots. *Addiction* **102**(5):740–51.

Rasmussen, N. 2008. *On speed: The many lives of amphetamine.* New York: New York University Press.

Sahakian, B. J., and S. Morein-Zamir. 2010. Neuroethical issues in cognitive enhancement. *Journal of Psychopharmacology* **25**(2):197–204.

Scheffler, R., S. Hinshaw, S. Modrek, and P. Levine. 2007. The global market for ADHD medications. *Health Tracking* **26**(2):450–7.

Spillane, J. F. 2000. *Cocaine: From medical marvel to modern menace in the United States, 1884–1920.* Baltimore: Johns Hopkins University Press.

Turner, D. C., T. W. Robbins, L. Clark, A. R. Aron, J. Dowson, and B. J. Sahakian. 2003a. Cognitive enhancing effects of modafinil in healthy volunteers. *Psychopharmacology* **165**(3):260–9.

Turner, D. C., T. W. Robbins, L. Clark, A. R. Aron, J. Dowson, and B. J. Sahakian. 2003b. Relative lack of cognitive effects of methylphenidate in elderly male volunteers. *Psychopharmacology* **168**(4):455–64.

Vastag, B. 2001. Pay attention: Ritalin acts much like cocaine. *Journal of the American Medical Association* **286**(8):905–6.

Volkow, N. D. 2006. Stimulant medications: How to minimize their reinforcing effects? *American Journal of Psychiatry* **163**(3):359–61.

Chapter 19

Challenges in regulating the use of stimulant drugs for cognitive enhancement in normal individuals

Wayne Hall and John Strang

19.1 Introduction

In principle, there are a number of approaches that could be taken to the regulation of cognitive-enhancing drugs (CED), and there are different classifications of these approaches (Babor et al. 2010; MacCoun and Reuter 2001; TDPF 2009). We use a modified version of the classification system of Strang et al. (2012) in their discussion of regulatory options for psychoactive drugs, that is, drugs that affect cognition, mood, and emotion, and that include CED as a special case. We outline the characteristics of each option, beginning with the least restrictive, a free market in which there is a minimum of regulation of producers, sellers, and users of CED. We then outline a set of progressively more restrictive regulatory options that gradually move toward total prohibition of all nonmedical use. We then discuss the challenges that arise in using these options to regulate the use of CED.

19.2 Options for regulation

19.2.1 A free market for CED

Liberal ethicists follow John Stuart Mill (1998) in giving absolute priority to the autonomy of adults. They propose that any adult should be free to use any psychoactive substance that they like, provided that their drug use does not harm others (e.g. Sandberg 2013; Savulescu 2013). Liberals therefore favor a laissez-faire free market approach to the regulation of CED that would allow these drugs to be freely sold to adults (Sandberg 2013; Savulescu and Bostrom 2009). The only regulatory constraints on the sale of these drugs, in their view, should be prohibitions on their sale to minors. Businesses that sold CED would have to meet basic consumer standards, and would potentially be legally liable for any harm arising from their use, as adjudicated by the courts.

Caffeine is one of the few drugs regulated in this way. It is a mild stimulant that is usually consumed in the form of tea and coffee and it is widely available in all developed and most developing societies. Those who favor free markets for CED argue that there is no ethically and empirically substantive difference between using stimulant drugs for cognitive enhancement (CE) and consuming caffeinated drinks. They seem to believe that stimulant drugs are as safe to use as coffee and often used for reasons similar to that for coffee. The historical experience with stimulant drugs such as amphetamines indicates that these drugs have substantial abuse potential, are capable of producing dependence, and in high doses can produce serious adverse health effects, such as fatal overdose, cardiovascular deaths, stroke, and psychoses (Rasmussen 2008). It is also of interest that recent marketing of high-dose caffeinated beverages to young people, particularly in combination with alcohol, have led to calls for more restrictive regulation of these drugs (Miller 2013).

Liberal ethical arguments for a free market for CED are often supported by the claim (rarely referenced) that the prohibition of alcohol use in the USA was an abject failure and hence that the same will be true of the prohibition of stimulant drugs (Savulescu 2013; Savulescu et al. 2004). Liberal views about the failures of drug prohibition are not shared by drug policy analysts (Strang et al. 2012). These analysts criticize draconian criminal penalties against drug users (e.g. Room and Reuter 2012), but their proposals for reform do not include a free market for all currently illicit drugs. They most often advocate a more liberal regulatory policy toward cannabis use (Room et al. 2010) but typically favor less radical forms of liberalization for other illicit drugs of the types discussed in sections 19.2.3 and 19.2.4. In the case of opioids, for example, they would not necessarily restrict medical use for pain relief and other approved medical indications while nevertheless still retaining the prohibition on nonmedical opioid use. They would also increase access to agonist maintenance treatment (e.g. methadone, buprenorphine, and heroin) for opioid addicts; use prescription monitoring systems to minimize the diversion and abuse of pharmaceutical opioids (Strang et al. 2012); and fund harm reduction measures such as needle and syringe programs, and supervised injecting rooms to reduce harms to heroin users (Babor et al. 2010).

Laissez-faire approaches to regulation receive minority support in the ethical debate about CED. More commonly advocated approaches include some type of regulation that restricts access to adults and ensures that adults who use CED are well informed about the potential risks. A range of regulatory approaches that vary in the restrictions imposed and in the extent to which they discourage use have been suggested (Dubljević 2013).

19.2.2 A regulated consumer market

In this model, adults could be legally able to use CED if the CED were regulated in much the same way as tobacco or alcohol. The use of alcohol and tobacco is not prohibited to adults; users do not require a prescription; and both products are legally sold in a wide range of sales outlets. However, governments impose taxes to discourage the heavy use of alcohol and encourage smokers to quit. They also minimally regulate the quality of both products to protect consumers.

The methods of regulating alcohol and tobacco have moved in opposing directions over the past 30 years in many developed countries (Room and Hall 2013). Alcohol regulation has been progressively liberalized. Taxes have not increased in line with inflation, effectively making alcoholic beverages much cheaper to purchase. Trading hours have typically been extended and there are few if any restrictions on how alcohol may be promoted. In the case of cigarettes, by contrast, public concern about the adverse health effects of smoking has led to much more restrictive forms of regulation. There have been large increases in tobacco taxes, and limits have been placed on how cigarettes can be displayed at point of sale, how they can be promoted, how they are packaged, and when and where they may be used.

The regulatory levers available to governments under a legal market can be used to reduce heavy use and the harm arising from it. For example, increasing taxes and limiting the availability of alcohol can reduce harmful alcohol consumption and alcohol-related problems (Babor et al. 2010; Room and Hall 2012; Wagenaar et al. 2010). Increased taxes on cigarettes, bans on advertising, and bans on smoking in public places, workplaces, bars, and restaurants have contributed to major reductions in the prevalence in smoking in many developed countries over the past several decades (Room and Hall 2012; Scollo and Winstanley 2012; Wakefield et al. 2014).

A major problem with these regulatory systems is that companies that market alcohol and tobacco have a strong interest in undermining their effectiveness by liberalizing them. They may also use the regulations to protect their market positions by discouraging new entrants. The effectiveness of regulatory systems depends on the preparedness of governments to operate the system in the public interest (Room and Hall 2012). This cannot be assumed when alcohol and tobacco interests have the resources to lobby governments and make substantial political donations, and when government obtain substantial tax revenue from the sale of these products. The alcohol industry, for example, opposes more restrictive regulations by arguing that its products are safe when used in moderation. It argues that the problems arising from alcohol use only occur to a small minority of people whose behavior is best managed by treating or legally restraining them. They oppose policies that will reduce heavy use of their product, such as increasing taxes and restraining availability and promotion, by claiming that these policies will punish the majority of drinkers for the sins of the minority of problem drinkers (Room and Hall 2012). They also seek to "capture" initiatives intended to reduce alcohol-related harms. They do so with offers of sponsorship or direct funding to advocacy groups that promote initiatives with less evidence of efficacy (McCambridge et al. 2014).

19.2.3 A prescription system

Prescription systems allow the medically supervised use of drugs that have beneficial health effects when used in appropriate doses. Availability is restricted to "prescription only" because these drugs can cause harm when used in larger than recommended doses or in combination with other drugs, including alcohol (Babor et al. 2010), and many can produce dependence. The opioids exemplify this regulatory approach: only licensed physicians can prescribe opioids, which must be dispensed by a pharmacist, often in limited quantities to minimize diversion for nonmedical use, and reduce the risk of dependence.

Prescription regimens do not eliminate nonmedical use of drugs because of potential diversion in the manufacture and supply processes and also because prescribed drugs may be diverted to nonmedical use, e.g., by giving drugs to family members or friends, or selling them on the black market (Strang et al. 2012). Prescription monitoring systems can reduce irregular prescribing and patient utilization while allowing patients to access these drugs for appropriate medical purposes (Strang et al. 2012). Systems can be devised to prevent "doctor shopping" and drug diversion by allowing physicians and pharmacists to access patient records at the time of writing a prescription or dispensing a drug (Holmes 2012).

Off-label use occurs when a prescription is provided by a physician for a purpose other than the medical indication for which the drug has been licensed. The American Academy of Neurology guidelines on prescribing CED argue that off-label use is warranted under the proviso that physicians are using the drug to treat a patient (Larriviere et al. 2009). The Academy of Neurology guidelines have been criticized for not providing helpful guidance to health professionals that is consistent with commitments to evidence-based medicine and socially responsible medical practice (Boot et al. 2012; Racine and Forlini 2010).

19.2.4 **Prohibition of nonmedical stimulant use**

Under a policy of prohibition, any nonmedical use of a drug by an adult, including its use for CE, is prohibited by state criminal law. Criminal penalties, including imprisonment, may apply to those who are found using the drug without a medical prescription. The prohibition of a drug is usually justified by an argument that it is too addictive or dangerous, even when used by informed adults, so its use by adults should be banned to protect the public health. This was the justification for the global ban on the recreational use of cannabis, heroin, and cocaine under the 1961 UN Single Convention on Narcotic Drugs (Room and Reuter 2012).

Liberal bioethicists often seem unaware that there is prohibition of stimulant use for CE under the international drug control treaties of 1961 and 1971 that have been signed by the majority of UN member states (UNODC 2015). Under these treaties, signatory states allow the use of controlled substances (such as stimulants) for medical and scientific purposes only. Recreational use is prohibited. These treaties preclude the option of legalizing stimulant drugs for CE use, for example, and so preclude the use of taxes to discourage their heavy use.

These treaties have been interpreted in creative ways to allow some recreational use of prohibited drugs. In the 1970s and 1980s, for example, the Netherlands used a constitutional exemption allowed in the Single Convention to achieve *de facto* decriminalization of the recreational use and small scale retail sales of cannabis (Bewley-Taylor et al. 2014; Room et al. 2010). In 1976 state prosecutors issued a statement that they would not prosecute persons for the possession and use of up to 30 g of cannabis, effectively making possession and use of these small quantities of cannabis legal *de facto* (Grund and Breekesma 2013; Korf 2002). The Dutch authorities later extended this policy to the sale of small amounts of cannabis in coffee shops, thereby creating a *de facto* legal retail cannabis market (Room et al. 2010).

The Dutch government argued that their policy complied with the international drug control treaties because cannabis use and its sale remained illegal; the authorities used the expediency principle within the Netherlands' constitution to give a low priority to enforcing these laws if certain conditions were met in order to give a greater priority to enforcing laws against "hard drugs" like heroin (Room et al. 2010). This argument was initially accepted by the guardian of the drug control treaties, namely the International Narcotic Control Board (INCB), until the late 1990s but the Netherlands has since been strongly criticized for failing to comply with the treaties.

Greely (2013) has questioned the relevance of these international control treaties. He asks: what would be the consequences of any state failing to comply with their provisions? The major consequence would be a chorus of disapproval from the majority of member states, including the USA, the architect of the treaties and the leading defender of their provisions (Bewley-Taylor 2012). The provisions of the international drug control treaties do allow member states to renounce the treaties and then re-accede with specific reservations. Bolivia has recently done this in order to allow the production of cocoa leaf for traditional use. But these provisions have rarely been exercised and their use has brought strong international disapproval on countries that have made such changes (Room 2012).

19.2.5 Proposals for regulating CED

Bioethicists of a liberal persuasion often write as if we can design regulations for CED by starting with a blank sheet of paper, and deducing an ideal regulatory scheme from broad ethical principles, such as respect for adult autonomy. The most commonly suggested regulatory approaches have been permissive or at least tolerant of CED use (e.g., Greely et al. 2008, Larriviere et al. 2009). This probably reflects the greater priority that North American bioethicists give to respecting adult autonomy than protecting public health.

For those who concede that CED may be used in ways that can cause harm, the usual preference is for an extension or adaptation of the pharmaceutical regulatory system (e.g., Dubljević 2013; Greely et al. 2008). Under this type of scheme a government agency would license pharmaceutical companies to develop CED on condition that these drugs were also used to treat cognitive disorders. Dubljević (2013) suggests that pharmaceutical companies would be interested in the large potential market for CED.

The government of New Zealand has experimented with a similar proposal to license and regulate new psychoactive drugs that are used recreationally by young adults (Wilkins 2014; Wilkins et al. 2013). This scheme was intended to regulate rather than prohibit new psychoactive drugs by requiring manufacturers to demonstrate the safety of their products before they could market them. Uncertainty about the standards that should be required to demonstrate safety effectively prohibited these drugs because no manufacturer was able to provide evidence that their products were safe. The same challenges will be faced by any regulatory regime for CED that is based on a modified form of the pharmaceutical regulatory system.

Even if such a regulatory scheme were introduced, it is not clear that pharmaceutical companies would be interested in having their drugs registered. They already have large markets for their medicinal products and they may not be willing to take on the extra expense and risk involved in marketing neuropharmaceuticals to healthy people for the purposes of CE (Foresight UK 2005). Some plans to market stimulants for cognitive enhancement have been abandoned (Bell et al. 2012). It is unclear whether it would be legal to study the safety and efficacy of CED in normal individuals. Even if it is legal, it is uncertain whether pharmaceutical companies would be eager to fund studies that may require large samples followed up over periods of years. The legal liability for any harms arising from the use of these drugs by healthy people may be another deterrent to developing CED. There may be very low social tolerance for the sale of drugs that produce serious adverse events (such as deaths from strokes, cardiovascular events, or psychoses) even if these effects largely occur among young adults who used the drugs in excess of recommended doses, by nonapproved routes of administration (e.g., by injection or smoking) or in combination with other drugs.

19.2.6 A hybrid system: licensing producers and users of CED

Transform, a drug policy reform group in the United Kingdom, has proposed that adults should be licensed to use stimulants and dance party drugs (e.g., TDPF 2006). Some bioethicists have proposed similar forms of licensing of users of CED. Bostrom and Sandberg, for example, have suggested that "enhancement licenses" be issued to people who demonstrate an understanding of the risks and a capacity to use these drugs responsibly. They argue that this would ensure that users gave free and informed consent to use CED and that any adverse effects of use could be properly monitored (Bostrom and Sandberg 2009).

Dubljević (2013) has produced a detailed argument for using such a system to regulate the enhancement use of sustained-release forms of methylphenidate, i.e., formulations of the drug that allow it to be steadily released into the body over a day rather than as a single dose that is quickly absorbed. He has described this as an "economic disincentives model" because under his model there would be (1) tight regulation of companies that produced and marketed the drug for enhancement purposes, (2) the drug would be heavily taxed to discourage abuse, and (3) potential users would be required to receive a license to use the drug by undertaking educational courses on the risks posed by its use. Users would pay a fee to do a course about the effects and adverse side effects of using methylphenidate, prove that they had acquired this knowledge by passing an exam, become a registered enhancement user, and obtain additional medical insurance to cover the risks of using.

There are a number of problems with his proposal.

First, sustained-release forms of drugs may not be safer than immediate-release forms. The epidemic abuse of sustained release OxyContin in the USA provides a cautionary tale (Hall and Degenhardt 2007; Hall and Farrell 2011). These formulations were marketed as having a low abuse potential but users soon discovered how to extract the active ingredient. The result was an epidemic of pharmaceutical opioid abuse and pharmaceutical opioid overdose deaths outnumbering overdose deaths caused by heroin (Okie 2010; Van Zee 2009). The illicit production of methamphetamine from cold medicines containing pseudoephedrine indicates that the reprocessing of sustained-release pharmaceuticals is not confined to opioids (McKetin et al. 2011).

Second, there are doubts about the feasibility and effectiveness of the proposed regulatory system. Tight regulations and high taxes would be major disincentives to the participation of the pharmaceutical system in the production and marketing of this drug for enhancement use. It is also not clear whether health insurers would be prepared to insure registered users and if they did so, whether the premiums would be affordable.

Third, high taxes and the inconvenience of obtaining a license to use may deter users from registering for licenses. They may prefer to use diverted pharmaceutical stimulants or those illicitly produced rather than navigate the complicated system of regulation.

19.3 Barriers to more liberal forms of CED regulation

The major challenge in choosing any regulatory systems for CED is the lack of experience in any countries with any form of regulation apart from the existing combination of prohibition and prescription systems for stimulants. The regulatory systems that apply to alcohol and tobacco are historical anomalies in that no such regulatory system would be used to regulate any new drug that had pharmacological effects similar to those of alcohol or nicotine. These regulatory systems are, in any case, moving in very different directions: toward prohibition in the case of tobacco, and, across much of the world, toward greater liberalization in the case of alcohol.

The prohibition of CED reduces their use by making these drugs more difficult to obtain, and more expensive to purchase than would be the case if they could be legally sold to adults (MacCoun and Reuter 2001). This is one of the reasons why heroin use is so rare and the use of most prohibited drugs is a small fraction of the use of caffeine, alcohol, and tobacco. Contrary to widespread misconception, national alcohol prohibition in the USA also reduced alcohol use and alcohol-related harm between 1920 and 1932 (Hall 2010).

The benefits of prohibition in reducing use comes at a cost: the creation of black markets that meet consumer demand for the prohibited drug. If demand is large, and sales are profitable, those supplying black markets can generate large profits that can be used to corrupt public officials in order to reduce their risks of arrest and punishment. Prohibition also makes illegal drug use riskier in various ways: it inhibits the development of social norms to regulate such drug use; and it makes it more difficult to monitor rates of illegal drug use and to detect any adverse effects of their use. As libertarians point out, prohibition is a strongly paternalistic policy that deprives all adults of the right to use these drugs in order to protect their own health and that of the public.

19.4 Conclusion

Prohibition is the current regulatory approach in most developed countries for regulating the nonmedical use of stimulant drugs. The use of these drugs for CE is prohibited because it constitutes nonmedical use that is proscribed under the 1971 international drug control treaty. This is also the most probable regulatory response to any new neuropharmaceuticals that may have putatively cognitively enhancing effects, especially if they prove popular among young adults. In the absence of good evidence about their safety and efficacy when used for CE, prohibition is a precautionary response that is justified by the argument that it will minimize the risk of serious adverse health outcomes that could occur if these drugs were allowed to be used recreationally.

The neuropharmaceuticals most often advocated for CE are regulated through pharmaceutical prescription systems. Their use for CE purposes is facilitated by the diversion of pharmaceuticals from medical use. Examples include the fraudulent presentation of symptoms to a physician for the purpose of obtaining a prescription for a stimulant or obtaining neuropharmaceuticals without a prescription, either from a friend or through the internet. A physician who prescribes a neuropharmaceutical off-label for enhancement purposes to someone without a disorder is also arguably acting illegally in most developed countries.

International drug control treaties currently prohibit any form of regulation that would allow adults to use stimulant drugs for CE. These treaties are currently under challenge in the USA with four states legalizing recreational cannabis use. It remains to be seen whether any reconsideration of the way that the treaties regulate cannabis use in the USA will be extended to allowing the non-medical use of stimulant drugs for CED.

References

Babor, T., J. Caulkins, G. Edwards, B. Fischer, D. Foxcroft, K. Humphreys, I. Obot, J. Rehm, P. Reuter, R. Room, I. Rossow, and J. Strang. 2010. *Drug policy and the public good.* Oxford: Oxford University Press.

Bell, S., J. Lucke, and W. D. Hall. 2012. Lessons for enhancement from the history of cocaine and amphetamine use. *AJOB Neuroscience* **3**:24–9.

Bewley-Taylor, D. 2012. *International drug control: Consensus fractured.* Cambridge,UK: Cambridge University Press.

Bewley-Taylor, D., T. Blickman, and M. Jelsma. 2014. *The rise and decline of cannabis prohibition: the history of cannabis in the UN drug control system and options for reform.* Amsterdam/Swansea: Global Drug Policy Observatory/Transnational Institute.

Boot, B., B. Partridge, and W. D. Hall. 2012. Better evidence for safety and efficacy is needed before neurologists prescribe drugs for neuroenhancement to healthy people. *Neurocase* **18**:181–4.

Bostrom, N., and A. Sandberg. 2009. Cognitive enhancement: methods, ethics, regulatory challenges. *Science and Engineering Ethics* **15**:311–41.

Dubljević, V. 2013. Prohibition or coffee shops: regulation of amphetamine and methylphenidate for enhancement use by healthy adults. *American Journal of Bioethics* **13**:23–33.

Foresight UK. 2005. *Drugs futures 2025? Perspective of the pharmaceutical industry.* London: Department of Trade and Industry.

Greely, H. 2013. Some first steps toward responsible use of cognitive-enhancing drugs by the healthy. *American Journal of Bioethics* **13**:39–41.

Greely, H., B. Sahakian, J. Harris, R. C. Kessler, M. Gazzaniga, P. Campbell, and M. J. Farah. 2008. Towards responsible use of cognitive-enhancing drugs by the healthy. *Nature* **456**:702–5.

Grund, J., and J. Breekesma. 2013. Coffee shops and compromise: separated illicit drug markets in the Netherlands. New York: Global Drug Policy Program, Open Society Foundation.

Hall, W. D. 2010. What are the policy lessons of National Alcohol Prohibition in the United States, 1920–1933? *Addiction* **105**:1164–73.

Hall, W. D., and L. Degenhardt. 2007. Regulating opioid prescribing to provide access to effective treatment while minimising diversion: an overdue topic for research. *Addiction* **102**:1685–8.

Hall, W. D., and M. Farrell. 2011. Minimising the misuse of oxycodone and other pharmaceutical opioids in Australia. *Medical Journal of Australia* **195**:248–9.

Holmes, D. 2012. Prescription drug addiction: the treatment challenge. *Lancet* **379**:17–18.

Korf, D. 2002. Dutch coffee shops and trends in cannabis use. *Addictive Behavior* **27**:851–66.

Larriviere, D., M. A. Williams, M. Rizzo, and R. J. Bonnie. 2009. Responding to requests from adult patients for neuroenhancements: guidance of the Ethics, Law and Humanities Committee. *Neurology* **73**:1406–12.

McCambridge, J., K. Kypri, C. Drummond, and J. Strang. 2014. Alcohol harm reduction: corporate capture of a key concept. *PLoS Medicine* **11**:e1001767.

MacCoun, R., and P. Reuter. 2001. *Drug war heresies: Learning from other vices, times and places.* Cambridge, UK: Cambridge University Press.

McKetin, R., R. Sutherland, D. Bright, M. Norberg. 2011. A systematic review of methamphetamine precursor regulations. *Addiction* **106**:1911–24.

Mill, J. S. 1998. *On Liberty (1859) and Other Essays*, edited with an introduction and notes by John Gray. Oxford: Oxford University Press.

Miller, P. 2013. Energy drinks and alcohol: research supported by industry may be downplaying harms. *BMJ* **347**:f5345.

Okie, S. 2010. A flood of opioids, a rising tide of deaths. *New England Journal of Medicine* **363**:1981–5.

Racine, E., and C. Forlini. 2010. Cognitive enhancement, lifestyle choice or misuse of prescription drugs? *Neuroethics* **3**:1–4.

Rasmussen, N. 2008. *On speed: The many lives of amphetamine.* New York: New York University Press.

Room, R. 2012. Reform by subtraction: the path of denunciation of international drug treaties and reaccession with reservations. *International Journal of Drug Policy* **23**:401–6.

Room, R., B. Fischer, W. D. Hall, S. Lenton, and P. Reuter, eds. 2010. *Cannabis policy: Moving beyond stalemate.* Oxford: Oxford University Press.

Room, R., and W. D. Hall. 2012. Population approaches to alcohol, tobacco and drugs: effectiveness, ethics and interplay with addiction neuroscience. In *Addiction neuroethics: The ethics of addiction neuroscience research and treatment*, ed. A. Carter, W. D. Hall, and J. Illes, 247–60. Amsterdam: Elsevier.

Room, R., and W. D. Hall. 2013. A history of drug use in Australia. In *Drug use and Australian society*, ed. M. Hamilton, A. Ritter, and T. King, 51–66. London: Oxford University Press.

Room, R., and P. Reuter. 2012. How well do international drug conventions protect public health? *Lancet* **379**:84–91.

Sandberg, A. 2013. Enhancement policy and the value of information. *American Journal of Bioethics* **13**:34–5.

Savulescu, J. 2013. A liberal consequentialist approach to regulation of cognitive enhancers. *American Journal of Bioethics* **13**:53–5.

Savulescu, J., and J. Bostrom, eds. 2009. *Human enhancement.* Oxford: Oxford University Press.

Savulescu, J., B. Foddy, and M. Clayton. 2004. Why we should allow performance enhancing drugs in sport. *British Journal of Sports Medicine* **38**:666–70.

Scollo, M., and M. Winstanley. 2012. *Tobacco in Australia: Facts and issues*, 4th ed. Melbourne: Cancer Council Victoria.

Strang, J., T. Babor, J. Caulkins, B. Fischer, D. Foxcroft, and K. Humphreys. 2012. Drug policy and the public good: evidence for effective interventions. *Lancet* **379**:71–83.

Transform Drug Policy Foundation (TDPF). 2006. After the war on drugs: Options for control. Available at: http://www.tdpf.org.uk/. Accessed February 2, 2015.

Transform Drug Policy Foundation (TDPF). 2009. After the war on drugs: Blueprint for regulation. Available at: http://www.tdpf.org.uk/Transform_Drugs_Blueprint.pdf. Accessed February 2, 2015.

UNODC. 2015. Treaties: drug related treaties. Vienna: UN Office on Drugs and Crime. Available at: http://www.unodc.org/unodc/en/treaties/index.html#Drugrelated. Accessed February 2, 2015.

Van Zee, A. 2009. The promotion and marketing of OxyContin: commercial triumph, public health tragedy. *American Journal of Public Health* **99**:221–7.

Wagenaar, A., A. Tobler, and K. Komro. 2010. Effects of alcohol tax and price policies on morbidity and mortality: a systematic review. *American Journal of Public Health* **100**:2270–8.

Wakefield, M., K. Coomber, S. Durkin, M. Scollo, M. Bayly, M. J. Spittal, J. A. Simpson, and D. Hill. 2014. Time series analysis of the impact of tobacco control policies on smoking prevalence among Australian adults, 2001–2011. *Bulletin of the World Health Organization* **92**:413–22.

Wilkins, C. 2014. A critical first assessment of the new pre-market approval regime for new psychoactive substances (NPS) in New Zealand. *Addiction* **109**:1580–6.

Wilkins, C., J. Sheridan, P. Adams, B. Russell, S. Ram, and D. Newcombe. 2013. The new psychoactive substances regime in New Zealand: a different approach to regulation. *Journal of Psychopharmacology* **27**:584–9.

Index

Notes: Page numbers followed by *f* indicate material found in figures. *vs.* indicates a differential diagnosis or comparison.